THE HOLT PHYSIOLOGY PROGRAM

HUMAN PHYSIOLOGY   *Morrison, Cornett, Tether, and Gratz*

Supplementary Materials to HUMAN PHYSIOLOGY

EXPERIMENTS IN PHYSIOLOGY   *A laboratory manual designed to accompany the text*

TEACHERS GUIDE TO THE HUMAN PHYSIOLOGY PROGRAM
Professional materials on teaching the text and laboratory experiments. Answers to questions and suggested tests are included.

Other Supplementary Materials

MODERN HEALTH   *Otto, Julian, and Tether*

TECHNIQUES AND INVESTIGATIONS IN THE LIFE SCIENCES   *Feldman*

EXPERIMENTS IN BIOLOGICAL DESIGN   *Feldman*

SEX EDUCATION AND FAMILY LIFE   *Julian and Jackson*

Holt Library of Science

BIOMEDICAL ASPECTS OF SPACE FLIGHT   *Henry*

CANCER   *Woodburn*

EXPERIMENTS IN PSYCHOLOGY   *Blough and Blough*

HUMAN EVOLUTION   *Lasker*

LIFE AND THE PHYSICAL SCIENCES   *Morowitz*

RADIATION, GENES AND MAN   *Wallace and Dobzhansky*

A TRACER EXPERIMENT   *Kamen*

VIRUSES, CELLS AND HOSTS   *Sigel and Beasley*

HUMAN

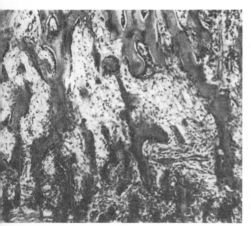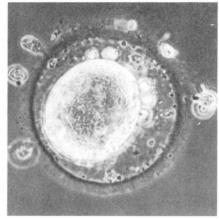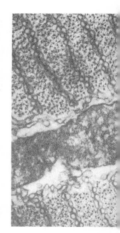

PHYSIOLOGY

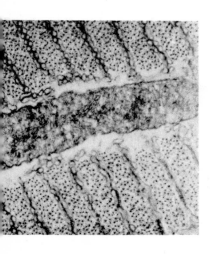

Thomas F Morrison

Frederick D Cornett

J Edward Tether

Pauline Gratz

Holt, Rinehart and Winston, Inc.

New York, Toronto, London

6

The Authors

THOMAS F MORRISON *was formerly a member of the Science Department and a teacher of physiology at Milton Academy, Milton, Massachusetts*

FREDERICK D CORNETT *is a member of the Science Department and a teacher of physiology at Santa Monica High School, Santa Monica, California*

J EDWARD TETHER, MD, *is Assistant Professor of Neurology, Indiana University Medical Center and School of Medicine, Indianapolis, Indiana, and a Fellow of the American College of Physicians*

PAULINE GRATZ, EdD, *is Assistant Professor of Natural Sciences in the Department of Science Education at Teachers College, Columbia University, New York*

PREFACE

The role of physiology in the general plan of education is to familiarize students with the functions of their own bodies and to prepare those who are interested for further specialized work in the field. In this text, the authors have had the former as their chief objective because they believe the more advanced and technical journals will give specially interested students the additional knowledge required for professional courses.

HUMAN PHYSIOLOGY is an introductory text written for students who wish more detailed explanations of some of the functions of the human body beyond those already covered in the standard secondary school biology or health courses. In this volume the authors have attempted to develop the basic knowledge of physiology as represented by the latest advances in the various areas of the subject, but have deliberately avoided the inclusion of findings of a highly theoretical nature or those supported by incomplete data not yet proved. Although a background knowledge of biology and chemistry may be helpful, students using this text can assimilate the material without either course because complete explanations are given for each physiological principle and all terms and concepts are defined and explained clearly. Thus, the book may be used as either a second course in the biological sciences or an introductory course in physiology without previous experience in this field. It may be used effectively as a text in those junior colleges or in schools of nursing where the study of physiology is an integral part of the curriculum.

In common with all introductory courses, there is a vocabulary load that must be mastered in order to understand physiological concepts clearly and concisely. To aid in this mastery, all physiological words and terms are printed in *italics* the first time they are used or at the point where they are defined. The more basic terms are collected at the end of each chapter in a matching test titled *Vocabulary Review*.

Also at the end of each chapter is a *Test Your Knowledge* section including multiple choice questions which provide an effective check of the assimilation of information and short essay questions which require organization of the material contained in the chapter.

At the end of the text is an *Appendix* containing various conversion factors which can be used to translate some of the data given in the text into more familiar terms. There is also a complete *Glossary* containing definitions of the important physiological terms occuring in the text.

EXPERIMENTS IN PHYSIOLOGY is a complete laboratory manual, following the sequence of topics in the text. The experiments are designed to use a minimum of specialized equipment. Thus, they are particularly adaptable to an elementary course in human physiology.

The text was written by Thomas F. Morrison, Frederick D. Cornett and J. Edward Tether. Pauline Gratz was mainly responsible for the preparation of the supplementary materials including the TEACHER'S GUIDE, and EXPERIMENTS IN PHYSIOLOGY.

Throughout HUMAN PHYSIOLOGY illustrations appear wherever they are required to help explain the topics under discussion. Some have been borrowed from other sources, but the majority are new illustrations drawn especially for this book. The authors appreciate the excellent work of the artists: Stephen Rogers Peck, Mrs. Pauline Burr-Thomas, Tom Morgan, Warren J. Chandler, Paul J. Singh-Roy, and Edward Lenaro whose drawings add greatly to the understanding of the text. Certain of the illustrations represent original drawings by Mr. Morrison, the senior author. The inclusion of the "Trans-Vision" insert of the human body should help in understanding the structure and functions of the more important organs.

The authors extend their sincere gratitude and appreciation to Miss Ruth E. Weimer, Washington High School, Massillon, Ohio and to Mr. Charles E. Herbst, Beverly Hills High School, Beverly Hills, California. These two classroom teachers of physiology read the entire manuscript of the first edition and offered many helpful comments and suggestions. The authors also wish to thank Dr. Alexander T. Ross, Professor and Chairman of the Department of Neurology, Indiana University Medical Center, for his excellent and constructive comments on the chapters of the first edition dealing with the nervous system, each of which he carefully analyzed in manuscript.

In preparation for this revision of HUMAN PHYSIOLOGY, the 1963 edition was read and criticized by Berl L. Huffman, Jr., Woodrow Wilson High School, San Francisco, California, by Antone Ara Bia, Monterey High School, Monterey, California, and by Dr. Lyon Hyams, Cornell University Medical College, New York, New York. Their comments and suggestions are gratefully acknowledged. In addition, the complete manuscript for EXPERIMENTS IN PHYSIOLOGY was read by Robert L. McMahon, Anaheim High School, Anaheim, California.

CONTENTS

The Body as a Whole

Bones and Muscles

The Skin, Temperature,
Metabolism, and Excretion

The Endocrine System

Reproduction and Heredity

Chapter 1   Basic Concepts

Everyone has a natural curiosity about living things but even more he has a thirst for knowledge about his own body—the form that it takes and the way it functions. He is curious to learn about the ceaseless activity of the heart, how it is controlled, and how it makes adjustment to bodily changes. He would like to investigate the sensitivity of the eye; the processes of digestion, respiration, and excretion; the functions of muscles and how they move the bones of the body or keep him warm; the chemical activity of hormones and their influence on individual organs or the body as a whole. As the student examines these areas he will realize that the human body reveals an almost unending complexity of structures as well as a diversity of physical and chemical activity. He will see how everything is interrelated and works together with a remarkable degree of harmony.

As is true of all fields of study, but especially of scientific disciplines, knowledge is added to, bit by bit, like a jigsaw puzzle. Complete understanding cannot be expected at first, but a piece here and there will fit together to make learning a rewarding experience. It will be necessary to make constant refresher trips back to earlier material and be satisfied with progress from general areas of understanding to specialized fields of concentration. Where, then, does the investigation of *human physiology* begin?

Since the biological sciences include all phases of knowledge dealing with living things, physiology is a biological science. *Botany,* the study of plants, and *zoology* the study of all animal life, are also biological sciences. The diversified fields related to these sciences may deal with the description of living things or with the detailed examination of relationships between living things and their environment and experimental research will often be involved. The ever changing pattern of life constantly presents a challenge to the student of the life sciences. Very few theories can be established to explain how living things will behave under all conditions, but scientists are continually discovering new facts about the life processes. Scientific advances have been most rapid in the last two decades but questions still remain unanswered. The techniques, tools, and knowledge gained in such studies as *biochemistry,*

1

the study of chemistry of living things, and *biophysics,* the study of energy as it is related to living things, can be utilized to find answers to these questions in the future. We should recognize that physiology is a branch of biological science that is concerned with all of the physical and chemical processes that occur within the body of a living organism. All of these processes involve transformation of energy from a quiescent *potential* form as it is found in the living material of the body to an active or *kinetic* state that results in the manifestations of life.

As shown in Fig. 1-1, the relation of physiology to some of the other biological sciences is threefold. On the one hand, there is a group of sciences dealing primarily with the structure of the body. These are grouped under the general heading of *morphology* which is the study of the form and structure of organisms. A part of the general area of morphology, is the science of *anatomy* which considers gross fea-

tures that can be seen with the unaided eye. Historically, anatomy is probably the oldest of the biological sciences. It is primarily descriptive in nature, although modern anatomical studies include some discussion of the function of an organ to make a knowledge of its role more complete and satisfying. With the development of the microscope, study of the finer details of structure became possible and branches of morphology appeared: *histology,* the study of tissues, and *cytology,* the study of cells.

A second side of our triangle is composed of the branches of the biological sciences that are primarily *functional* in nature. Included here is physiology itself and the contributing fields of biochemistry and biopyhsics. Here also we can include *psychology* which deals with behavior and emotions of the individual, both of which frequently have a profound effect on the physiological well-being of the person. Only recently has there been a satisfactory explanation of the influence emotions have on *visceral* activity. Under certain conditions, the emotional influence may lead to the pathological (pertaining to disease) condition resulting in a gastric ulcer. This is but one example of a group of conditions that are said to be *psychosomatic* in origin. The word psychosomatic is derived from the two Greek stems of *psyche,* meaning soul or mind, and *soma,* referring to the body. Man's relation to his fellow beings constitutes the field of *sociology.* This may be considered a functional aspect of the biological sciences because it frequently involves problems affecting the health and physical welfare of communities.

Fig. 1-1   The relationships of some of the biological sciences.

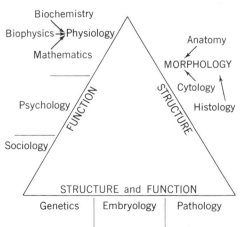

A third group of the biological sciences consists of subjects that combine descriptive and functional elements. One of these, *pathology,* is of primary concern in medicine as it applies to the study of disease. Two other branches are *embryology* and *genetics.* Embryology deals with development of living organisms before birth. Recent experimental work in this field has shed light on the processes occuring in the early stages of human development. Genetics, one of the most rapidly advancing sciences at the present time, is a study of the manner in which various characteristics are transmitted from one generation to the next. The pattern of inheritance exhibited by any person is established at the very beginning of his existence. Not only are physical features determined at this time, but certain physiological traits are also fixed, due to the activities of transmissible material.

Fig. 1-2 Hippocrates, the father of medicine, is responsible for formulating the Hippocratic oath of medical ethics. (Bettman Archive)

The Historical Picture

Physiology is a relative newcomer on the stage of the biological sciences. Much of what we know today of its origins is contained in the records of previous civilizations. These writings record things as they appeared to the inquisitive but superstitious minds of the healers and experimentalists of the day. Clay models of various human organs have been preserved dating back to Babylonian times. In early Egypt the technique of embalming the dead led to a knowledge of anatomy. Among the Greeks the science of physiology received its first great impetus. In the 4th century B.C., Hippocrates a Greek physician, became famous for his medical texts. His collection of writings from different sources, called the *Hippocratic Corpus,* was the first widely used book of medical practice in the Western world. The Hippocratic Code is still the criterion for medical ethics today.

The next great influence in the development of physiology, lasted for over one thousand years, and is attributed to Galen, a Greek physician with Roman citizenship (131–210? A.D.). His many experiments laid a foundation for further studies, although many of his followers misinterpreted his results. Galen lived at a period when animals were considered to be governed by laws quite different from those governing inanimate objects. It was thought that all living things were controlled by some mystical influence called "vital force." Although some of Galen's observations led him to the threshold of discovery of important physiological facts, the concept of "vital forces" among living things prevented him from grasping the true significance of his observations, For many years after Galen, the study of physiology continued to be shrouded in mysticism, although for a brief period there was a trend toward a more logical approach to the subject.

In the 16th century, a German named Theophrastus Hohenheim (1490–1541), who wrote under the name of Paracelsus, attempted to study the action of the body in much the same way as we do today. Paracelsus believed that if the true structure of the body was to be learned, direct observation was essential rather than "textbook anatomy." Although many of his experiments provided insight into the functioning of the body, he was unsuccessful because the field of chemistry was not sufficiently developed to furnish him with the necessary information or techniques.

One of the great names in the development of scientific thought in general, and physiology in particular, is that of the Englishman William Harvey (1578–1657). In 1628 he published a small book, only 72 pages long, written in Latin, the language used by all scientists at that time. It was called *Exercito Anatomica de Motu Cordis et Sanguinis in Animalibus* which may be freely translated as *An Anatomical Dissertation on the Movement of the Heart and Blood in Animals.* The importance of this work was that it demonstrated clearly that the flow of blood

around the body is really a circulatory movement in which the arteries, veins, and capillaries form the conducting channels. Due to the fact that microscopes had not reached a satisfactory stage in development at his time, Harvey had to imagine the presence of the smallest blood vessels, the capillaries. In 1632, however, their presence was demonstrated by the Italian anatomist, Marcello Malpighi. Harvey's methods were so painstakingly developed and his thinking so clear that this small report is considered to be one of the finest examples of scientific investigation. Prior to the appearance of Har-

Fig. 1-3   William Harvey, one of the early pioneers in the study of physiology. (Bettman Archive)

vey's book, the course the blood followed in passing around the body was unknown. Paracelsus, for example, believed that there were small openings in the wall which separates the two sides of the heart through which the blood passed. Others believed that the arteries were tubes which carried air around the body because after death the blood frequently leaves the major arteries to collect in the veins and tissues.

From Harvey's time onward, great strides were made in the biological sciences. As the microscope was perfected, more details of structure were observed and linked with the activities of various parts of the body.

One of the greatest physiologists of the 19th century was Claude Bernard. His keen insight into experimental physiology led to many discoveries concerning chemical and physical activities of living organisms. Included in his list of achievements was the concept of internal secretion and the discovery of the function performed by pancreatic juice. Bernard discovered the mechanism of carbon-monoxide poisoning in living cells. Finally, through his understanding of physiology, he presented to the world of science the concept of the constancy of the internal environment; that at all times the body maintains within itself a dynamic equilibrium.

The 20th century has truly been the century of scientific discovery. The excellence of physiological research has received world recognition in the presentation of numerous Nobel prizes. Pavlov (1904) studied digestive responses of animals to symbols, which resulted in the understanding of conditioned reflexes. In 1906, Cajal and Golgi developed a silver staining technique which made the detailed structure of nerve cells visible for the first time. The isolation of insulin and its application to the treatment of diabetes mellitus was achieved in 1922 by Banting, Best and Macleod. Landsteiner in 1930 developed a nomenclature for typing human blood into four groups: A, B, AB, and O. Heymans in 1938 showed how the rate of respiration is regulated by the chemistry of the blood leaving the heart. Kendall and Reichstein were able to produce cortisone in 1950 and in the same year Hench showed that cortisone could be used to treat a variety of diseases.

We have reached a point in history where it is beyond the scope of this text to present all of the challenging new concepts and discoveries. It will be the responsibility of today's student of science to keep himself informed of current advances. He should remember that the scientific discoveries of the past become the scientists's tools for future enlightenment.

Scientific Investigation

Historically, experimentation and scientific investigation had rather crude beginnings because of superstition, supposition, and acceptance of authoritarian ideas. Scientists have gradually thrown off these hindrances and through the years have developed an impressive array of methods and experimental techniques for objective investigation. The conscientious amassing of observational and experimental data, the correlating and tentative interpretation of this data, and its continual reinterpretation in the light of new evidence, these are basic activities of the research scientist looking for explanations of natural phenomena. The tentative explanations are called *hypotheses*. If testing shows them to be accurate they become incorporated into the body of scientific knowledge and they may form theories or parts of theories. The research scientist also has an obligation to publish his findings. Throughout the world other scientists can then incorporate these scientific facts into their research and thus make the growth of scientific knowledge possible on a world wide basis.

The Human Body

Before beginning a study of the details of the body's structure and function, it is well to consider the body briefly as a whole and learn certain terms that will be used from time to time throughout this text.

The basic structural unit of the human body is the *cell,* just as the structural unit of a brick wall, is a brick. Even though the body is made up of a variety of tissues and organs, the basic component of these rather complex structures is the cell. The cell is also the unit of function and since the activities of the cell rely upon its living content, *protoplasm,* this substance is considered the basic functional material of all living organisms. Both the structure and function of the cell will be considered in more detail in Chapter 3.

In the very early stages of embryonic development, cells begin to show changes in appearance; to *differentiate.* As this differentiation proceeds, groups of cells having a common origin become specialized for certain physi-ological functions. They also develop structural characteristics that are peculiar to their group. Cells with a common origin, appearance, and function form the various *tissues* of the body.

When two or more differing tissues are organized to perform a specific function, they constitute an *organ.* Each individual muscle of the body, for example, is considered an organ since each of the primary types of tissue are organized into a structure performing a highly specialized activity.

A *system* is a group of organs which serve a common functional cause. There is a division of labor among the organs resulting in independent activities, which however are all interrelated and collectively result in an integrated whole.

Man, like all of the vertebrates (animals with backbones) has a type of body symmetry that is known as *bilateral.* This arrangement is determined by dividing the body along a *midsagittal plane,* that is lengthwise from the middle of the skull in such a manner that it will bisect the vertebral column and the breast bone. By viewing such a division in a mirror, we see that each half of the body is essentially like the other. An example of this phenomenon is found in the shape of the two hands. Their general form is such that a left hand resembles a right hand when its reflection is viewed in a mirror. Internally, of course, the plane does not divide the body into exactly equal halves because the various organ systems have developed different growth patterns.

If a vertical plane is drawn through the human body at right angles to the midsagittal plane, the body is divided into *anterior* (ventral) and *posterior* (dorsal) portions along a *frontal* or *coronal plane*. Finally, if the body is divided horizontally, in other words at right angles to both the sagittal and frontal planes, it is then divided into upper and lower portions along a *transverse plane*.

When the human race began to walk upright, the relative position of the body surfaces changed. In a four-footed animal, the upper surface along which the backbone lies is called the dorsal side, while that along the belly of the animal is the ventral side. In man, the dorsal side has therefore become the posterior area, and the former ventral side, the anterior portion of the body. Since these terms are sometimes used interchangeably, their original use should be kept in mind.

The main axis of the body consists of three principal regions, the *head, neck,* and *trunk*. Attached to the axis are the extremities, which make up the appendages. In Fig. 1-4 are indicated the two principal parts of the trunk. The upper part is the *thorax*, a region that is usually spoken of as the chest. Its lower limit is marked by the course of the lowest rib that can be easily felt. Below the thoracic cavity lies the *abdomen*. Within its cavity lie

Fig. 1-4 Front and side views of the human body to show planes of references (red), general regions, and the orientation of parts.

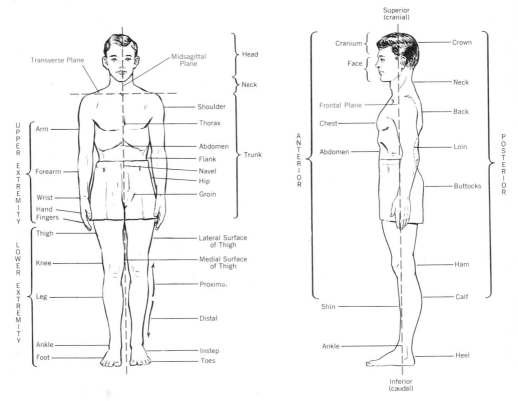

the various organ systems shown in the Trans-Vision insert following page 212.

Several terms are applied to an appendage to describe the position of its parts in relation to each other and to the main axis of the body. The *proximal* end of a limb is the region that lies toward the main part of the body. Thus, one can say that the upper arm is proximal to the forearm, or the shoulder proximal to the arm. Likewise, the region that lies toward the free end is spoken of as being *distal* to other parts that are nearer the point of attachment. In this case, the hand is said to be distal to the forearm.

Two other terms are used in connection with appendages. The surface that is toward the side of the body is known as the *lateral surface* to distinguish it from the *medial* (mesial) *surface* that lies toward the midsagittal line of the body. If the *palmar surface* of the hand is held upward as in the motion of asking for something, the hand is then in the *supine position*. The same term is used to describe the position of the body when lying on its back. If, however, the hand is extended with the palm facing downward, it is in the *prone position*, as is the body when lying on the anterior surface.

All of these terms will be found very useful in the study of physiology. One of the most important aspects of science training is exactness in describing the location of a structure or the nature of a process, hence a specialized vocabulary is necessary.

VOCABULARY REVIEW

Match the phrase in the left column with the correct word in the right column. *Do not write in this book.*

1	The person who first demonstrated that blood circulated in a closed system.	a	anatomy
		b	Bernard
2	The science dealing primarily with the thought processes of the mind.	c	bilateral symmetry
		d	cell
3	A body form in which the right and left sides show a mirror image of each other.	e	Galen
		f	genetics
4	A preliminary supposition requiring verification.	g	Harvey
5	The "structural unit" of all living things.	h	hypothesis
6	Sciences concerned mainly with form and structure of living organisms are grouped under this heading.	i	morphology
		j	pathology
		k	psychology
7	One of the great experimental physiologists of the 19th century.	l	psychosomatic
		m	tissue
8	A group of cells which have a common origin and function.	n	Paracelsus
		o	organ
9	The study of disease.		
10	The study of inheritance of characteristics.		

11 Probably the oldest of all biological sciences concerned with the normal animal.

12 An emotional influence resulting in physiological and anatomical changes.

14 An early physician, anatomist and physiologist whose influence lasted for more than one thousand years.

TEST YOUR KNOWLEDGE

Group A

Select the correct statement. *Do not write in this book.*

1 The type of symmetry shown by the human body is (a) radial (b) bilateral (c) asymmetric (d) unilateral.

2 The chest region of the human body is called the (a) abdomen (b) thorax (c) pelvis (d) cervix.

3 The study of anatomy is most closely linked with the science of (a) chemistry (b) histology (c) physiology (d) sociology.

4 The science primarily concerned with the study of disease is (a) embryology (b) histology (c) physiology (d) pathology.

5 Physiology is mainly concerned with (a) growth of the body (b) activity of the body (c) structure of the body (d) diseases of the body.

6 The term describing a person lying on his back is (a) prone (b) supine (c) palmar (d) lateral.

7 A branch of the biological sciences that is primarily functional is (a) anatomy (b) morphology (c) pathology (d) histology.

8 The end of a limb that lies toward the main part of the body is called (a) distal (b) proximal (c) dorsal (d) ventral.

9 The surface of an appendage that is toward the outside of the body is said to be (a) medial (b) lateral (c) anterior (d) posterior.

10 Parts of the body toward the head end are said to be (a) posterior (b) anterior (c) dorsal (d) ventral.

Group B

1 Describe how physiology is related to the various other sciences, both physical and biological.

2 Who were some of the great scientists who contributed to our present-day knowledge of physiology? What was the great achievement of each?

3 Briefly describe each of the following planes: sagittal, midsagittal, frontal, and transverse.

4 List the primary organ or organs found in each of the following body cavities: cranial (head), thoracic, abdominal.

Chapter 2   Energy and Matter

Energy

Living material is in a constant state of activity. This fact is readily seen in the mechanical action of walking, the electrical phenomenon of stimulated nerve endings, in the osmotic work of the kidneys or sweat glands, in chemical activity which causes changes in body temperature. In all activity, *energy* is involved.

Energy is a fundamental component of our universe and may be defined in one way as *the capacity for doing work or causing a chemical or physical change*. It is identified by what it does or what it is capable of doing. If one considers an object's capacity to cause a change because of its position or form as one kind of energy, called *potential energy,* then the resulting change, involving motion of some sort, is energy of a *kinetic* form. The former is often considered stored or bound energy as seen in various chemicals; the latter is seen in the activity of electrical current, heat, light, a chemical reaction, or the motion of an object.

Regardless of the form energy takes, it cannot be created nor destroyed. However, it can be converted from one form of energy to another. This is the *Law of Conservation of Energy*. When you weigh yourself on a set of bathroom scales, you are transferring energy from one form to another many times. The potential energy of your muscles is transformed into kinetic energy as you step upon the scales. The weight of your body pressing down upon the platform causes tension on the resisting spring, and you read your weight in pounds on the dial. This represents yet another transfer of energy from kinetic to potential. The coiled spring now begins to uncoil as you step off and the dial moves back to zero. Again potential energy has been changed into energy of motion.

Similarly radiant energy of the sun is converted by plants to the potential energy of complex chemical compounds which form our food. Some of the chemical potential energy of the food we eat is converted to kinetic energy of muscular activity for locomotion or movement of our internal organs. It may be changed back to potential chemical energy in the form of foods stored within our bodies or to potential mechanical energy, as when we climb a flight of stairs or a ladder.

Matter

The other essential component of our universe is *matter*. It may be defined as *anything that has mass and occupies space*. Matter can be identified in one of three forms: *gas, liquid,* or *solid.* If you heat water to the boiling point (100°C at sea level) the continued heating causes the production of water vapor (a gas). Water remains in a fluid state (a liquid), between 100°C and 0°C and below 0°C water turns to ice (a solid). In each instance, the total *mass* or quantity of matter remains the same. This principle is known as the *Law of Conservation of Matter.* In ordinary chemical activity, the quantity of matter present before and after a reaction or change remains the same.

At the present time it is known that matter and energy are related. It is also known that energy can be transformed into matter and matter converted to energy, but only under special conditions. This fact has been established during the last 30 years through nuclear reactions and a knowledge of the fusion taking place in the sun. It is now believed that matter and energy, rather than being separate physical quantities, are the same physical quantity in different forms.

Structure of the Atom

The basic physical unit of structure of all matter is the *atom.* Though individual atoms are exceedingly small, they are quite complex. In recent years many component parts of an atom, known as *subatomic particles,* have been discovered and identified. Three of these particles are of interest to us in our brief consideration of the chemical composition of all living things.

In the center of the atom is the part known as the *nucleus.* This very small dense center is usually composed of two different particles: positively charged *protons* that have a definite mass, and varying numbers of *neutrons,* which are electrically neutral but have approximately the same mass as the proton. The *atomic number* of an atom refers to the number of protons, in the nucleus. This is one way scientists can identify different atoms. Atoms identified in this way have names and are called *elements.* If the nucleus contains one proton, we know it is an atom of the element hydrogen. If there are two protons, it is an atom of helium. For each additional proton, a new element is formed, each with its own name and structure, until we reach an atom containing 103 protons in its nucleus, and this element is known as lawrencium. 103 is the total number of different elements presently known to man. Each is assigned a symbol, the scientists' shorthand, to indicate one atom of the element. Usually the first letter or the first two letters of the element's name is used as the symbol; thus, H stands for one atom of hydrogen, He stands for one atom of helium, C for carbon, and O for oxygen.

Negatively charged particles, that for all practical purposes shall be considered massless, are called *electrons.* They move about the nucleus in a number of fairly definite regions called *shells* or *energy levels.* The electrons orbit around the nucleus of the atom at tremendous speeds and nuclear scientists describe the atom as having a *spherical electron cloud* appearance as seen in Fig. 2-1. Note that in the illustrated examples of different atoms, the

Hydrogen   Helium   Lithium

Fig. 2-1 *Examples of atomic structure. A hydrogen atom has one proton in its nucleus; one electron moves about this nucleus. A helium atom has a nucleus consisting of two protons and two neutrons; two electrons move about this nucleus. A lithium atom has a nucleus consisting of three protons and four neutrons; two electrons occupy the first shell and one is in the second shell.*

total number of electrons equals the total number of protons in the nucleus. Atoms are, thus, electrically neutral.

The chemical activity or reaction of one atom with another atom is determined by the number and arrangement of electrons in the outer shell. The nucleus is not involved in this. The atoms of most of the 90 different kinds of naturally occurring elements, and of the 13 which have been artificially made, have incomplete outer shells. Such atoms are unstable and will try to share electrons with other atoms so that each can attain greater stability. This sharing establishes a *chemical bond* between the two atoms. Because of the energy present within each atom, changes in these bonds during chemical reactions will either absorb or expend energy. Thus, all of the functions of life involve a transfer of energy arising from chemical reaction—the metabolism of the living organism.

Isotopes. The same element may exist in several different forms, called *isotopes.* In these the number of protons and electrons is always equal and the same, but the number of neutrons varies. An example of this is seen in the atomic structure of hydrogen. The common hydrogen atom has one proton in its nucleus and one electron in its single shell. One rare isotope of hydrogen, called heavy hydrogen or deuterium, has a neutron and a proton in the nucleus. Another, tritium, has two neutrons and one proton (Fig. 2-2).

Since elements have different numbers of isotopes, a system of relative

Fig. 2-2 *Isotopes of hydrogen. Each isotope has one proton in the nucleus and one orbiting electron. Hydrogen has no neutrons in its nucleus, deuterium has one and tritium two.*

Hydrogen   Deuterium   Tritium

weights has been devised. This scale or the *atomic weight* of an atom, utilizes the average weight of the atoms of the naturally occurring isotopes of an element compared to that isotope of carbon having exactly a weight of $12(C^{12})$. In this way scientists have a standard for comparison of various kinds of atoms.

In some heavy atoms the nuclei are unstable. These atoms break down spontaneously to produce atomic particles of varying sizes. They also release penetrating rays at the same time. Such atoms are said to be naturally radioactive. It is possible to make some atoms artificially radioactive by bombarding them with atomic particles from nuclear reactors, and by the use of other complex machines, such as the cyclotron. The Oak Ridge and Brookhaven Laboratories produce a radioactive isotope of iodine with atomic weight of $131(I^{131})$, for use in the treatment of diseases of the thyroid gland, and the calcium isotope (Ca^{45}) for the study of bone and muscle activity. A naturally occurring isotope of carbon (C^{14}) is now being widely used to determine the age of ancient organic materials such as bones and wood.

Ions and molecules. When an atom gains or loses electrons, it becomes negatively or positively charged

Fig. 2-3   *Two types of apparatus that can be used to decompose water into hydrogen and oxygen.*

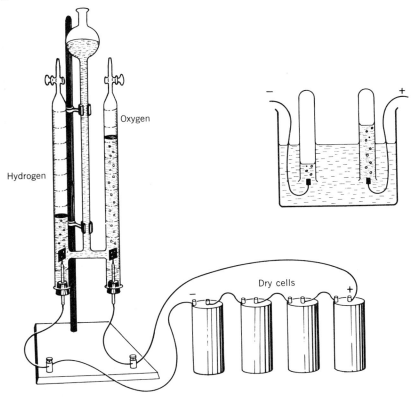

and is known as an *ion*. For example, table salt (sodium chloride) is composed of one atom of sodium combined chemically with one atom of chlorine. In this combination the sodium atom has lost one electron to form the sodium ion (Na$^+$), and the chlorine atom has gained one electron to form the chloride ion (Cl$^-$). If a direct current of electricity is passed through melted sodium chloride, the salt is decomposed into sodium and chlorine. The sodium ions are attracted to the negative electrode where they gain one electron each to convert them into sodium atoms. In the opposite way, chloride ions are converted to chlorine atoms as they lose electrons at the positive electrode. The decomposition of chemical compounds by electricity is known as *electrolysis*.

In a similar way, electrolysis can be used to decompose water (H$_2$O) into hydrogen atoms and oxygen atoms (Fig. 2-3). As the ions are discharged at the electrodes, the hydrogen atoms become bonded with each other to form hydrogen gas (H$_2$) and in the same way the oxygen atoms bond to form oxygen gas (O$_2$) (see Fig. 2-4). Since water is composed of two atoms of hydrogen for every one of oxygen, two volumes of hydrogen gas are evolved for every one volume of oxygen.

Fig. 2-4 Diagram showing how two atoms of oxygen sharing electrons form a molecule of oxygen.

Fig. 2-5 Diagram of a molecule of water formed by the hydrogen and oxygen atoms sharing electrons. This is an example of a chemical compound.

A *molecule* is formed when two or more atoms unite by sharing or transferring electrons. Molecules are electrically neutral. Thus, in the electrolysis of water, all of the gas appearing at one electrode is composed of molecules of the element hydrogen, while that appearing at the other is composed of molecules of the element oxygen.

Chemical compounds. The water from which the hydrogen and the oxygen were obtained is an example of a chemical compound. In a compound, the atoms are held together by their electrical charges (Fig. 2-5) and cannot be separated by means that do not involve an expenditure of energy. The hydrogen and oxygen in water will not separate on standing; electrical energy is required in this case to separate the bonds that hold these elements together.

A second characteristic of a compound is that the number and relationship of the atoms that compose its molecules must always be the same. In a molecule of water, there are always twice as many hydrogen atoms as oxygen atoms, so that the *molecular formula* for water is H$_2$O. Under certain conditions a compound of hydrogen and oxygen can be made in which the number of oxygen atoms in one molecule equals the number of hydrogen atoms. This is a very unstable liquid, hydrogen peroxide (H_2O_2), which has properties quite different from those of water. Therefore, a change in the relative number of atoms in a compound produces a new compound.

The relative position of the atoms within a molecule of a compound will affect the properties of that compound. This is well shown by two simple sugars that play important parts in our body's chemistry. Glucose, a very common sugar manufactured by green plants, is composed of 6 atoms of carbon, 12 of hydrogen, and 6 of oxygen ($C_6H_{12}O_6$). Within its molecule, the atoms are arranged as in Fig. 2-6, left. If these same atoms are arranged in a slightly different manner, another sugar is formed that has properties quite different from glucose. This is fructose,

the molecular formula for which may also be written as $C_6H_{12}O_6$. Since these two sugars are so nearly alike, ordinary chemical tests will not distinguish between them. However, if each is exposed to a beam of polarized light, the glucose will rotate the light to the right and the fructose to the left. Glucose is sometimes called dextrose (Latin, *dexter*, right) and fructose is known as levulose (Latin, *laevus*, left).

A fourth characteristic of a compound is that its properties are quite different from those of the substances that produce it. Water is a liquid composed of two elements that appear as gases in nature. Similarly, common table salt has properties quite different from those of the sodium and chlorine which compose it. Sodium is a soft, silvery, highly active metal that will react violently with water. Chlorine, on the other hand, is a yellow gas which

Fig. 2-6   Structural formulas for glucose (left) and fructose (right). Above are the open-chain forms and below are the ring forms. Note that although both compounds have the same general formula ($C_6H_{12}O_6$) the arrangement of the atoms in the molecules is different.

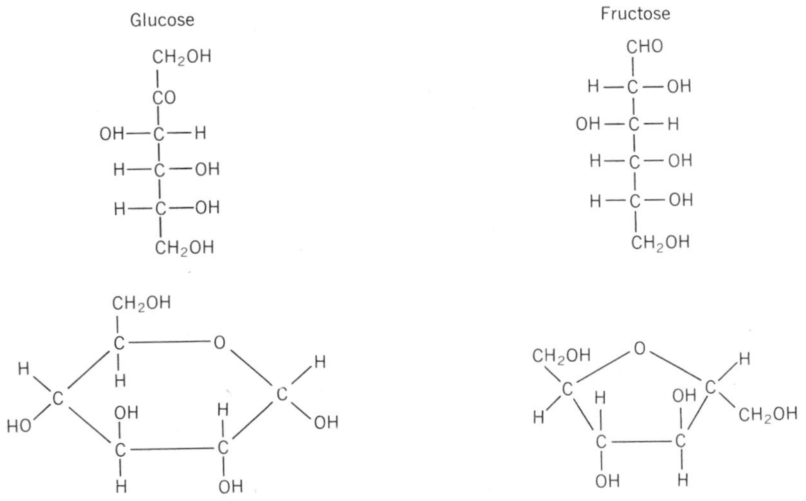

was used during World War I as a poison gas. However, if these two elements are chemically combined by transfer of electrons, the resulting compound is table salt, a white crystalline substance which is easily soluble in water and is essential for all animal life.

Suspensions

Colloidal suspensions. Living matter is vastly complex in its composition and structure. Its basic nature is that of a mixture of extremely minute particles of solid matter suspended in a liquid. This is known as a *colloidal suspension* (Greek, *kollos,* glue). The

particles may range in size from 0.000001 millimeter to 0.0002 millimeter. These size limits are a rather arbitrary way of defining a colloidal suspension because larger particles tend to settle out from the liquid in which they were suspended, whereas colloidal suspensions rarely settle out. The colloidal particles are sometimes single huge molecules, as in the case of some of the proteins, while at other times they represent groups of molecules that stick to each other in clumps.

If a colloidal suspension is viewed under the ordinary microscope or is seen in direct light, it appears to be

Fig. 2-7 Tyndall effect. A pair of battery jars one containing a suspension of gelatin and the other containing water. When these liquids are viewed under ordinary light (above) they appear similar in transparency and color but when a spotlight is shone through them (below) its path shows up clearly only in the colloidal suspension. (Fundamental Photographs)

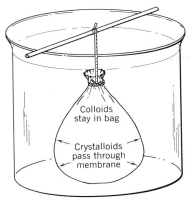

Fig. 2-8 Apparatus for demonstrating dialysis. The colloidal particles are too large to pass through the membrane.

transparent. But if a beam of light is passed through it at right angles to the viewing angle, it will appear turbid because the particles scatter the light. This is called the Tyndall effect (Fig. 2-7). The size of the particles in a colloidal suspension also prevents them from passing easily through a parchment or an animal membrane. Thus it is possible to separate a colloidal suspension from a true solution (solution of crystalloids) by a process known as *dialysis* (Fig. 2-8).

The relation between the suspension and the liquid in which the particles are suspended can change from time to time. A well-known example of this property is seen when gelatin is mixed with water. Gelatin, a simple type of protein, will form what appears to be a clear solution when it is heated with water. In this condition, the colloidal particles of the gelatin are in a greatly folded or contracted condition and are quite widely scattered among the water molecules. In this state it is called a *sol*. As the gelatin cools, the particles straighten out or expand to form a firm network that traps the water within its meshes and forms a *gel* (Fig. 2-9).

The physiological significance of colloidal particles lies in the fact that, being extremely small, they expose a tremendous amount of surface over which chemical reactions can occur. It has been estimated that if a one-centimeter cube of a substance were divided into particles of colloidal size, its total

Fig. 2-9 The colloidal state. Left: the colloidal particles are shrunken and folded among the molecules of water (small dots). Right: the particles have straightened out and now form a network that traps the water. As the arrow indicates, the sol state can be changed into the gel state, or the process can be reversed. (From Marsland, Principles of Modern Biology, Holt 1964)

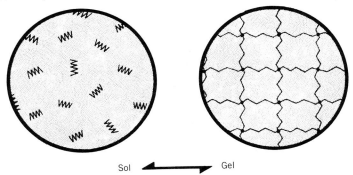

Sol ⬅➡ Gel

original surface area would be increased from 6 square centimeters to over 6 million square centimeters. Since living material is colloidal in nature, this means that within each cell there is a vast amount of surface area over which chemical reactions can occur.

Emulsions. An *emulsion* is a type of colloidal suspension in which a liquid is suspended in a liquid. In milk, for example, cream will separate out and rise to the top if the milk is allowed to stand. If the milk is homogenized, the fat is broken down into particles of colloidal dimensions that remain suspended in the rest of the liquid. In this latter state, milk is an emulsion. We are all familiar with what happens when we try to mix oil and water. The oil becomes divided into fine droplets if we shake it vigorously with the water, but soon the two will again separate into distinct layers. If soap is present, the oil remains suspended in the water and gives it a milky appearance. The soap surrounds the fat droplets and prevents them from recombining, thus forming an emulsion. The soap acts as an *emulsifying agent* (Fig. 2-10). In our discussion of the digestive system in Chapter 17 we will see how bile acts as an emulsifying agent in the digestion of fats.

Motion of Molecules

Matter, as we have seen, is composed of atoms and combinations of atoms (molecules). Atoms are in a constant state of motion, due to the movement of the electrons. This atomic motion is transmitted to the molecules so that there is a constant minute motion in all substances. It may be rather difficult to imagine this occuring in a bar of lead or a piece of gold—both very solid substances. But if a piece of pure gold is fused to the end of a bar of pure lead, atoms of gold will slowly migrate into the lead and atoms of lead into the gold. This movement can only be explained on the basis of the motion of molecules. In a solid the spaces between the molecules are infinitely small, and the distances permitted the

Fig. 2-10   The action of an emulsifying agent. In A, when an emulsion made by shaking oil and water together is allowed to stand, the oil particles tend to run together. In B, the emulsifying agent surrounds the particles and a stable emulsion is formed. (After Marsland, Principles of Modern Biology, *Holt 1964)*

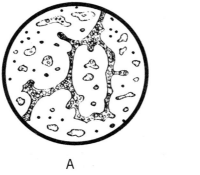

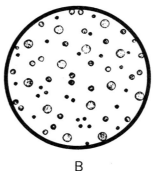

A   B

11 A positively or negatively charged atom.
12 Energy involving motion.
13 A subatomic particle that has no electrical charge and may be found in the nucleus of an atom.
14 The negatively charged particle of an atom.

TEST YOUR KNOWLEDGE

Group A

Select the correct statement. *Do not write in this book.*

1 Energy of motion is said to be (a) potential (b) kinetic (c) mechanical (d) dynamic.
2 Nearly all the weight of an atom is concentrated in its (a) protons (b) neutrons (c) electrons (d) nucleus.
3 The number of protons in an atom determines its (a) atomic weight (b) atomic number (c) atomic mass (d) atomic energy.
4 The chemical activity of an element is determined by its (a) nucleus (b) protons (c) electron shells (d) neutrons.
5 A chemical element that breaks down spontaneously is said to be (a) metallic (b) nonmetallic (c) radioactive (d) isotopic.
6 A substance made up of atoms that are all alike is a(n) (a) element (b) compound (c) molecule (d) isotope.
7 A colloidal suspension can be separated from a true solution by (a) the Tyndall effect (b) dialysis (c) evaporation (d) osmosis.
8 An example of an emulsion is (a) sugar solution (b) salt solution (c) homogenized milk (d) muddy water.
9 Diffusion occurs most rapidly in a (a) gas (b) liquid (c) solid (d) solution.
10 A fluid with pH greater than 7.5 is (a) blood (b) intestinal juice (c) gastric juice (d) hydrochloric acid.

Group B

1 Explain what is meant by the statement, "Living material is in a constant state of activity."
2 Explain the difference between potential energy and kinetic energy. Give examples of each type of energy in your answer.
3 Explain how matter and energy are related.
4 Briefly describe the structural form of an atom.
5 Briefly define each of the following: atom, molecule, compound, element, isotope, ion, mixture, and colloid.
6 Explain why two different molecules can have the same molecular formula but different structural formulas.
7 Explain the basic differences between diffusion and osmosis.

Chapter 3 The Cell

Those features which one usually associates with the condition of being alive are actually manifestations of the activities of protoplasm. The protoplasm of the body of any animal is always contained in unit packets, the *cells*. An adult human body of average size is estimated to contain about 26,500,000,000,000 cells. A protozoan, such as the ameba, has only one. In any case, the cell contains a quantity of protoplasm which carries on those various processes, the sum total of which is called the state of living. Thus we find that all the different parts of the human body are composed of cells and the products of cells.

Not only is protoplasm responsible for the functions of the various cells, but it is also the material from which the parts have been formed. In certain cells in the bodies of all animals, the protoplasm is replaced by nonliving material which either gives strength to the region or serves as a protective layer for underlying parts. In man, parts of the nails, hair, and skin are examples. It must be remembered, however, that in all cases these are modified cells that have once been living and filled with protoplasm; they are the products of

highly specialized action within the living cell.

The Cell Theory

The study of the minute structure of living matter was dependent on the invention of the compound microscope. It was not until this instrument had been developed to a point where it could be used for the examination of these parts that man was able to see the fine structures that made up the human body. This invention came during the seventeenth century. In 1665, Robert Hooke, an English scientist, cut very thin sections of cork and examined them under this new instrument. He saw that the sections were composed of numerous walled structures that he called "cells." Hooke's use of this term for what he saw differed from our use of the same term. What he actually saw were the dead cell walls of the cork and not the "stuff of life" which they had at one time contained.

During the next fifty years, many basic discoveries were made by such men as Marcello Malpighi, Jan Swammerdam, and Anton van Leeuwenhoek, who opened up the field of microscopic anatomy.

24

In 1839, a concept was presented by Matthias J. Schleiden, a botanist and Theodor Schwann, a zoologist, which summarized many ideas and observations of their predecessors. These two German scientists postulated that the cell is the basic unit of life, both in structure and function. This *cell theory* is one of the cornerstones of modern life science. For the next 100 years or so, great strides were made in our knowledge of these structural units. However, scientists were limited by the magnifying power of the microscopes then in existence.

Today, with the help of microdissection techniques, the electron microscope, and improved methods of chemical analysis, new facts are being discovered about the details of cell structure. Although much has been learned about these structural units during the 300 years that have elapsed since Hooke's time, there remain many unanswered questions.

The cells of the human body differ greatly in shape and size. Some are minute disk-shaped objects, like the red blood corpuscles which are so small that between 5 and $5\frac{1}{2}$ millions of the cells are contained in $\frac{1}{25}$ of a drop of blood. Others, the nerve cells, have a very irregular outline with exceedingly thin projections, several feet in length.

Fig. 3-1   A generalized cell (right) diagrammatically drawn from a number of electron micrographs. Some of the structures are also shown in the electron micrograph of part of a human lung cell (left). (Micrograph: Drs. K. A. Siegesmund and C. A. Fox, Marquette University School of Medicine)

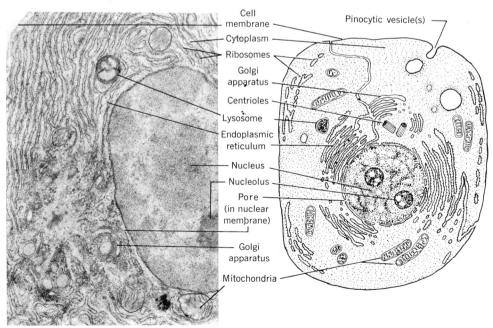

Cell membrane
Cytoplasm
Ribosomes
Golgi apparatus
Centrioles
Lysosome
Endoplasmic reticulum
Nucleus
Nucleolus
Pore (in nuclear membrane)
Golgi apparatus
Mitochondria
Pinocytic vesicle(s)

Still other types of cells are spindle-shaped with cross-markings that distinguish them as muscle cells attached to the bones of the skeleton. As we shall see in the next chapter, the shape and size of the cells are usually associated with the particular function they perform.

Not only do cells serve as "building blocks" for the body, but they also determine its various activities. This is because the functions of the body as a whole are the result of the combined activities of the individual cells. Furthermore, each cell contains the materials that control the inheritance of the individual, although, only those in reproductive cells actually determine the offspring's characteristics. To accomplish all these activities, the internal structure of a cell must be highly complex although superficially it may appear to be quite simple in structure. No cell of the human body, regardless of its apparent simplicity or complexity, is self-sufficient. A cell cannot supply its own needs without help from other cells. All of the body's millions of cells are highly specialized for specific functions; hence, they are not capable of meeting all of the problems that arise.

The diagram given in Fig. 3-1 may be considered a composite of a large number of electron micrographs of many cells. Few cells contain all of the structures shown here, but all are present in one or another type of cell.

Fig. 3-2 Electron micrograph showing the plasma membranes of two adjacent cells. Note that each membrane is composed of two protein layers (dark lines) between which lies a layer of lipid (light colored). The double membrane of each cell measures about 80 Ångstroms across. (Dr. J. D. Robertson, Harvard Medical School)

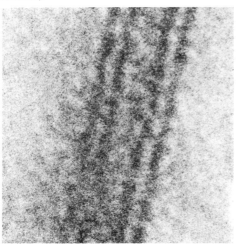

CELL STRUCTURE

Cell or Plasma Membrane

A membrane forms the outer covering of all animal cells (plants have a wall of cellulose outside this). Seen through a light microscope, it appears as an exceedingly thin boundary line to the cellular contents. But the nature of the membrane—its thickness and its true structure—requires the use of an electron microscope. See Fig. 3-2. Observing electron micrographs of this surface reveals two dense outer lines and a less dense inner area. Evidence from biochemical analysis suggests that the outer dark lines are protein molecules and the inner light area is composed of a double layer of lipid (fat-like) molecules composed of *phospholipids* and cholesterol. The total thickness of these layers of molecules measures between 75 and 100 Ångstroms (1 Å or Ångstrom unit equals 1×10^{-8} centimeter).

Because of the protein surface molecules, the membrane shows a remarkable flexibility. This feature is highly protective to the cell, which must withstand many bumps and abrasions throughout its life. Yet another feature of the plasma membrane that makes it unique is the method of formation. Although it is generally considered a living part of the cell, under certain conditions it behaves as though it were a product of chemical reactions that occur when the contents of the cell come in contact with the surrounding fluid. For example, if, during a microdissection experiment, a plasma membrane is torn and the contents allowed to escape, a new membrane will form if calcium ions are present in the surrounding fluid. If no such ions are present, the contents of the cell will continue to flow through the break. However, in the case of a small puncture of the membrane, the gap is healed spontaneously.

Cytoplasm

When viewing cells under the light microscope, everything inside the plasma membrane, except for the nucleus, appears as a slightly granular semifluid. This material is called *cytoplasm.* (*Cyte,* cell + *plasma,* something moulded—refers to a fluid substance). The electron microscope reveals that the cytoplasm is actually highly differentiated into various "little organs," the *organelles.* These structures are specialized for specific activities pertaining to the life of the cell and, in turn, the organism as a whole.

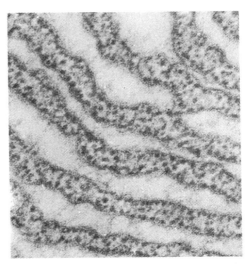

Fig. 3-3 Electron micrograph of a portion of cat pancreas cell. Note the parallel rows of endoplasmic reticulum and the ribosomes along the inner walls. (Drs. K. A. Siegesmund and C. A. Fox, Marquette University School of Medicine)

At present, we are concerned with their structure, location, and composition. Later on in this chapter we will consider in more detail their function and chemical activity.

Endoplasmic reticulum. In Fig. 3-1 and in Fig. 3-3, numerous clear membranous tubes or small canals, running parallel to the external surface of the cell, occur throughout the cytoplasm. This is known as the *endoplasmic reticulum.* Some of these canals appear to form an intricate connecting link between the plasma membrane and the membrane that surrounds the nucleus of the cell.

Ribosomes. On the inner surface of some of the canals that form the endoplasmic reticulum, there are numerous small granules. See Fig. 3-3. These particles contain a large amount of an extremely important protein called *ribonucleic acid* (RNA), and are consequently known as *ribosomes*. In addition to their RNA content, ribosomes possess certain chemicals which aid in the production or synthesis of various kinds of proteins necessary for cellular life.

Mitochondria. *Mitochondria* number among the few cytoplasmic organelles visible in the light microscope. They are small rod-shaped granules in the cytoplasm. Fig. 3-4 shows a diagramatic cutaway drawing of an electron micrograph of one mictochondrion. It appears as a fluid-filled vesicle, bounded by a double membrane. The inner membrane displays numerous infoldings called *cristae*. Both membranes consist of alternating layers of protein molecules and phospholipids. Mitochondria have often been called the "power plants" of the cell, since they are responsible for cellular respiration during which energy is released for cellular activities. Fig. 3-5 shows two mitochondria associated with the endoplasmic reticulum and the nucleus of a cell.

Lysosomes. These small bodies may be concerned with digestive activities in the cell (intracellular digestion). Their shape and size are determined largely by the type of material they contain.

Golgi apparatus. A network of minute cytoplasmic structures seen as groups of parallel, double-layered membranes is often found near the nucleus of the cell. These membranes comprise the *Golgi body* or *apparatus*. See Fig. 3-6. They often appear to be associated with droplets and with the endoplasmic

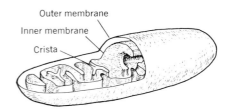

Fig. 3-4   Structure of a mitochondrion.

reticulum. There is some evidence that they may be related to the secretory activities of the cell.

The centrosome. This structure lies quite close to the nucleus and Golgi apparatus. It consists of two units, the *centrioles,* which are seen as a pair of

Fig. 3-5   Electron micrograph of a cat liver cell showing two mitochondria (upper center) in relation to the endoplasmic reticulum and nucleus. (Drs. K. A. Siegesmund and C. A. Fox, Marquette University School of Medicine)

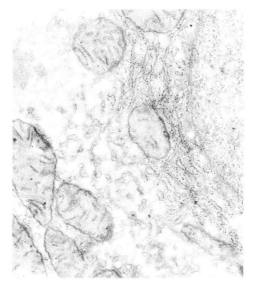

cylinders situated at right angles to each other. The wall of each cylinder is composed of nine groups of three fibers each. As we will find later in this chapter the centrioles play an active part in the division of the cell.

Vacuoles. Just as solids may be stored in the cytoplasm, liquids may also be present as minute droplets of fluid within vacuoles. Each is surrounded by a very thin membrane, the vacuolar membrane, which separates it from the rest of the cytoplasm. Some of the liquids are stored material, such as fat, others are manufactured substances which the cell may liberate through the cell membrane for use elsewhere and still others are waste materials.

Fig. 3-6 Electron micrograph of Golgi apparatus in a rat liver cell. Note the characteristic parallel nature of the layers of the double membranes (lower center). (Carlo Bruni M.D.)

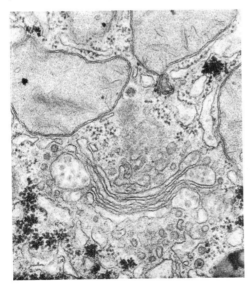

The cytoplasmic organelles just described are found in most cells but it should be remembered that some normal cells may lack one or more of these structures.

The Nucleus

The *nucleus* of a cell functions in two important ways. It controls and regulates the metabolic activities of the cell as a whole, and it plays a leading role in the processes of cell division and heredity. The protoplasm within the nucleus is called *nucleoplasm*. A nuclear membrane separates it from the cytoplasm. This thin double-layered membrane behaves in a manner similar to that of the plasma membrane. The nuclear membrane has small openings or pores (Fig. 3-7) which open to the cytoplasm or make connections with canals of the endoplasmic reticulum.

Chromatin is a material capable of absorbing a variety of basic dyes, hence its name which is taken from the Greek word *chroma* meaning "color." From a chemical standpoint chromatin is composed of a type of protein known as a *nucleoprotein*. Nucleoproteins are combinations of a protein and a nucleic acid. The nucleic acid typical of all living cells is *deoxyribonucleic acid* (DNA). We shall have more to say about DNA when we discuss the subject of heredity (Chapter 34).

When a cell is not actively dividing (reproducing), the chromatin is usually in the form of long threads so thin and interwoven that they give a netlike appearance to the contents of the nucleus. However, when a cell is in the process of dividing, this network becomes aggregated, forming thick, rodlike bodies, the *chromosomes.*

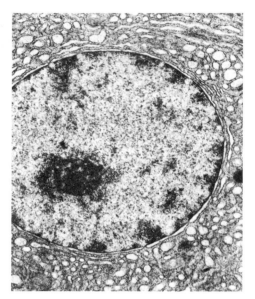

Fig. 3-7   Electron micrograph of the nucleus and surrounding structures of a cat pancreas cell. Note the nucleolus as a dark mass within the nucleus, the chromatin as more diffuse dark staining matter, and the pores in the double nuclear membrane. (Drs. K. A. Siegesmund and C. A. Fox, Marquette University School of Medicine)

Nucleolus. One or more *nucleoli* (Fig. 3-7) may be present in the nucleus of some cells while other cells seem to lack them. In fact, they may be present during some of the stages of the life of a cell and absent at other times, or they may divide into two or more pieces. During the process of cell division, the nucleolus disappears entirely, only to reappear after the new cells have been formed. The exact function of these bodies is not clearly understood, but they are rich in ribonucleic acid (RNA) and there seems to be increasing evidence that they are associated with the formation of proteins.

CELL FUNCTION

We have considered the parts of the cell as they comprise the structural unit of all living things. However, each cell carries on certain physiological activities which make it a functional unit as well.

Protoplasm

As mentioned earlier, the fundamental material of which all living organisms are composed is protoplasm. This highly complex material is in a constant state of change, and undergoes a continuous cycle of destruction and rebuilding which constitutes cell metabolism.

There are approximately twenty common different chemical elements present in the living material of the body. Identification of these elements is rather easy, but the part each plays in the living process is sometimes difficult to determine. The reason lies in the fact that the property of life depends on a very delicate balance between chemical compounds and reactions. These conditions are so easily upset that as soon as an attempt is made to analyze protoplasm, the material is so altered that it loses the living quality. When that occurs, we are no longer working with living material in which vital chemical reactions are taking place.

Many analyses have been made of protoplasm to determine its chemical contents and to measure the percentage by weight of each of its elements. These percentages will vary among different individuals, depending on what elements each is able to obtain from his environment, but they will approximate to those in the table.

Chemical Constituents of Protoplasm

Oxygen	65.00%
Carbon	18.30%
Hydrogen	10.00%
Nitrogen	2.65%
Calcium	1.40%
Phosphorus	0.80%
Potassium	0.30%
Sodium	0.30%
Chlorine	0.30%
Sulfur	0.20%
Magnesium	0.04%
Iron	0.004%
Iodine	0.00004%
Silicon, Zinc, Fluorine, Cobalt, Copper, Manganese, Bromine, Aluminum, and others	Minute traces

One must remember that these elements are not found in their free states in the protoplasm of cells. That is, one does not find pure carbon or hydrogen or potassium in the body. Instead, the elements are combined into compounds. The following table gives the approximate distribution of these:

Water	80%
Proteins	15%
Fats, oils, and other fatty substances	3%
Inorganic salts	1%
Other compounds including carbohydrates	1%

Protoplasm has a slightly grayish-green appearance when viewed under a microscope. Within the material can be seen small masses of a granular nature, which may be either cell structures or products of the cell's activities. The background material is colloidal in nature.

Protoplasm is the basis for all traits which distinguish a living from a non-living object. These characteristic activities will now be discussed.

Metabolism

Metabolism includes all of the chemical and physical processes that occur in the body—how the cells use different types of food, how they produce energy from these substances, and how they manufacture the many types of organic materials for which they are responsible. It consists of continuous reactions that go in two directions, but in each the transfer of energy is primary. In one direction, *catabolism,* large molecules are broken down into smaller ones with the liberation of energy. In reactions in the other direction, *anabolism,* small molecules are united to form larger ones and, in so doing, consume energy. Catabolism furnishes the energy necessary for all cellular activity. Some of the energy is used as heat and some is used to maintain the mechanical work of the cell but above all this energy activates anabolism. Anabolism utilizes energy in the production or synthesis of cellular products for growth, repair, and reproduction. In each of these instances the synthesis of protein is paramount.

The diagram in Fig. 3-8 summarizes the primary catabolic and anabolic activities involved in a protein cycle. Included in each step is the part each cytoplasmic organelle contributes toward the cycle and a brief explanation of what is taking place. It should be remembered, however, that although each step has its own function, it is really a part of a continuous process.

Without food no living organism can maintain life. In animal cells, food is the ultimate source of energy. Since cells cannot create matter, food is the only source of matter which can maintain or produce new protoplasm. In Step 1, Fig. 3-8, the principal organic substances are taken into the cell by *absorption*. This is the process by which food substances pass from the digestive cavity into the body proper or into cells. These organic substances are the end products of carbohydrate digestion (mainly glucose), fat digestion (glycerol and fatty acids), and of protein digestion (various amino acids).

Fig. 3-8   Chart showing how the different metabolic processes are related.

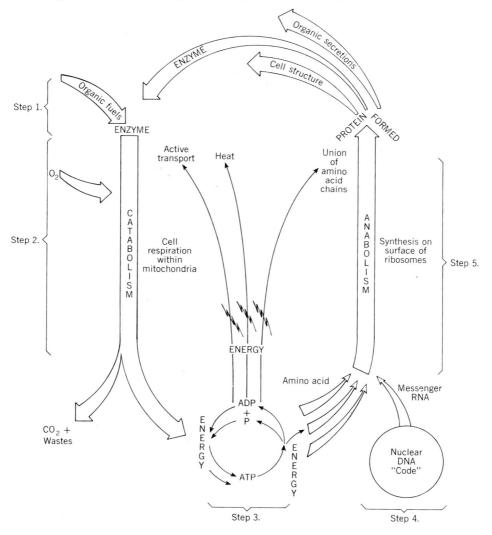

Most absorption takes place by the diffusion of materials through the plasma membrane. Large molecules, however, cannot pass through it but electron micrographs show that the plasma membrane forms an indentation containing these large molecules. It then seals off behind the molecules so that they are held in a vacuole within the cell. This process of indentation is called *pinocytosis* and is considered a type of *active transport* since energy is involved. An example of this process is seen in Fig. 3-9. In contrast, diffusion and osmosis are considered *passive transport* since no cellular energy is involved.

In the second step we are primarily concerned with the oxidation of organic molecules, mainly *glucose,* and the liberation of energy from chemical bonds. This is known as cellular respiration. While many steps are involved in the oxidation of glucose, it can be simplified in the following equation:

$$\underset{\substack{\text{glucose}}}{C_6H_{12}O_6} + \underset{\substack{\text{oxygen}}}{6O_2} \xrightarrow{\text{enzymes}} \underset{\substack{\text{carbon}\\\text{dioxide}}}{6CO_2} + \underset{\substack{\text{water}}}{6H_2O} + \text{Energy}$$

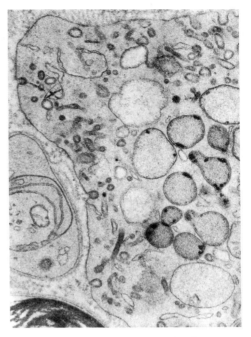

Fig. 3-9   Electron micrograph of an interstitial cell in the sciatic nerve of a mouse. Several pinocytotic vesicles can be seen as indentations in the portion of cell membrane running vertically just left of center of the micrograph. (Dr. J. D. Robertson, Harvard Medical School)

Note that in this form of oxidation, oxygen must be supplied and the end products are carbon dioxide and water. As these end products accumulate, they are removed as wastes.

The step-by-step breakdown of organic molecules is made possible only through the presence of enzymes. An *enzyme* is a type of protein classified as an *organic catalyst.* A catalyst is an agent which affects the speed of a chemical reaction without itself being permanently changed by the reaction. For each chemical activity that takes place within a cell, there is a specific enzyme. There are thousands of chemical reactions that occur within cells, so it is understandable when scientists say that each cell may contain several thousand different enzymes. In certain chemical reactions, enzymes require the assistance of other substances known as *coenzymes* to complete the chemical change. Many of the vitamins, especially the B-complex group, act as coenzymes in cell metabolism.

The energy resulting from this activity and which is available for cellular work is called *free energy.* The remaining energy which is not available

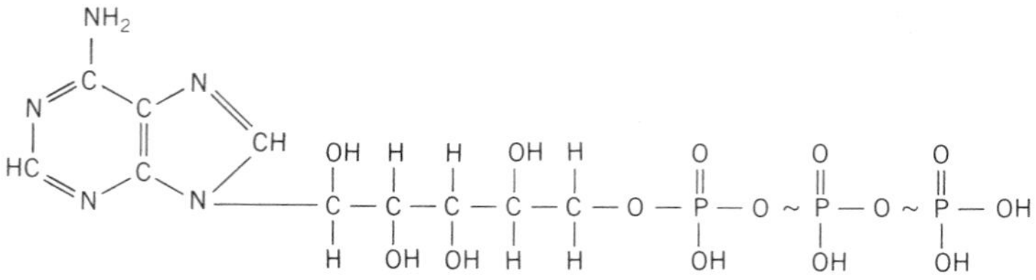

Fig. 3-10 One of the most important substances in all living cells is adenosine triphosphate (ATP). Note the high energy bonds which transmit all of the energy. This bond has a special symbol (~).

for cellular work is lost as heat. Approximately 50% of the energy resulting from the oxidation of glucose is conserved as free energy for the next step in the cycle.

Most of the activities mentioned in Step 2 of Fig. 3-8 occur in the mitochondria of the cell. Small particles located on the inner folds (cristae) are believed to be responsible for carrying out the oxidation reactions that supply energy.

Step 3 shows how the utilization of energy within living cells involves a remarkable molecule known as *adenosine triphosphate*—ATP for short. ATP is believed to be present in the cells of all living things. Its unique molecular structure makes this substance of prime importance in energy transfer. Two of the ATP molecule's three phosphate units are attached to adenosine by *high-energy phosphate bonds* (Fig. 3-10). When some part of the cell needs energy, it is immediately made available by splitting the outermost high energy bond. This removes one of the phosphate units, and adenosine triphosphate (ATP) becomes adenosine diphosphate (ADP). The free energy released is utilized by the cell. Thus, as ATP is used up by the energy requirements of the cell, ADP

accumulates. It now becomes necessary to re-attach the phosphate unit to ADP to reform ATP. The reversable reaction can be written as follows:

$$ADP + P + Energy \rightleftarrows ATP$$

The chemical energy required for this reformation comes from the breakdown of glucose, described in Step 2. Again, all of these activities take place in the mitochondria of the cell. It is no wonder that we call these minute cytoplasmic organelles the "power plants" of the cell.

The processes outlined above provide the energy necessary for all metabolic processes particularly the synthesis of protein. The type of protein must be determined by the activities in Step 4. Protein molecules are composed of long chains of various subunits called *amino acids*. About 25 different amino acids are known, and the sequence of these various amino acids determines the kind of protein. Since the length of each protein chain is variable, and the sequence and number of amino acids used is different for each protein, the possible number of different kinds of protein is tremendous. Some of the proteins are used for structural parts of the cell. Others make up the thousands of different enzymes which are used inside and outside the cell. Still other proteins are hormones—the products of various endocrine glands. It is quite obvious that, as seen from these few examples, proteins are indispensible for life. Since proteins are a product of cellular anabolism, how is their "blueprint" directed and controlled? The answer lies within the cell itself in the DNA. A more complete discussion of this substance is found in Chapter 34. It is believed that the basic instructions directing the production of proteins are coded in the DNA molecules found in the nucleus. There is a separate coded area on the DNA molecule for each kind of protein produced. Since the site of protein synthesis is on the ribosomes in the cytoplasm, somehow the instructions coded in DNA must be carried into the cyto-

plasm to the ribosomes where the work of protein production shown in Step 5 goes on.

The messenger, called *ribonucleic acid* or RNA, which carries this coded information is a molecule quite similar to the DNA molecule. As explained in Chapter 34, the DNA molecule takes the shape of a ladder that is twisted to form a double helix. Scientists have found that it takes three rungs of the ladder, see Chapter 34, to specify a single amino acid unit in a protein molecule. The messenger RNA takes its pattern from the DNA and moves out of the nucleus to the surface of a ribosome located along the endoplasmic reticulum. There it forms a template for the synthesis of a protein molecule.

Fig. 3-11 Protein synthesis. A single strand of DNA forms a strand of messenger RNA having complimentary bases (left). Messenger RNA is deposited on a ribosome where it acts as a template and accepts specific amino acid molecules brought by transfer RNA (middle). As the amino acids are laid down on the template, a protein is synthesized (right). (Redrawn from Otto and Towle, Modern Biology, *Holt, Rinehart and Winston 1965)*

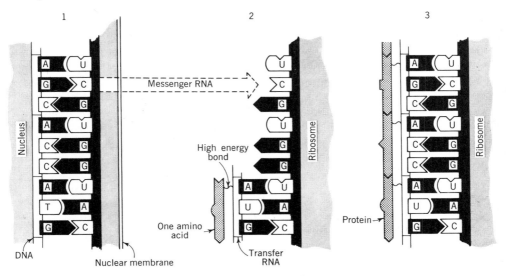

Yet another kind of RNA must bring the various amino acid units to the ribosome so that they will fit into the template that has been prepared for them. This material is called *transfer RNA*. It is thought that there are many transfer RNAs each containing a coded triplet and responsible for one specific amino acid (supplied originally by the absorption of proteins—see Step 1). The transfer RNA, carrying its amino acid, matches its counterpart in the RNA attached to the ribosome. As the RNA triplets fit into position, the amino acids join by chemical energy supplied by ATP and form the chain typical of protein molecules (Fig. 3-11).

The type of protein, originally determined by DNA, now goes to work for a specific function. Possibly it will become a part of a cytoplasmic organelle, or a specific enzyme or hormone, or some other part of the living organism. The cycle is complete when some of the proteins produced play their respective role in cellular respiration (Step 2).

Irritability

Another characteristic of protoplasm is its power to respond to environmental changes. A single-celled animal like the ameba is quite sensitive to differences in its surroundings although it lacks a definite nervous system or specialized organs to register these changes. This is an example of how protoplasm, without benefit of any highly differentiated parts, can respond to variations which may affect it either adversely or beneficially. An increase in size makes it necessary to develop highly specialized organs which can receive stimuli from outside

the body. Other parts must be able to transmit these messages to relatively distant regions, which set up a chain of events that constitute the reaction to the original stumulus. Thus, in the more advanced forms, certain cells of the body have developed to a high degree the function of irritability, and we find a nervous system appearing as an essential part of the animal's organization. Man's real claim to superiority lies in the fact that he has the most highly specialized nervous system found among animals. This permits him to respond to stimuli in a variety of ways and also gives him power to originate within his body other stimuli which appear as the thought patterns known as human intelligence.

Growth of Cells

The growth of a living body is dependent on two factors which occur in sequence: the increase in the volume of the individual cells and the increase in their numbers.

Individual cells grow in volume as a result of a process known as *assimilation*. During this process the cell takes in food and, under the influence of chemical materials produced by specialized parts of the cytoplasm, is able to form new living material. This increase in size occurs within the confines of the cell membrane and thus differs from the so-called "growth" of nonliving objects, such as icicles and stalactites, which increase by the addition of new material to the outside.

A result of this type of cell growth is that the volume increases more rapidly than the surface area. Thus, if a newly formed cubical cell measures 0.01 of a millimeter on each edge, growth may double its dimensions to 0.02 of a millimeter on each edge. This would mean that its volume would be increased eightfold, while the surface area had increased only four times. Since cells depend on the amount of food and oxygen they can absorb through their cell membranes, this difference between volume and surface makes continued growth impossible. If the cell grows larger, it will die because it cannot absorb enough of the necessary materials to supply its ever-increasing volume, nor can it get rid of all the wastes it produces. Cells, therefore, are limited in size by this surface area-volume relationship. When an individual cell reaches the maximum volume its surface area can support, it divides in half to restore the balance between area and volume.

The rate of normal division of individual cells is obviously a direct result of their rate of growth. In the human body, cells divide at varying rates of speed, depending on their functions. The result is that some parts of the body may grow very rapidly for a period of time and then slow down, while other parts grow slowly or not at all, once they have been formed. Good examples of how different kinds of cells grow at varying rates of speed are the cells of the middle and inner ear, and those in the outer layers of the skin. The tiny bones of the ear and the hearing apparatus have reached their maximum size some time before birth and do not grow after that time. The outer layers of the skin, however, are in a constant state of active growth because they are being constantly shed and replaced by new cells formed by the underlying skin layers.

Tissues are formed by cell divisions (Chapter 4), and when a sufficient number of cells have been formed to meet the requirements of the individual, they stop dividing. In the majority of tissues, even replacement of lost or injured cells is not possible once the "check" has come into play. Occasionally, this check disappears for some currently unknown reason, and some of the cells in a tissue may begin to divide rapidly and in an abnormal manner. When this occurs, these abnormal cells invade regions occupied by normal cells and upset their metabolic activities. The result is a tumor or a cancer. When a more complete understanding of the processes occuring in normal cells has been reached, the problems concerned with their abnormalities may possibly be solved. Basically, the control of cancer involves the riddle of cell growth and division and an understanding of why cells behave as they do.

Reproduction

The continuation of any species of living organism is dependent on its ability to produce others of a like nature. One-celled animals simply divide the material of the cell into two halves, each of which then develops into a new individual. The greater complexity of the higher animals has necessitated the development of specialized cells for this purpose. These cells contain all the information necessary for the development of a new individual and insure that the characteristics of the race shall be transmitted in an essentially unaltered form from one generation to the next. Thus, the inheritance pattern of the individual is closely associated with the process of *reproduction,* and among highly complex organisms this responsibility is shared by two sexes. Hereditary information from each sex is transmitted to the offspring so that new combinations of heritable characters occur.

Cell Division

The process of *cell division* results in the growth of the body as a whole, for it is by this means that the total number of cells is increased. As a general rule, those cells that grow slowly divide with a corresponding slowness, so that it may take days for them to go through the same process that more rapidly growing cells accomplish within twenty minutes.

The most common process by which cells divide is called *mitosis* (Greek, *mitos,* thread). During this process, the cell passes through a series of quite complicated stages which result in the formation of two new cells. Each of these daughter cells contains an amount of chromatin that equals that found in the other, not only in quantity but also in quality. Thus all the information encoded on the DNA is passed on to each new cell.

The process of mitosis is a continuous one and is usually of fairly short duration. For the convenience of study, the various stages through which the nucleus passes in its division have been grouped into five principal phases. It should be remembered, however, that although each phase has its own peculiarities, these are really parts of a continuous process. The five phases of mitosis (Fig. 3-12) are named: *interphase, prophase, metaphase, anaphase,* and *telophase.*

Fig. 3-12   Phases of mitosis as they occur in animal cells. (Redrawn from Otto and Towle, Modern Biology, *Holt, Rinehart and Winston, 1965)*

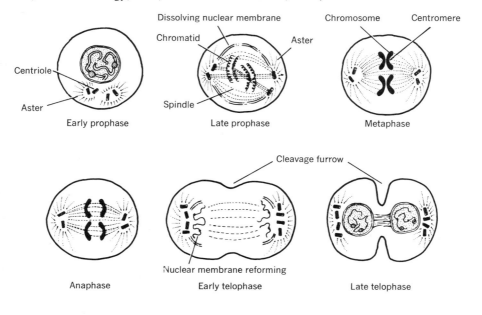

Interphase. This, strictly speaking, is not a stage in the process of mitosis but represents that growth period in the cell's life when it is not actively dividing. However, the nucleus, during the latter part of this period, is duplicating the DNA molecules in preparation for cell division. This self-duplication of the DNA molecules is referred to as *replication* and results in paired "daughter" chromosomes.

Prophase (Fig. 3-13). One of the first indications that a cell is about to divide is given by the *centrioles*. These, if you remember our discussion of the cytoplasmic organelles, are two short, rod-like cylinders. During the interphase these structures lie within the centrosome, but as the time of division approaches they replicate forming "daughter" pairs of centrioles. The two pairs of centrioles move slowly away from each other toward opposite poles of the cell. As the centrioles separate, lines of granules, the *spindle,* appear between them. Around each centriole similar fibers, *asters,* radiate into the cytoplasm. The exact function of these granules is in doubt, but that they are very definite structures can be demonstrated by microdissection. If a fine hooked needle is thrust into the cell, the spindle can be stretched and drawn out of position, indicating its fiberlike consistency.

Within the nucleus, changes occur in the chromatin. The netlike appearance of the chromatin is lost, and for the first time its true threadlike character is seen. The chromatin now appears as long double threads which are loosely coiled. These double threads become shorter, thicker, and appear as separate bodies, the chromosomes.

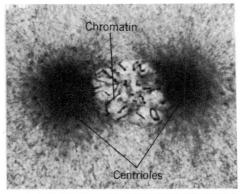

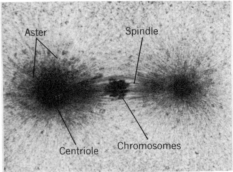

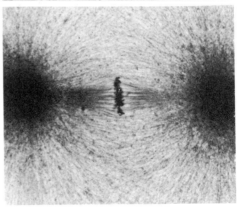

Fig. 3-13   Top, early prophase: the chromatin appears. The centrioles have moved to opposite poles of the cell. Middle, late prophase: the chromosomes and spindle appear. Bottom, metaphase: the chromosomes are arranged along the equator of the cell. (General Biological Supply House)

For many years the chromosomes were thought to be single structures, but recent advances in the techniques of staining cell parts show that they are

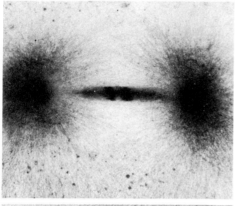

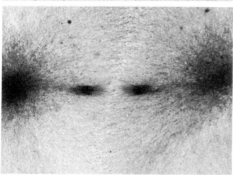

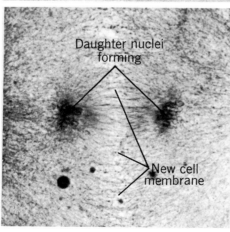

Daughter nuclei forming

New cell membrane

not. In fact, what appears to be a single rodlike body turns out to be a double thread surrounded by a sheath of material that is held to it by a very thin layer of protein. It is this surrounding substance that is stained by ordinary dyes and gives the impression that the chromosome is single. If special stains are used, the inner structures can be seen to consist of two daughter chromosomes, each one called a *chromatid*. The chromatids are joined to each other at a single point by a very small piece of chromatic substance, the *centromere*. This double nature of the chromosome explains how some of the later processes occurring in mitosis can take place. Both the nuclear membrane and the nucleolus disappear during the latter stages of this phase.

Metaphase. Metaphase is characterized by the separation of the chromatids. Up to this time the chromatids have been attached to each other by the centromere, but in this phase they break apart and the halves separate. Prior to their actual separation, under the influence of the spindle fibers, the chromosomes have become neatly arranged along the center (equator) of the cell.

Anaphase. Following their separation, the chromatids move towards opposite poles of the cell. Just how this

Fig. 3-14 Top, early anaphase: the halves of the chromosomes separate and move toward opposite poles of the cell. Middle, late anaphase: the chromosomes move farther apart and the spindle begins to break down. Bottom, telophase: a new cell wall appears and two distinct daughter nuclei are formed. (General Biological Supply House)

movement is accomplished is not clear, but the spindle fibers appear to attach to the centromere of the chromatids and contract, drawing the two members of the chromatid pair to opposite poles. As the halves separate, an indentation appears in the cell membrane at a point approximately in the middle of the cell. This indentation begins to divide the cytoplasm of the cell in half.

Telophase. The final stage of mitosis completes the division of the parent cell into two daughter cells. This is accomplished when the new cell membranes have completely divided the old cell into two new and separate cells. The chromatin material of each daughter cell then rapidly goes through a threadlike stage and eventually assumes the network appearance typical of the interphase. The nuclear membrane forms around the chromatin of each new cell, and nucleoli may reappear.

VOCABULARY REVIEW

Match the phrase in the left column with the correct word in the right column. *Do not write in this book.*

1 An organic catalyst.
2 A maze of small canals that appear to connect the surface of a cell to its nucleus.
3 The sum total of all chemical activity which takes place in a living organism.
4 The ability to respond to environmental stimuli.
5 That part of metabolism which results in the liberation of energy.
6 All the minute structures found in cytoplasm.
7 The basic "functional material" in all living organisms.
8 The outer covering of all animal cells.
9 That part of metabolism which uses energy in the construction of new protoplasm, etc.
10 A cytoplasmic organelle which is considered the "power plant" of the cell.
11 The oxidation of organic molecules with the liberation of chemically bound energy.
12 Networks of nucleoproteins found in the nucleus of non-dividing cells.
13 Small cavities or pockets found in the cytoplasm which may contain water, nutrients or waste materials.

a anabolism
b catabolism
c cellular respiration
d chromatin
e cytoplasmic organelles
f endoplasmic reticulum
g enzyme
h irritability
i metabolism
j cytoplasm
k mitochrondrion
l ribosomes
m plasma membrane
n protoplasm
o vacuoles

TEST YOUR KNOWLEDGE

Group A

Select the correct statement. *Do not write in this book.*

1   A scientist whose name is closely associated with the formulation of the cell theory is (a) Schwann (b) Hooke (c) Malpighi (d) van Leeuwenhoek.
2   The general name for all of the living material in a cell is (a) cytoplasm (b) protoplasm (c) nucleoplasm (d) endoplasm.
3   The nucleus of a cell contains (a) ribosomes (b) mitochondria (c) chromosomes (d) centrosomes.
4   An outer covering found on all cells is the (a) nuclear membrane (b) plasma membrane (c) permeable membrane (d) cell wall.
5   The cytoplasm of a cell normally contains (a) endoplasmic reticulum (b) cellulose (c) DNA (d) chromatin granules.
6   The most abundant chemical element in protoplasm is (a) nitrogen (b) carbon (c) hydrogen (d) oxygen.
7   The most abundant compounds in living cells are (a) proteins (b) fats (c) inorganic salts (d) carbohydrates.
8   During the interphase DNA is found mainly in the (a) nucleus (b) cytoplasm (c) endoplasm (d) nucleolus.
9   Oxidation reactions within a cell occur chiefly in the (a) mitochondria (b) centrioles (c) ribosomes (d) vacuoles.
10   Protein molecules are composed of long chains of (a) carbohydrates (b) ribonucleic acids (c) fatty acids (d) amino acids.

Group B

1   Briefly describe the historical significance of the "Cell Theory."
2   Outline the structure and function of the various parts that make up the cytoplasm and nucleus of a cell.
3   Briefly define each of the following: irritability, metabolism, growth of cells, reproduction.
4   Summarize the various steps in a "protein cycle."
5   What part do organic catalysts play in cellular activity? What are *coenzymes* and what is their function?
6   Describe the various stages the nucleus of a cell passes through while dividing by mitosis. Stress the part played by the chromatin.

Chapter 4 **Cell Specialization**

The very simplest animals have bodies composed of a single cell or, at best, a loose grouping of more or less similar cells. These few cells suffice for primitive forms of life. However, in those animals which are more complex in structure, the need for specialized parts to meet the demands of the larger body size becomes increasingly apparent. In human societies no single individual can be simultaneously a successful plumber, lawyer, artist, farmer, and merchant. Likewise, a group of similar cells cannot possible carry on all of the functions of a complex body. The work of society is conducted by groups of specialists, and the same is true of the work of the human body. *Division of labor* in the body is accomplished by the specialization of its cells, and of its various tissues and organs.

The cells which are the basic units of structure become adapted and changed for specific purposes and are grouped together to work with greater efficiency. Each of these groups of similar cells is called a *tissue*. When various tissues are combined to carry on a particular function or activity, the structure is called an *organ*. Organs which act together in the performance of some major vital role constitute an *organ system*.

TISSUES

The four primary tissues of the body are *epithelium* which protects surfaces and absorbs or secretes materials, *muscular tissue* which is highly contractile, *connective tissue* which provides support and holds parts together, and *nervous tissue* which is irritable and conducts impulses.

Epithelium

This is the general name given to all those tissues which cover the body surfaces, both externally and internally. The cells, found close together with little intercellular material, are arranged in one or more layers and serve as a protection against invasion by bacteria or as a buffer against mechanical injury. Some types of epithelium also produce materials useful to the body in localized areas, such as the *serous fluid* which lubricates adjoining surfaces. Other types form the secreting portion and the ducts of glands. As a rule, there are no blood vessels in epithelial tissues.

43

Nutrients and other required materials reach these and most other cells by diffusion through the fluid which fills the thin intercellular spaces. These tissues may be classified on the basis of the general appearance of the cells that comprise them.

Squamous epithelium. *Simple squamous epithelium* is composed of cells that are flat and slightly irregular in outline—almost scalelike (Fig. 4-1). All tissues composed of such cells are extremely thin, the average depth being approximately 0.0025 millimeters. Since they are so thin, they serve admirably as the lining of blood vessels and the heart. They are also present in the covering of the lungs (pleurae) and heart (pericardium). In the case of the coverings of the heart and lungs, the cells secrete fluid which reduces fric-

Fig. 4-1 Squamous epithelium from the skin of a frog. (Ward's Natural Science Establishment)

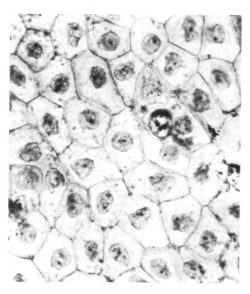

tion between the actively moving organ and its covering. This is the serous fluid which, in the case of the covering of the heart, is called the pericardial fluid and when found between the layers of the pleura, is known as the pleural fluid. The layer of cells forming the thin inner walls of all blood vessels is a specialized type of squamous epithelium called *endothelium*.

Endothelium. A type of squamous epithelium, highly specialized as to structure and function, is the tissue known as endothelium. The cells comprising it are flat and quite irregular in outline, resembling in shape the epithelial cells of the lining of the cheeks. An outstanding characteristic of these cells is that they are held together by a cementlike substance that changes in permeability as the occasion demands. The walls of the smallest blood vessels, the capillaries, are composed of endothelial cells. The diffusion of substances through the capillary walls to the surrounding tissues is thus regulated, since some part of the substances passes between the cells rather than through the cell membranes. The endothelial cells also line the walls of the arteries and veins, as well as the cavities between bones at the joints. This tissue will be discussed more fully later in connection with the circulatory system.

Stratified epithelium (Fig. 4-2). This is a highly complex type of epithelial tissue. Its primary function is the protection of some body surfaces that may be subject to mechanical injury. This tissue is composed of several layers of cells for added strength. The lowest layer is composed of cells that are tall and cylindrical in shape. Above these are irregularly shaped cells which are gradually transformed into flat cells that cover the surface. In the skin, for example, the outermost cells may eventually become nonliving because their protoplasm is replaced by a harder and more resistant material *(keratin)* that protects the more delicate lower layers. If we gently scrape the inside of the cheek with the blunt end of a tooth-pick and then examine the scrapings under a microscope, we will see the general form of some of these outer-most layers. The cells appear irregular in outline with gently folded surfaces. They lack the keratin that is present in the corresponding layers in the skin because they are not subjected to the same degree of mechanical abrasion.

Cuboidal epithelium. The simple cuboidal epithelial cells are somewhat heavier and stouter than the squamous type. The cells are not real cubes, as will be noted from Fig. 4-3, but the name has been applied to them because of their appearance when seen in a section cut through the middle of the cell.

Cuboidal epithelium is not a common type of tissue. The secreting parts of some glands, tubules in the kidneys, tissue covering the ovary are the best examples of this type of epithelium.

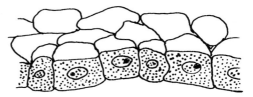

Fig. 4-3 Cuboidal epithelium.

Columnar epithelium. Simple columnar epithelium is composed of cells that are much taller than they are broad. They are packed close together to form a protective covering for the inner surface of an organ (Fig. 4-4). The lining of the stomach is composed of cells of this type. In some regions of the body these cells are modified so that they are able to produce materials (secretions) which aid in a particular body function. This is true of those found in the walls of the stomach, where many of the cells manufacture the various components of the gastric juice. In some tissues, cells of the columnar type have been modified to produce *mucus,* a lubricating material. These are the so-called *goblet cells* (Fig. 4-5).

Fig. 4-4 Columnar epithelium. Note the height of cells in this type of tissue.

Fig. 4-2 Stratified epithelium.

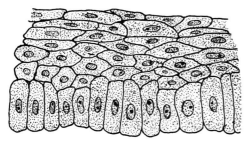

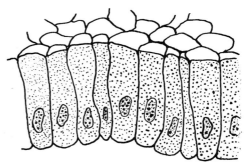

Ciliated epithelium. Fig. 4-5 also shows a group of ciliated epithelial cells. These cells are modified columnar epithelium that has the function of moving small particles of debris or individual cells along a surface or through a tube. The free end of each cell is equipped with numerous small hairlike projections, the *cilia.* By the continuous movement of the cilia, particles of dust and dirt are moved up the trachea (windpipe) from the lungs, or eggs and sperm are swept through the oviducts or seminal tubules from the ovary and testis respectively.

Muscle tissue

The movements of the body organs are due to the activities of muscle tissue. There are certain types of movement that are so slight that they are only visible with the aid of a microscope. In the ordinary sense, however, the movement of any part of the body, whether it is rapid or slow, depends on the type of cells that comprise the muscle tissue. Muscle tissue can be classified on the basis of the types of cells that compose it.

Striated, skeletal, or voluntary muscle. The majority of the striated muscles require conscious effort to make them contract. They are therefore often called *voluntary muscles.* If a piece of lean meat is carefully torn apart with needles so that its fibers are separated, it will be seen under the microscope that the cells making up the tissue are crossed by very delicate lines. Fig. 4-6 is typical of muscles that are attached to the skeleton of an animal. Each individual line is known as a *striation.* In human muscles, these cells may vary from 1 millimeter to over 40 millimeters in length and from 10 to 40 microns in width (1 micron = 0.001 millimeter). Muscle cells are relatively large and very active. They require several centers (nuclei) to regulate their activities. It is not unusual, therefore, to find cells containing twenty or more nuclei which lie in the protoplasm outside of the main body

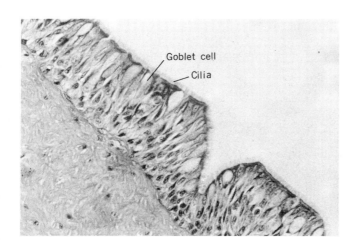

Goblet cell

Cilia

Fig. 4-5 Ciliated epithelium with goblet cells. (Ward's Natural Science Establishment)

of the cell but surrounded by a very thin membrane, the *sarcolemma*. Internally, the muscle cell is made up of many small fibers which lie parallel to each other and run lengthwise in the cell. These in turn are composed of still finer fibers, called *filaments*. When a muscle contracts, the filaments shorten due to chemical changes occurring in the proteins they contain. More will be said about these changes in Chapter 6.

Smooth muscle. This tissue is also called *involuntary muscle* since its action, unlike that of striated muscle, is not consciously controlled. Striated muscle cells may be over 40 millimeters in length, whereas the length of the smooth muscle cells is measured in thousandths of a millimeter. Their appearance also is quite different because they lack the cross-markings which are characteristic of striated muscles (Fig. 4-7). Another difference is that each smooth muscle has a single nucleus which lies in the approximate center of the cell. Smooth muscle cells are found in the walls of the organs of the digestive system. Their slow movements pass food along the canal during the process of digestion. They are also present in the walls of arteries and veins. Here their contraction and relaxation controls the flow of blood to various parts of the body.

Cardiac muscle. Cardiac muscle contracts rhythmically. The word "cardiac" always refers to the heart and it is only here that cardiac muscle tissue is found. The cells of the heart (Fig. 4-8) are unlike those of either smooth or skeletal muscles. They are greatly branched and join each other to form a protoplasmic network. The muscle

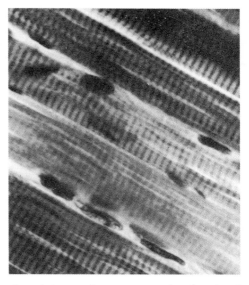

Fig. 4-6 A photomicrograph of striated muscle. Note the numerous nuclei. (Ward's Natural Science Establishment)

Fig. 4-7 Photomicrograph of smooth muscle. Note the tapering cells and the absence of striations. (General Biological Supply House)

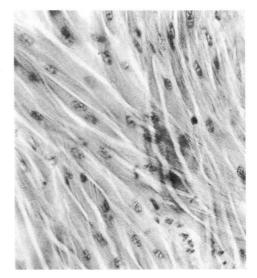

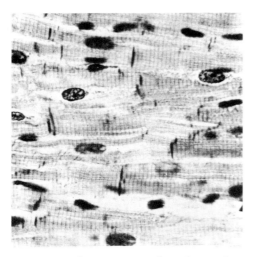

Fig. 4-8 Photomicrograph of cardiac muscle. Note interlacing cells and location of the nuclei. The disks are clearly shown. (General Biological Supply House)

Fig. 4-9 Photomicrograph of loose connective tissue from a mammal. Note the cobweb-like arrangement of white fibers (in bundles) and yellow fibers (single), the cell nuclei (black dots), and the large amount of intercellular space. (Carolina Biological Supply Co.)

cells show some striation, but it is not as distinct as that found on skeletal muscles. Also, disks are present which cross the muscle cells at more or less regular intervals. The nuclei are found in the middle of the cells. Functionally, the most important characteristic of this type of tissue is its ability to contract without being stimulated by a nerve-borne impulse. No other type of muscle tissue will do this normally. It is possible to remove the heart from a freshly killed animal and have it continue to beat rythmically when completely separated from the body.

Connective Tissue

The most widely distributed tissue throughout the human body is *connective tissue*. It supports and joins the various other tissues together. If, for example, the skin and subdermal tissues from the surface of a muscle are separated, it will be found that there is a sheet of tough thin cells that hold these two layers together. Under the microscope, this layer of connective tissue will show more intercellular material than cells. This intercellular substance is called *matrix* and may exist as a fluid, semi-fluid, gelatinous, or ridged material. You may find numerous nuclei, but the cell membranes may be indistinct. These cells may appear as disorganized threads without any apparent cellular structure. This is a primary feature which distinguishes connective tissue from the closely packed cells of most other tissue. Connective tissue is highly vascular. In addition to supporting and binding parts of the body together, it aids in the distribution of nutrients. Connective tissue may be classified as follows.

Loose connective tissue. This tissue is composed of relatively few inconspicuous cells in a semifluid matrix. Found scattered throughout the matrix are bundles of flexible but strong white fibers composed of a protein called *collagen.* Also present are single interlacing fibers of great elasticity. They are composed of the protein *elastin.* The spaces between the fibers is filled with tissue fluid, Fig. 4-9. Loose connective tissue is found in the epidermis of the skin and in the subcutaneous layer along with fat cells. It surrounds various organs, and supports the nerves of the body and the network of blood vessels that bring nutrients and carry away wastes from various structures.

Adipose tissue (Fig. 4-10). Fat tissue is quite widely distributed through the body. The cells that make up this type of tissue are relatively large and have a single vacuole containing a droplet of fat. Fat accounts for a sizable percentage of the total body weight; in males it is approximately 18 percent of the total weight and in females about 28 percent. Fat plays three important roles in the body. It serves as a reserve supply of energy-producing materials. It also serves as padding to absorb the jolts and jars to which the body is being constantly subjected. Fat tissue is deposited in quantities around the eyes, in the palms of the hands, in the soles of the feet, and between the joints. The third function of fat is that it serves as an insulator. Deposits of fat beneath the skin help to maintain the normal body temperature by preventing the loss of heat generated within the body tissues.

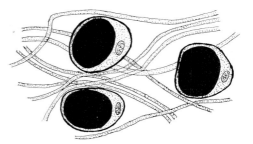

Fig. 4-10 *Adipose tissue. Three fat cells are shown among strands of connective tissue.*

Liquid or circulating tissue. All the cells of the body must be supplied with food and oxygen and have waste products removed from them. To accomplish this end there is a transportation system which contains blood; a liquid connective tissue.

The blood is composed of two distinct parts: a fluid portion, or *plasma,* which forms the matrix, and the solid elements, the *corpuscles.* The particular characteristics of each of these will be dealt with in later chapters. For the time being, let us note that the solid parts of the blood can be divided into three main types of cells. These are the red corpuscles (erythrocytes), the white corpuscles (leukocytes), and the platelets (thrombocytes), Fig. 4-11. We should also note that the fluid which is present in the cells, and which bathes the cells and tissues, is formed from the plasma.

Fibrous connective tissue. This is a dense tissue sometimes referred to as *white fibrous tissue* because of the closely packed white collagen fibers. These unyielding fibers run parallel giving strength to the length of the tissue. Although these bundles of fibers

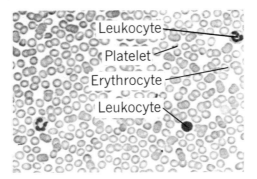

Fig. 4-11 *Photomicrograph showing human blood cells. This photomicrograph of a thin smear of blood shows several types of cells. When in the body, the cells float in plasma. (Carolina Biological Supply Co.)*

are flexible, they show little elasticity. Fibrous connective tissue forms the major part of some *ligaments* which hold bones firmly in place at the various joints. *Tendons* are also made of this tissue and serve to attach the skeletal muscles to bones of the body. Sheets of fibrous connective tissue cover the muscles and keep the muscle bundles in place.

Cartilage. Cartilage, commonly called gristle, is a flexible but firm material. In the early stages of development before birth, it forms the model for the future bones of the body. In the adult, cartilage is present in those parts where flexibility is a desired condition. Thus we find the ribs attached to the sternum (breast bone) by cartilage which permits their movement during breathing. It is also present in the tip of the nose and the external ear. Cartilage is composed of a nonliving material, the matrix, through which are scattered groups of cartilage cells. These

cells may be in groups of two, four, etc., each group surrounded by a capsule of semitransparent material. Microscopically, it is possible to distinguish several different types of cartilage. Two of the more common forms are the *hyaline cartilage,* with a uniformly clear matrix, and the *elastic cartilage* (Fig. 4-12), in which there are strands of denser substances between the cells.

Bone or osseous tissue. The framework of the body consists of two types of supporting tissues: cartilage and bone. Although other types of tissues play some role in the general support of the body, these two are the principal ones, since they make up the skeleton. Most of the bones of the skeleton first appear as cartilage structures, although in some instances special membranes form the bases into which bone cells migrate. This process of bone formation (ossification) by the replacement of cartilage or the invasion of membranes proceeds slowly

Fig. 4-12 *Photomicrograph of elastic cartilage showing the relation of the cartilage cells to the matrix. (General Biological Supply House)*

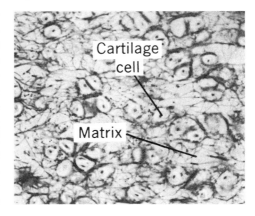

throughout life. In a very young child the skeleton is characterized by the presence of a relatively large number of living cartilage cells and their products. In an elderly person most of the cartilage has been replaced by nonliving calcium and phosphorus salts that have been laid down by the bone-forming cells. Thus, a young person has quite flexible bones, but those of an older individual are brittle because they contain more of the inorganic salts.

Bone tissue is composed of two general types of cells, one that forms the bony material, the *osteoblasts,* and another that reabsorbs it, the *osteoclasts.* By the action of these two opposing types, the skeleton undergoes continual changes that result in the building up of new bone and the sculpturing and reabsorption of old material. The formation and fine structure of bone is discussed in more detail in the next chapter.

Nervous Tissue

The cells that comprise this type of tissue are adapted for the function of relating the individual to his surroundings. Some of the cells receive stimuli. These are the receptors. Others transmit the nervous impulse from one part of the body to another, while still other types bring about a response to the stimulus.

The basic unit of structure in the nervous system is the nerve cell, or *neuron.* This type of cell shows many variations in form and size depending on the part of the body in which it is located and on its particular function. In the brain, for example, the cells may be relatively short but extremely branched, while in other parts of the

body some of the extensions of the nerve cell may be several feet long. One of the characteristics of nerve cells is their branching nature. By means of these branches the various parts of the body are connected and their activities coordinated. The details of the structure of the neuron are found in Chapter 8.

ORGANS AND SYSTEMS

Organs

An organ can be defined as a group of tissues which function together in the performance of some vital activity. The tissues in an organ are not necessarily all alike in structure or function, but by the coordination of their individual activities they form a distinct part of the body.

The hand may be taken as an example of an organ. Here we have a collection of different tissues, all of which work together to give us a very useful part of the body. The outside of the hand is covered by a highly complex type of stratified epithelium which in certain areas has become folded into ridges (the fingerprints). These ridges supply a slightly roughened surface to aid in grasping. Embedded in the skin are highly sensitive nerve receptors. Also in the skin and throughout the entire hand are blood vessels that carry the blood to supply the tissues with needed food and oxygen. In the walls of the blood vessels are smooth muscle fibers, while the fingers are moved by the action of striated muscles, many of which are located in the forearm, although there are some in the hand to perform its more delicate motions.

Bone and cartilage serve as the supporting elements of the hand, and there is adipose tissue in the palm and elsewhere to absorb shocks. All of these various types of tissues are held together by bands of connective tissue. This is an example of a group of tissues functioning as a coordinated organ.

Organ Systems	Principal Types of Tissues						
	Epithelium	Muscle	Nerve	Blood	Bone	Cartilage	General connec-tive
Digestive							
Mouth	X	X	X	X	X	X	X
Esophagus	X	X	X	X			X
Stomach	X	X	X	X			X
Intestines	X	X	X	X			X
Circulatory							
Heart	X	X	X	X			X
Arteries	X	X	X	X			X
Veins	X	X	X	X			X
Capillaries	X		X	X			X
Lymphatics	X		X	X			X
Skeletal							
Skeleton		X	X	X	X		
Respiratory							
Nasopharyngeal area	X	X	X	X	X	X	X
Trachea	X	X	X	X		X	X
Lungs	X	X	X	X			X
Pleurae	X		X	X			X
Diaphragm	X	X	X	X			X
Intercostal muscles	X	X	X	X		X	X
Excretory							
Kidneys	X	X	X	X			X
Ureters	X	X	X	X			X
Kidney tubules	X			X			X
Urinary bladder	X	X	X	X			X
Urethra	X	X	X	X			X
Reproductive							
Ovaries and testes	X		X	X			X
Associated organs	X	X	X	X			X
Nervous							
Brain and spinal cord	X		X	X			X
Sense organs	X	X	X	X		X	X
Muscular							
Muscles	X	X	X	X			X
Endocrine							
Endocrine glands	X		X	X		X	X

Organic Systems

These are groupings of several organs which function together in carrying out some major bodily activity. The division of labor in the body is accomplished not only by tissues and organs, but on a more elaborate scale by the organ systems. In the table on page 52 the various organ systems are listed. These systems are further subdivided to show the organs and tissues that comprise them.

VOCABULARY REVIEW

Match the phrase in the left column with the correct word in the right column. *Do not write in this book.*

1	A group of cells having the same origin and function.	a	adipose
2	A group of different tissues having a common function.	b	cilia
3	A tissue in which fat is stored.	c	collagen
4	Connective tissue which joins the bones together at a joint.	d	endothelium
		e	epithelium
5	Attaches a muscle to a bone.	f	ligament
6	Groups of different organs working together with a common function.	g	matrix
		h	mucus
7	The inner lining of all blood vessels.	i	organ
8	A lubricating secretion.	j	system
9	The intercellular material in connective tissue.	k	tendon
10	A protein found in non-elastic connective tissue fibers.	l	tissue
11	Small hair-like projections on certain cells.		

TEST YOUR KNOWLEDGE

Group A

Select the correct statement. *Do not write in this book.*

1 Cartilage (a) can be transformed into bone during ossification (b) binds muscles to bones (c) is part of epithelial tissue (d) covers muscles.
2 A group of cells similar in structure and function is a(n) (a) organ (b) system (c) tissue (d) organism.
3 Smooth muscle tissue (a) is attached to the skeleton (b) contains many nuclei in each cell (c) forms the walls of the heart (d) is in the walls of veins.
4 Adipose tissue is (a) a storage tissue (b) a muscle tissue (c) held together by cartilage (d) an epithelial tissue.
5 A leukocyte is a type of (a) muscle cell (b) blood cell (c) receptor (d) cellular inclusion.

6   Epithelial cells (a) line body cavities (b) cover bone surfaces (c) form connective tissue (d) have many nuclei in a single cell.

7   Voluntary muscle cells are (a) smooth (b) striated (c) ciliated (d) branched.

8   Cardiac muscle is found in the (a) stomach (b) intestine (c) heart (d) veins.

9   Red blood corpuscles are called (a) lymphocytes (b) erythrocytes (c) thrombocytes (d) leukocytes.

10   One type of bone cell is a(n) (a) osteocyte (b) neuron (c) thrombocyte (d) goblet cell.

Group B

1   Briefly describe the importance of division of labor in the animal body.

2   Describe briefly the various types of cells that are found in each of the four types of tissues.

3   What significant differences are there between epithelial and connective tissues?

4   Briefly describe the activity of each of the following: osteoblasts, osteoclasts and osteocytes.

5   Classify the following as a tissue or an organ: skin, lining of the stomach, finger nails, teeth, heart. State the basis for your decision in each case.

6   How can you distinguish between voluntary muscle, involuntary muscle and cardiac muscle by their appearance? By their function?

Chapter 5   **The Skeleton**

The bony connective tissue that makes up the general framework of the body constitutes the body's *skeleton* (see Fig. 5-14, pages 68 and 69). There are 206 named bones. The bones that make up the skeleton vary greatly in size and shape; some are extremely small, such as those of the middle ear, others are large and heavy, such as the thigh bones. Their shapes range from long and cylindrical to thin and curved, such as those of the skull. Regardless of their size and shape, all of the bones are joined together by tough, fibrous bands of tissue, the *ligaments*. This is true even in the case of the bones of the skull, although here the softer ligamentous tissue may be replaced by bone as individual parts fuse.

The functions of the skeleton are to afford a supporting framework for the soft parts of the body, to protect delicate inner structures, to serve as a place for attachment of muscles, to supply the body with certain types of blood cells, and to provide a storehouse of minerals that the body can draw on in times of necessity.

Not all bones can perform all of the functions listed above; some are more highly adapted for specific activities

than others. The function of protection is illustrated by the bones of the skull that enclose the brain and by those of the thorax (the ribs, breast bone, and part of the vertebral column) that enclose the heart, lungs, and other internal organs. Attached to the outer surfaces of bones are muscles that supply the force that enables us to use our bones as a series of levers in such motions as walking, lifting, and sitting (see Chapter 7). The long bones of the body, such as those of the upper arms, the ribs, and the breast bone, supply many of the solid parts of the blood. The skeleton as a whole can supply minerals to other parts of the body in times of need. This function is especially evident during pregnancy. At this time the mother furnishes the developing child with the calcium and phosphorus salts it needs for the formation of its skeleton.

BONE

The hardness of bone is due to its high mineral content. Approximately ninety percent of all of the calcium in the body is found in bone, combined with phosphorus in the form of phosphate (PO_4). The general formula for this material is $3Ca_3(PO_4)_2 \cdot Ca(OH)_2$ which is known as *hydroxyapatite*. This forms the crystalline structure, but there are also other compounds present which contain small amounts of soluble calcium, sodium, citrate, potassium, magnesium, and chlorine. These materials lie between the masses of the crystals, bearing very much the same relation to them as mortar does to the bricks in a wall. Fibers of collagen are also found in this intercrystalline substance. Collagen normally furnishes many of the minerals that are required

Fig. 5-1   A section through the humerus to show the internal structure of bone.

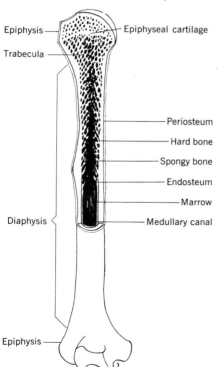

Epiphysis — Epiphyseal cartilage

Trabecula —

— Periosteum

— Hard bone

— Spongy bone

— Endosteum

— Marrow

Diaphysis —

— Medullary canal

Epiphysis —

by the body since they are in forms that are quite easily soluble in the fluid part of the blood.

Structure of Bone

There are two types of bone tissue: *spongy* and *compact*. Spongy bone, which is found in the interior of a bone, is more porous and contains more blood vessels than does the compact bone of the surfaces. In a long bone, such as that of the upper arm (Fig. 5-1), the *diaphysis* (shaft) is composed almost entirely of compact tissue, while each *epiphysis* (end) contains mostly spongy bone. The compact bone of the diaphysis encloses the long cylindrical *medullary canal* which contains the marrow. Most bone marrow (yellow marrow) contains blood vessels and fat cells, but some (red marrow) contains blood forming tissue which is discussed later (see Chapter 22). Lining the medullary canal is a membrane called the *endosteum*. Covering the bone is the *periosteum,* a thick double-layered membrane containing blood vessels, nerves, and the bone-forming cells. The growth and development of bone is initiated from this membrane.

Bone Formation

The general form that the adult skeleton will take is determined long before birth. In the early embryo the potential skeletal framework is laid down in two forms: flexible hyaline cartilage and fibrous membranes. As many as seven months before birth, bone cells begin to replace these with bone tissue. This process of ossification continues throughout life but the major activity occurs in the first twenty-five years. Thereafter, bone tissue is capable of

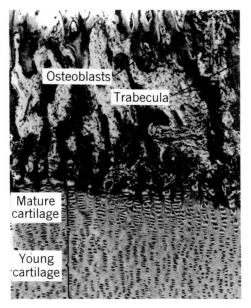

Fig. 5-2 Developing bone. The hyaline cartilage is being rapidly transformed into bone. Masses of solid bone, the trabeculae, may be seen toward the top of this photomicrograph. (General Biological Supply House)

matrix dissolves. Simultaneously, bone-forming cells *(osteoblasts)* move rapidly from the periosteum into the area and begin to manufacture their own ground substance, the bone matrix, in long columns called *trabeculae*. It is interesting to note that though bone matrix also becomes calcified, the bone cells do not die, as do the cartilage cells. This is because the osteoblasts have long processes that come in contact with the branches of other bone cells and with the capillaries. During the formation of the bone matrix these provide channels for the passage of nourishment to the cells. After the matrix is laid down and calcification

local growth, as in the healing of a fracture. Most of the bones of the body are formed from cartilage. The cartilage of the embryo is not strong enough to bear the weight of the human body, though it serves as an excellent model for the laying down of the strong bony tissue that characterizes the mature skeleton.

Cartilage consists of cartilage cells in a large amount of gelatinous matrix (Fig. 5-2). As the young cells mature, they produce a chemical substance that brings about the calcification of the matrix. The cells are surrounded by this hardened material and are cut off from their source of nutrition. They eventually die and most of the calcified

Fig. 5-3 Longitudinal sections showing stages in ossification of a typical long bone. Left: the primary center of ossification replaces the calcified cartilage at the center of the diaphysis. Note that the periosteal bone collar is also well established at this stage. Right: ossification centers develop in the ephiphyses (top) and eventually eliminate the epiphyseal cartilage plate as the bone ceases to grow in length (bottom). Bone tissue is colored, cartilage is stippled, and calcified cartilage is black.

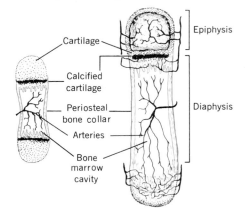

begins, the processes are drawn back into the cell, leaving *canaliculi* (small canals). These are filled with fluids, thus continuing the flow of life-sustaining fluid between the osteoblasts and the capillaries. By this unique mechanism bone cells are able to give to bone the quality of living tissue, in spite of the fact that about 70 percent of bone is composed of a mixture of mineral salts of an extremely hard consistency.

Ossification (Fig. 5-3) commences in the center of the diaphysis and proceeds rapidly toward the epiphyses. After this primary *center of ossification* has its work well under way, a similar ossification center arises in each epiphysis and begins its slower replacement of cartilage. In the X-ray photograph (Fig. 5-4) of an infant's hand, the epiphyses appear to be separated from the bone. Actually there is a band of cartilage connecting them to the shaft. In the adult hand (Fig. 5-5), the epiphyses have become fused to the diaphysis, at the *epiphyseal line*.

The age at which the epiphyses first appear and then later become firmly attached to the shaft varies somewhat with different individuals as well as with the bones themselves. As a general rule, in females the epiphyses appear earlier and fuse to the diaphyses earlier in life than in males. Referring to the photograph of the child's hand, Fig. 5-4, we find that the epiphysis on the distal end of the metacarpal bone will appear between the ages of ten months and two years. It becomes solidly attached to the shaft between the ages of fourteen and twenty-one. Likewise, the epiphysis at the end of the radius makes its appearance between three months and one and one-half years. In girls,

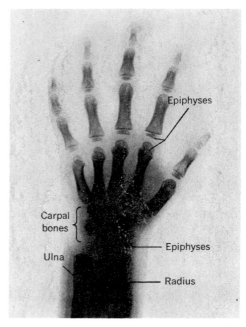

Fig. 5-4 *Hand of a young child. Note the absence of solid bone formation in the wrist and the separation of the epiphyses. (F. E. Wheatley, Jr., M.D.)*

this bone becomes attached to the radius at an average age of seventeen, but in boys at an average age of nineteen. This indicates that the skeletal age of an individual as shown by X-ray photographs corresponds roughly to the chronological age, but does not exactly equal it. Since the epiphyseal cartilage is a weak link in bone structure, a violent blow may cause the bone to separate here instead of fracturing. This often occurs in children, but the separation usually heals together quickly.

While replacement of cartilage continues in the interior centers of ossification, osteoblasts between the periosteal membranes are depositing layer after layer of thin bone, forming a collar around the shaft. This type of ossification differs from that of replacing cartilage. The bone matrix is laid down by osteoblasts between the membranes and is calcified directly. A similar process occurs in the formation of the skull bones, for which membrane rather than cartilage serves as the model for the future bone.

Because all bone tissue, when first formed, is of the spongy type, these external layers of the long bones and the first bony formations of the skull are only temporary. They must be replaced by compact bone by the processes of destruction and rebuilding. Destruction is probably the work of the *osteoclasts,* large cells with many nuclei that are thought to dissolve bone tissue. In the interior of the long bones these cells carve out a large cylindrical space, the medullary canal, which fills with marrow. In the flat bones of membranous origin, such as those of the skull, the spaces hollowed out by the osteoclasts contain blood vessels and nerves. At this stage, newly formed bone is very porous, with many spaces between the trabeculae. It must now be converted to more compact bone by building the layers into a more closely woven network with fewer and smaller spaces.

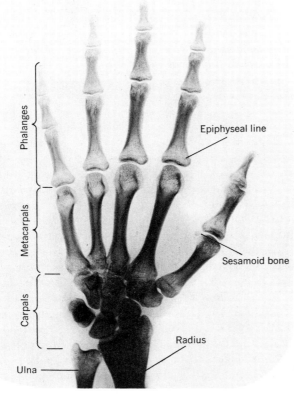

Fig. 5-5   Adult hand. Compare the development of bone tissue with that shown in Fig. 5-4. (F. E. Wheatley, Jr., M.D.)

To do this, concentric layers *(lamellae)* of bone are laid down on the inside surface of the channels between the trabeculae. As these layers are added, the channel is gradually narrowed to form the Haversian canals, which contain blood vessels and lymphatics. The osteoblasts (bone-forming cells) come to lie in cavities *(lacunae)* between the lamellae and are then called *osteocytes* (bone-maintaining cells).

The unit of structure of compact bone, then, is the Haversian system (Fig. 5-6) with its lamellae in concentric layers around the *Haversian canal,* the osteocytes lying in the lacunae between the layers of bone, and the tiny canaliculi that channel the tissue fluid between the cells and the blood vessels in the canal. Because of this intricate,

interconnecting system of canals there is an extensive blood supply throughout the bone.

Joints

The junction of two or more bones is a *joint.* Such an *articulation* (joining) may be freely movable, as the shoulder joint; slightly movable, as the joint formed by the rib in its articulation with the spine; or immovable, as the bones of the skull. The degree of flexibility is determined by how closely the bones are joined together by ligaments and the amount of freedom permitted them by nearby structures.

There are several *types of movable joints,* described according to the type of articulation and the range of movement (Fig. 5-7). The *ball and socket*

Fig. 5-6 Photomicrograph of bone tissue. Note the arrangement of the bone cells around the Haversian canal. (Carolina Biological Supply Co.)

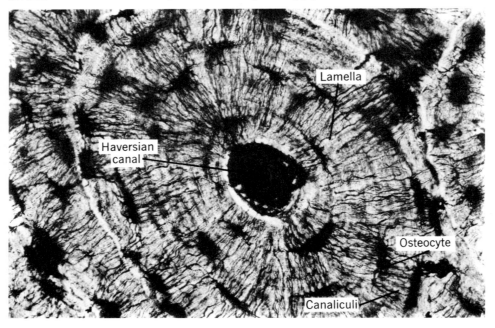

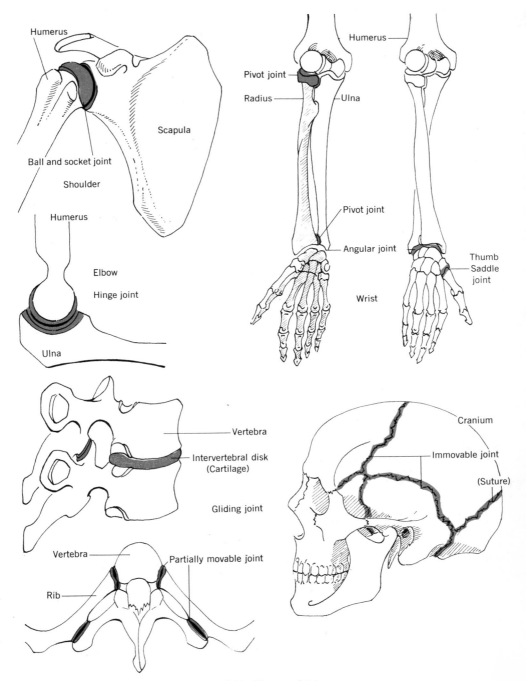

Fig. 5-7 Types of joints.

joint provides the freest movement. One bone with a rounded head moves in a cuplike cavity of the other, as the humerus in its articulation with the scapula. *Hinge joints,* such as the elbow, allow movement largely in one plane. *Pivot joints* provide a rotary movement in which a bone rotates on a ring or a ring of bone rotates around a central axis, as is seen in the turning of the skull on the spine. The vertebrae of the spine are good examples of *gliding joints* in which the articulating parts of one vertebra slide on those of another. An *angular joint* is formed by the articulation of an oval-shaped surface with a concave cavity, as in the wrist. It permits movement in two directions. A *saddle joint* is similar to the angular joint in its range of move-ment, but the bones that form the joint each present a concavo-convex articulating surface. The thumb joint is an example.

There are several terms which describe the movement of joints, and these terms will also be used in Chapter 7 which deals with the muscles and their action. *Flexion* is bending or decreasing the angle between the parts, as when the leg bends back toward the thigh. *Extension* is the opposite action, stretching out, as when the leg is straightened. *Rotation* is turning on an axis, much as the earth turns on its axis, except that in the body complete rotation is impossible because blood vessels, nerves, and other tissue would be torn. *Abduction* is drawing away from the middle line of the body, as the

Fig. 5-8   The arrangement of ligaments, cartilages, and bursae in the region of the knee. (Left, back view; right, side view.)

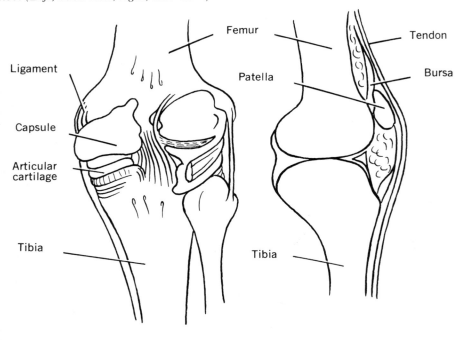

lifting of the arm away from the body. *Adduction,* the reverse, is turning toward the midline of the body, as when the arm is brought toward the trunk. These last two examples describe the action of the shoulder joint.

The movable joints of the skeleton are enclosed by a *fibrous capsule,* and the end of each bone is covered by the *articular cartilage* (Fig. 5-8). In the space between these cartilages is a small amount of lubricating fluid, the *synovial fluid,* that is secreted by the endothelial membranes lining the capsule. The fibrous capsule is strengthened by bands of ligaments. Another type of lubricating and cushioning device, a *bursa,* is found between such bones as those of the elbow, knee, hip, shoulder, and ankle joints. Bursae are found in those joints where pressure may be exerted or where the attachment of a tendon rides over a bone.

Tendons and Ligaments

Muscles are attached to bones by tough cords called *tendons.* These are composed of fibrous connective tissue to which muscle fibers are fused, and they are attached to the surface of the periosteum. Through the tendon, the muscle exerts traction on the bone to which it is attached, and is thus able to control the movement of the bone.

A *ligament* is a strong band of connective tissue which connects the bones of a joint. Its function is to support the joint by holding the bones in place.

Bones as Levers

The movements of the muscles and joints provide action in which the bones act as levers. From the standpoint of

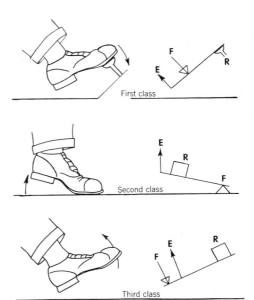

Fig. 5-9 The three classes of simple levers illustrated by different motions of the same part of the body. (From Miller and Haub, General Zoology, *Holt, 1956)*

elementary physics, a *simple lever* is a rod that can be moved about a fixed point, the *fulcrum,* to exert force for the movement of an object. We may consider the bones as levers and the joints as the fulcra. Muscle action supplies the force. Simple levers are classified in one of three groups depending on the relation of the effort (E) and the load (R for resistance) to the fulcrum (F). In Fig. 5-9 these three classes are illustrated by the motion of the foot. In this case, the same bones are involved in all three instances, but as the effort is applied in different directions the type of lever changes. These simple levers are more useful in moving small loads rapidly than they are in moving a heavy load because their mechanical advantage is low. When the body is required to move heavy loads. it uses another type of lever that is seldom found among man-made machines because of its complex structure. This is the *end-loaded compound lever.* This consists of two or more long levers (bones) connected by joints in such a manner that the load comes at the end of the levers when they are almost straightened. Examples of this type of lever are found in the upward thrust of the leg when it is almost straight or the out-thrust arm of a boxer at the end of a jab. In these levers the mechanical advantage is great and accounts for the fact that considerable weight can be moved or force exerted.

THE SKELETON

The skeleton as a whole can be divided into two main divisions: the *axial* portion and the *appendicular* portion. As the name implies, the former makes up the main axis of the body, while the appendicular portion makes up the appendages and those structures which attach them to the axial part. In the axial part we find the skull, the vertebral column, the breast bone, and the ribs. The appendicular portion consists of the shoulder girdle and arms and the hip girdle and legs. See Fig. 5-14.

The Axial Skeleton

The skull. The skull is composed of two distinct regions: the cranium, which houses the brain, and the facial bones, those irregularly shaped bones which support and guard the mouth, nose, eyes, and ears. The bones of the cranium are generally membranous in origin, while those of the face are largely cartilaginous in origin. Some of the facial cavities are composed of both bone and cartilage. For example,

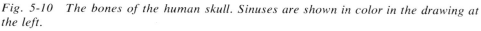

Fig. 5-10   The bones of the human skull. Sinuses are shown in color in the drawing at the left.

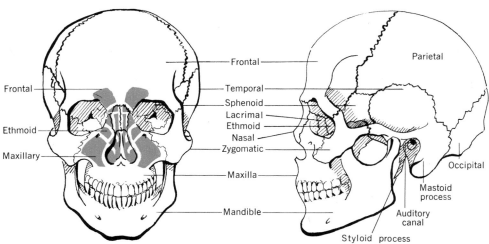

in the nasal cavity the *nasal bones* form the upper part of the bridge of the nose, while the lower part is cartilage.

The *cranium* consists of relatively thin, slightly curved bones. In infancy, the edges of these are held snugly together by an irregular line of connective tissue called a *suture*. This tissue is later replaced by hard bone. The result is a solid, dome-shaped structure that is a highly efficient protective mechanism for the soft brain that lies beneath it. From a mechanical standpoint, a rounded surface, such as that of the upper part of the skull, is a better protection against injury than is a flat surface. Not only do blows glance off more easily, but the archlike structure serves to strengthen it. A severe blow sometimes results in a fractured skull. However, if any part of the bone is pushed inward, the underlying brain tissue may be injured. A depressed fracture of this type usually requires surgical intervention to relieve the pressure on the brain.

The eight bones of the cranium are the *frontal, right and left parietal, occipital, right and left temporal, sphenoid,* and *ethmoid.* The remainder of the skull is composed of *facial bones* that house the organs of smell, taste, sight, and hearing. The cheek bone may be seen in Fig. 5-10 as the *zygomatic bone.* It joins with the temporal bone to complete the *zygomatic arch.* Certain bones aid in the process of chewing, such as the *mandible,* or lower jaw,

Fig. 5-11 Vertebral column as seen from the left side. The intervertebral disks occupy the spaces between the vertebrae.

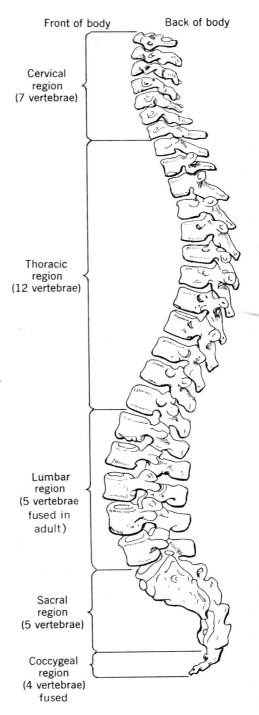

Front of body   Back of body

Cervical region (7 vertebrae)

Thoracic region (12 vertebrae)

Lumbar region (5 vertebrae fused in adult)

Sacral region (5 vertebrae)

Coccygeal region (4 vertebrae) fused

and the *palatine bone* and part of the *maxilla* that together form the hard palate in the roof of the mouth.

Within the bones of the skull are several large cavities in the facial region. These are the *paranasal sinuses* (Fig. 5-10). They are lined by ciliated columnar epithelium and normally drain through small openings into the nasal cavity.

Certain bones of the skull contain small cavities that are sufficiently numerous to give the bone a spongy appearance. This condition is well illustrated by the *mastoid process* of the temporal bone. This lies just behind the ear and may be identified as a slightly raised area.

The vertebral column. The vertebral column (Fig. 5-11) is made up of a series of separate bones, the *vertebrae,* linked by cartilage and ligaments in such a manner as to permit flexibility of the trunk. It is divided into five regions from top to bottom: *cervical*

(neck), *thoracic* (chest), *lumbar* (small of the back), *sacral* (hip), and *coccygeal* (tail). There are 33 separate vertebrae in the embryo but fusion of some reduces this number to 26 by the time of birth. It will be noted that these bones are not in a straight line as seen from the side. The thoracic and sacral curves appear before birth, but the others develop later. The lumbar curvature enables an infant to sit up without support, and the cervical curve appears when he begins to walk. These changes in the curvature of the column result in shifts in the body's center of gravity.

An individual vertebra, as represented by one from the lumbar region of the column (Fig. 5-12), is composed of three main regions of bone with a centrally placed opening. The large, solid part of the vertebra is the body, from the top of which winglike transverse processes project to either side. Behind the body of the vertebra is an opening, the vertebral foramen, through

Fig. 5-12   *Lumbar vertebra as aeen from the side (left) and from the top (right). Articulating surfaces are shown in color.*

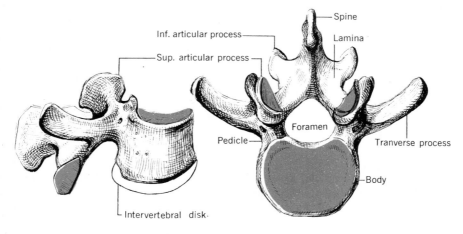

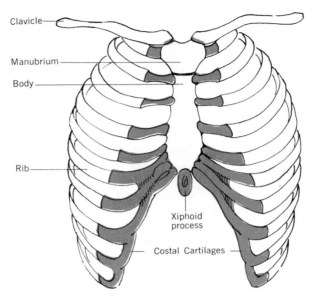

Clavicle

Manubrium

Body

Rib

Xiphoid process

Costal Cartilages

Fig. 5-13 The sternum from the front. Costal cartilages (color) connect the ribs.

which the spinal cord passes. The roof of this arch is then completed by the spinous and auricular processes.

Between the individual vertebrae are the *intervertebral disks* of cartilage which serve as the principal connecting bonds between the vertebrae. Bands of ligaments secure the vertebrae to each other.

The sternum and ribs. The chest region is supported and protected in back by the thoracic vertebrae and in front and along the sides by the *sternum* (breast bone) and the *ribs*.

The sternum consists of a single bone that is divided into three general regions (Fig. 5-13): the upper part, or *manubrium;* the *body;* and a cartilaginous portion, the *xiphoid process*. The general shape of the bone suggested the form of a Roman thrusting-sword, hence the name manubrium (Latin, *manus,* hand), for the upper part, which

has been likened to the hilt region of the sword. While the two upper parts are solid, the lowest is soft and flexible due to the presence of cartilage. Ligaments attach a clavicle (collar bone) to each side of the top of the sternum. To the sides of the breast bone are attached seven pairs of *costal cartilages* (Latin, *costa,* rib), by means of which the ribs are joined to the sternum.

There are normally 24 *ribs* in all human beings. These are curved bones arranged in pairs that articulate with the thoracic vertebrae in back and the costal cartilages in front. Seven of the twelve pairs are known as *true ribs* because each has a direct cartilage connection with the sternum. The next three pairs are *false ribs* since their cartilage connections join with that of the seventh pair. The last two pairs of ribs have no connection with the sternum and are called the *floating ribs*.

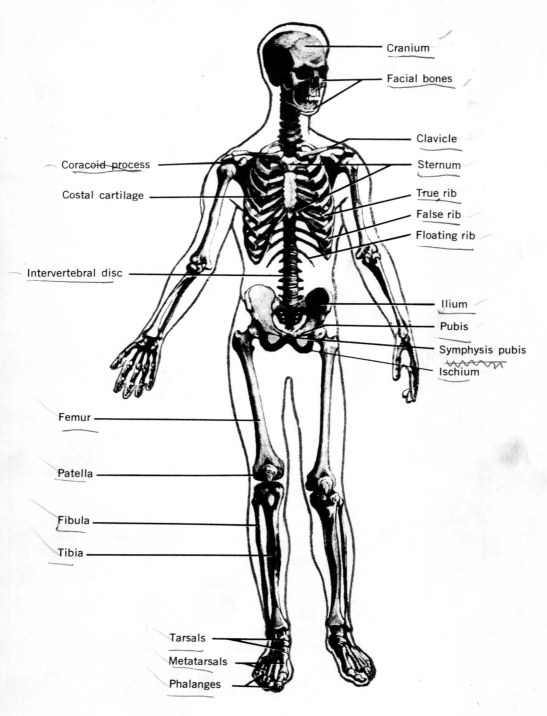

Cranium

Facial bones

Clavicle

Coracoid process

Sternum

Costal cartilage

True rib

False rib

Floating rib

Intervertebral disc

Ilium

Pubis

Symphysis pubis

Ischium

Femur

Patella

Fibula

Tibia

Tarsals

Metatarsals

Phalanges

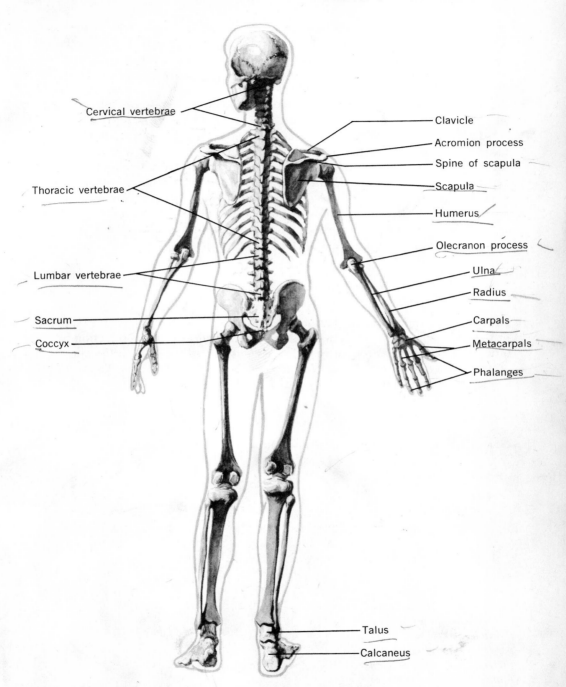

Cervical vertebrae

Clavicle

Acromion process

Spine of scapula

Scapula

Thoracic vertebrae

Humerus

Olecranon process

Lumbar vertebrae

Ulna

Radius

Sacrum

Carpals

Coccyx

Metacarpals

Phalanges

Talus

Calcaneus

Fig. 5-14   Front view (left) and back view (right) of the human skeleton.

Fig. 5-15 shows the different parts of a typical rib. It will be noted that the vertebral end is equipped with smooth surfaces, the head and the tubercle, which fit into the articular processes of the thoracic vertebra and permit the rib a slight gliding motion when a person breathes. Between these two surfaces is the neck region, where ligaments are attached to hold the rib securely to the vertebra. The shafts of adjacent ribs are joined by the *intercostal muscles,* the action of which serves to increase and decrease the capacity of the chest cavity during the act of breathing.

The Appendicular Skeleton

The word "appendicular" refers to a structure one end of which is attached to the axial part of the body while the other is free. Such an organ (e.g., an arm or a leg) is an *appendage.* In the skeleton, these appendages are at- tached to the axial skeleton by means of a series of bones that form the *girdles*—the shoulder girdle (pectoral girdle) and the hip girdle (pelvic girdle).

The shoulder girdle. Each half consists of a large, roughly triangular *scapula* (shoulder blade) and a thin, slightly curved *clavicle* (collar bone) (Fig. 5-16).

The broad, flat surfaces of each scapula permit the attachment of muscles which aid in the movement of the arm and which also serve to hold the upper arm bone firmly in place. There is a prominent shelflike ridge running obliquely across the upper third of the bone on the posterior (back) side. This is the spine of the scapula, and it ends in a flattened projection, the *acromion process,* that overhangs the shoulder joint. Beneath this projection is the *glenoid cavity* which articulates with the head of the humerus. The acromion can be felt as a slight bony projection

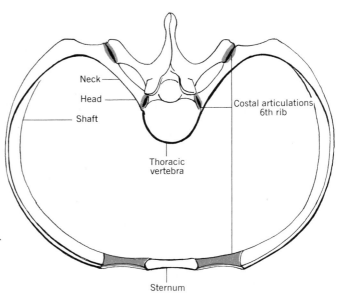

Neck

Head

Shaft

Costal articulations, 6th rib

Thoracic vertebra

Sternum

Fig. 5-15   Sixth rib, showing its articulation with a thoracic vertebra. Compare the general outline of this type of vertebra with that of the lumbar vertebra shown in Fig. 5-12.

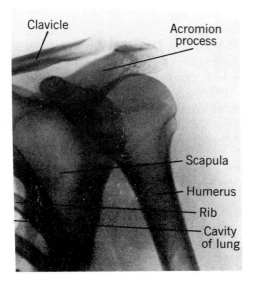

Clavicle

Acromion process

Scapula

Humerus

Rib

Cavity of lung

Fig. 5-16 X-ray of the left shoulder showing the relation of the bones in this ball and socket joint.

ligaments are attached to it. These help to hold the humerus in the socket. Also, along the shaft of the bone we find ridges and depressions that show where various other muscles are attached to it. (The muscles cannot be seen in the X-ray.) At the lower end of the humerus are the areas which serve as the points of articulation with the bones of the forearm. On the posterior surface of the bone just above one of these articular surfaces is a notch or depression into which the head of the ulna fits when the arm is extended at the elbow. This depression is called the *olecranon cavity.*

The skeleton of the forearm is composed of two bones, the *radius* and the *ulna.* The name radius is derived from the fact that this bone can be rotated around the ulna, since both of its ends are flattened and the bone can be moved freely through an arc. This method of articulation permits the hand to be rotated and greatly increases its flexibility. The ulna lacks this ability to rotate. At the upper end of the ulna there is a projection, the *olecranon process,* which fits into the olecranon cavity of the humerus when the arm is extended. This not only prevents a side-to-side motion, but it also serves to stop the backward motion of the arm beyond an approximately straight line. At its lower end, the position of the radius is indicated by a projection just above the thumb side of the wrist, and the corresponding end of the ulna forms an easily felt prominence on the opposite side.

The *wrist* (carpus) contains eight small bones which are held together by ligaments. These *carpal* bones are roughly arranged in two rows and can

on the upper surface of the shoulder. Articulating with the acromion is one end of the clavicle, the other end of which rests against the top of the sternum. The two collar bones thus serve to brace the shoulders and prevent excessive forward motion. The *coracoid process,* which projects forward, is an attachment for several muscles and ligaments.

The arm. Each arm is divided into four general regions by the principal joints which permit action of these parts. The upper arm extends from the shoulder to the elbow and its skeletal support is supplied by the *humerus.*

The head of the humerus (Fig. 5-16) is smoothly rounded and articulates with the scapula at the glenoid cavity. The opposite surface is somewhat roughened, indicating that muscles and

move on each other in such a manner as to allow flexion of the wrist, but they do not permit a wide lateral (side) movement. We can consider these a complex form of gliding joint. On their inner (palmar) surface are attached some of the short muscles that move the thumb and little finger.

The *hand* can be divided into two main sections: the palm of the hand with its five *metacarpal bones* and the five fingers supported by 14 *phalanges* (singular, phalanx). Four of the fingers have three phalanges in each; the thumb has only two. Due to the way the muscles and ligaments are attached to these bones, there is very little motion in the palm of the hand. The fingers, however, have great flexibility in a vertical plane. The joints between each phalanx form a hinge that permits the parts of the fingers to be bent. The thumb is more flexible than the other four fingers because the end of its metacarpal bone is more rounded and because it is supplied with muscles that lie within the hand itself. These permit the thumb to be brought across the palm of the hand, a condition that does not exist in any other group of mammals except that to which man

belongs, the primates. Occasionally in the hand, and elsewhere, small nodules of bone are formed in the tendons at the region of a joint. These *sesamoid bones* may remain completely separated from the adjoining bony structures or may become attached to them. The kneecap, discussed later, is an important sesamoid bone.

The pelvis (Fig. 5-17). In the course of its development, the pelvic girdle is first formed as six separate bones, three on each side of the middle line of the body. These are *ilium, ischium,* and *pubis.* As growth continues, the three bones on each side fuse into a single structure and with the sacrum form a bowl-shaped cavity, the *pelvis,* that supports the viscera (soft organs) of the lower abdomen. Eventually the two sets of bones form a joint with the pubic bones in front, the *symphysis pubis,* and with the sacrum in back, the *sacroiliac joint.*

One of the more obvious differences between the male and female skeletons is evident in the region of the pelvis (Fig. 5-17). In the male, the angle made by the two pubic bones (*pubic arch*) is more acute than in the female. Also, the height of the male pelvis is some-

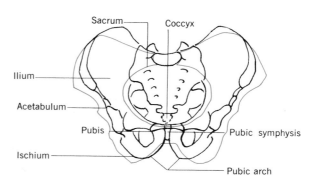

Fig. 5-17 Human pelvis. Male, black outline; female, colored outline.

what greater than that of the female, although it is not as broad. As a result, the heads of the thigh bones are more widely separated in women than in men. Since the thighs grow toward the center line of the body as they approach the knees, this outward flare at the hips has a tendency to bring the knees of a woman somewhat closer together than is the case in a man.

The leg. The *thigh bone* (femur) (Fig. 5-14) is the longest and strongest member of the skeleton. The head of the femur is rounded and smooth and articulates with the ilium at a pronounced indentation in the latter, the acetabulum, forming a strong ball and socket joint. The head is set off at an angle of about 125° from the main shaft of the bone by a short region, the neck. The weight of the body is thus supported at a point slightly inside of the main axis of the limb. The neck region of the femur is a fairly common site of fracture, especially in older persons, in whom it tends to become rather porous and thus weaker.

The internal architecture of the femur follows a very definite pattern. The shaft of the bone is filled with marrow and is almost completely cylindrical in outline. The walls are composed of very compact bone. At the lower end the femur expands to a broad surface to permit articulation with the tibia. In both the lower end of the bone and head and neck regions, the femur shows excellent examples of the application of mechanical engineering principles to body structures. The trabeculae in the upper end of the bone grow in curved lines so that they offer the maximum strength to the bone. In this respect they may be likened to girders in a structure that is designed to carry a heavy load. At the lower end of the bone the trabeculae are arranged in parallel, longitudinal walls, thus affording support for a direct downward thrust. While some measurements have been made on the amount of compression the femur can stand before breaking, these have not given uniform results due to the variability in the specimens used and to the methods employed. The results indicate, however, that if a direct compressing force is put on top of the femur, a pressure of between 15,000 and 19,000 pounds per square inch is needed to break the bone. If cylinders are cut from the shaft and subjected to similar tests, fresh bone has been found to withstand over 27,000 pounds pressure per square inch. Fracture of the bone occurs when it has been subjected to a twisting motion or receives a blow from the side. In these circumstances, a few hundred pounds pressure per square inch will break the bone.

The lower end of the femur is expanded into a large flattened region, the two sides of which can be felt at the knee joint (Fig. 5-8). This is the point of articulation with the larger of the two bones of the leg, the *tibia*. The *patella* (kneecap) is located just in front of this joint. This large sesamoid bone develops in the tendon of the principal muscle of the front of the thigh (quadriceps femoris). It usually appears between the second and third year of life in girls, but may be delayed until about the age of six in boys. Ossification is completed about the age of puberty. The patella is flat and roughly triangular in outline, with the apex of the triangle pointing downward. A ligament attaches it to the tibia. Surrounding the patella are four bursae which serve to cushion the joint.

From the standpoint of comparative anatomy, the bones of the lower leg are comparable to those of the forearm. The *fibula* would correspond to the radius of the arm in position and mode of development, while the *tibia* can be compared to the ulna. There, however, the similarity stops. Both the tibia and fibula are immovable as far as rotation is concerned. This results in a lower leg that cannot be rotated from side to side in the same manner as the forearm. It increases the stability of the leg as a whole and makes it a more suitable organ of locomotion. The lower end of the tibia forms the projecting ankle bone on the inside of the leg, while the end of the fibula can be felt on the outer side.

The *ankle* (tarsus) is composed of seven bones and is comparable to the wrist, in that it serves as the connecting link between the bones of the leg and the foot. Both the tibia and fibula articulate with a broad-topped tarsal bone, the *talus*. As in the wrist, the movement of the ankle is a sliding motion which permits the foot to be extended and flexed each time a step is taken. The largest bone of the ankle is the heel bone, the *calcaneus*.

The *foot* contains five *metatarsal bones* that correspond to the metacarpals of the palm of the hand. Each of these bones articulates with at least one of the tarsal bones, and in some cases they articulate with other metatarsals. There is one important difference between the palm of the hand and the metatarsal region of the foot: the foot has two main curvatures in its

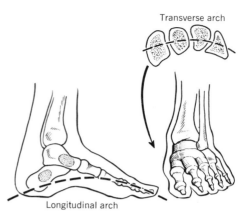

Fig. 5-18   Arches of the human foot.

architecture that are lacking in the palm (Fig. 5-18). These curves are the result of the arching of the metatarsals as they join the tarsus. One of these arches is longitudinal in its direction and the other lies perpendicular to it and is transverse. Together they strengthen the foot and give a certain amount of springiness to the stride. Due to any one of a variety of causes, including faulty prenatal nutrition, improper posture, excessive weight, fatigue, or improperly fitting shoes, these arches may be lowered. The result is a "flat-footed" condition which throws unnatural stresses and strains on the muscles involved in walking and may lead to fatigue and pain.

The phalanges of the toes are similar in number and relative position to the corresponding bones in the fingers. Four of the toes have three phalanges each; the great toe has only two. Unlike the thumb, the great toe is nonopposable, that is, it cannot be brought across the sole of the foot, although in a young infant it is more flexible than in an adult.

SKELETAL DISORDERS

The skeleton, like any other part of the body, is subject to various diseases and injuries. Bone infections may result from invasion by bacteria from some other region where these organisms are the cause of a difficulty quite unconnected with the skeleton. Thus bone infections may develop as a result of badly infected teeth, pneumonia, typhoid fever, or other diseases. Injury to a bone as a result of a hard blow may also serve as a predisposing condition that will be followed by an infection. The general name given to several conditions involving infection of bones is *osteomyelitis*.

One of the more common types of skeletal disorders is *arthritis*. This term is used to include a large number of different conditions in which joints are affected. Arthritis is usually thought of as a disease of the aged, but only certain types, particularly hypertrophic arthritis, occur as a result of aging. The most serious forms of arthritis usually develop between the ages of 25 and 50. Rheumatoid arthritis is a particularly devastating form of the disease. It affects about three times more women than men. The cause is unknown, though emotional factors, infections, metabolic and endocrine abnormalities, and diseases of the nervous system are suspect. In this condition the joints become badly swollen and painful (Fig. 5-19). The pain causes muscle spasm which may result in deformity. As the disease progresses, the cartilage that separates the bones of the joints is gradually destroyed and hard calcium bands fill the spaces. These calcified bands restrict the movement of the joint so that crippling and wasting of the muscles result.

There are several different types of *fractures* that have been named according to the way the broken ends of the bone behave. The simplest type is known as a *green-stick fracture* because the broken bone does not separate completely but acts like a sap-filled stick, the fibers of which separate

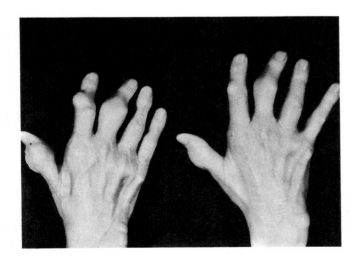

Fig. 5-19 Two right hands showing the effects of rheumatoid arthritis. Note the swollen joints and the muscular contractions of the fingers. (National Institutes of Health)

lengthwise when it is bent. Green-stick fractures are usually found in young children whose bones still contain flexible cartilage. A *simple fracture* is one in which a bone is broken but the ends do not push outward through the skin. As the X-ray photograph in Fig. 5-20 shows, both of the bones in the left forearm have been broken just above the wrist and the ends have over-ridden each other. This slipping of the bones over each other may have been the result of the blow or of a contraction of the muscles that pulled the broken bones into that position. The most serious type of fracture is the *compound fracture,* in which the end of the bone protrudes through the skin. This may lead to an infection of the bone or of the surrounding tissues.

The periosteum and the endosteum play important roles in healing broken bones. After a fracture has occurred, these membranes immediately begin to produce bone-forming cells. In bones of membranous origin, healing occurs as a direct extension of new growth from each side of the break. However, in long bones there is the formation of a fibrous cartilage model to fill the gap, and this model is re-

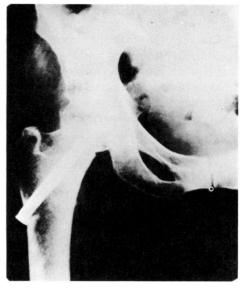

Fig. 5-21 In this X-ray photograph of a hip fracture, a metal pin has been used to hold the broken bone until it can knit together normally. (Jerry Harrity)

placed by bone, as in the initial ossification of long bones. The fibrous connective tissue, cartilage, and new bone form a *callus* or bridge between the broken ends of the bone until healing is complete. In some cases, surgeons use metal pins to hold fractured bones

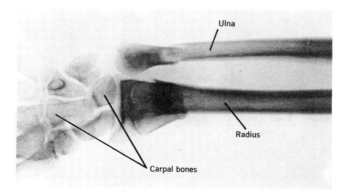

Fig. 5-20 A simple type of fracture of both the radius and ulna just above the region of the wrist.

Ulna

Radius

Carpal bones

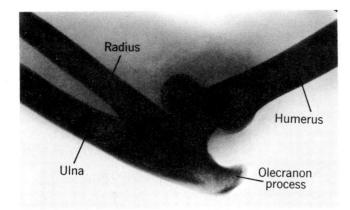

Fig. 5-22 Dislocation of the elbow. Note the olec-ranon process projecting away from the bottom of the humerus.

together until they can heal properly (see Fig. 5-21).

A *dislocation* (Fig. 5-22) occurs when a bone is forced out of its proper position in a joint. In the case shown here, the top of the ulna has been pushed out of position. Note the displaced olecranon process, the sharply curved upper part of the ulna.

When a joint is subjected to a sudden, unnatural motion, a *sprain* may result, because of the tearing or straining of the tendons that hold the muscle to the bones, or of the ligaments that connect the bones. This is usually accompanied by acute pain and rapid swelling in the region which has been injured.

VOCABULARY REVIEW

Match the phrase in the left column with the correct word in the right column. *Do not write in this book.*

1	Attaches a bone to a bone.	a	bursa
2	Is constantly remodeling bone by changing its internal structure.	b	cartilage
		c	epiphysis
3	Internal columns of bony tissue.	d	epiphyseal cartilage
4	The patella is an example of this type of bone.	e	ligament
5	The type of cell that forms bone.	f	osetoblast
6	A small bony structure found at the ends of a long bone.	g	osteocyte
		h	sesamoid bone
7	The forerunner of bone in the majority of the skeleton.	i	suture
		j	tendon
8	An immovable type of joint.	k	trabeculae
9	A fluid-filled cushion between bones at a joint.	l	osteoclast
		m	membrane bone

TEST YOUR KNOWLEDGE

Group A

Select the correct statement. *Do not write in this book.*

1 A structure associated with a Haversian canal is a (a) lamella (b) bursa (c) tendon (d) ligament.
2 Straightening the arm at the elbow is an example of (a) flexion (b) distortion (c) extension (d) abduction.
3 A bursa (a) acts as a joint (b) forms part of a bone (c) is a type of lever (d) serves as a cushion.
4 The sacrum is made up of (a) tarsals (b) carpals (c) vertebrae (d) phalanges.
5 Sutures are found in the (a) knee (b) skull (c) elbow (d) wrist.
6 The cerbical region of the spinal column is found in the (a) thorax (b) trunk (c) neck (d) pelvis.
7 Moveable bones are attached to each other by (a) ligaments (b) tendons (c) epiphyses (d) joints.
8 Parts of the pelvis include the (a) humerus and pubis (b) ilium and ischium (c) sacrum and coccyx (d) femur and tibia.
9 The shaft of a long bone is called the (a) epiphysis (b) diaphysis (c) trabecula (d) periostium.
10 The axial skeleton includes the (a) humerus (b) femur (c) tarsus (d) cranium.

Group B

1 What are the functions of the skeleton, and how are these exemplified by the ribs?
2 What types of joints permit the following actions: (a) bending the arm at the elbow (b) rotating the hand (c) swinging the arm in a circle (d) lifting the ribs when taking a breath (e) bending over to pick an object off the floor.
3 Briefly describe the formation of bony tissue.
4 What are the principal parts of the skeleton, and what bones, or groups of bones, are found in each?
5 Identify each of the following: olecrenon process, fibula, scapula, ilium, false rib, femur, humerus, phalanx, clavicle, cervical vertebra.

Chapter 6 Muscle Physiology

The ability of any part of the body to move is the result of the specialized ability of certain cells to contract in response to a stimulus. Just as in other parts of the body cells have been developed which are modified for a particular function, so the muscle cells have been developed specifically for the conversion of chemical energy into mechanical energy. As a result of this we have the ability to move, speak, breathe, and carry on all of those other activities which involve motion.

In any device that converts energy from one form to another so that work can be done, the energy is changed from an inactive, *potential* form into an active, *kinetic* form. Muscle tissue is able to convert about 40 percent of the potential energy provided by food into motion, compared with a gasoline motor which is about half as efficient. Heat is a waste product of both.

Muscle Structure

Striated muscle makes up about 40 percent of the weight of the body. It is the most common single type of tissue. As was pointed out in Chapter 4, striated muscle tissue is composed of elongated, highly specialized cells, the *muscle fibers*. Each fiber is apparently crossed by numerous alternately light and dark bands, the *striations*. Under the ordinary light microscope, these bands appear to be continuous from one side of the fiber to the other. With the aid of an electron microscope, the bands are seen to be features of the extremely small *myofibrils* which are very closely packed together and run lengthwise through the fiber. As shown in Fig. 6-1, each fiber is surrounded by a thin layer of connective tissue, the *sarcolemma*.

When an exceptionally well-prepared section of striated muscle tissue is viewed with a light microscope, a series of dark bands are seen. These are called the *A bands,* and in the middle of each is found a narrow light area, the *H band*. On either side of the A band is another light area known as the *I band* which is crossed by a very thin line of darker material, the *Z band*. The material lying between the two Z bands makes up what is now considered to be the unit of structure of a muscle fiber, the *sarcomere*. The banding of the myofibril is determined by the distribution of the materials composing it. See Fig. 6-2.

79

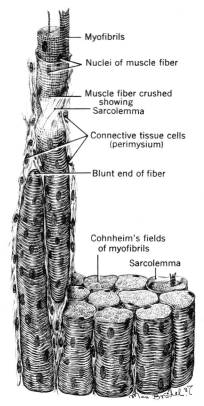

Myofibrils

Nuclei of muscle fiber

Muscle fiber crushed
showing
Sarcolemma

Connective tissue cells
(perimysium)

Blunt end of fiber

Cohnheim's fields
of myofibrils

Sarcolemma

Fig. 6-1 A three-dimensional diagram of the structure of a muscle. Note the relation of the various parts to each other. (Courtesy of Department of Art as Applied to Medicine, John Hopkins Medical School)

It has been known for some time that striated muscle contains two principal types of proteins, *actin* and *myosin*. These were thought to be connected with the contraction of the fibers. Various theories were advanced to explain how they acted and what their relation was to the shortening of a fiber. All of these theories had certain basic defects, and it was not until 1957 that the actual relation of actin and myosin to

the process of contraction was satisfactorily explained. In that year two unrelated English researchers by the name of Huxley independently published results which provide the best explanation to date for the banded appearance of the fibers and also for the process of contraction. A combination of X-ray studies, electron microscope photographs, and microchemical analyses led to the present widely accepted explanation of the process.

The myofibrils have now been found to contain the two proteins, actin and myosin, in the form of filaments. The myosin filaments are about 100 Ångstroms in diameter, while the actin filaments are only half as thick. (An Ångstrom $= 1 \times 10^{-8}$ centimeter). The striated appearance of the fiber is caused by variations in its actin and myosin content at different positions. In the A band both actin and myosin filaments are present, in the H band there are only myosin filaments, and in the I band only actin filaments. The structure of the Z band is currently unknown, but the actin filaments of one sarcomere pass through the Z band and join similar filaments in the adjoining sarcomere. The relationship of these two types of filaments is shown in the electron micrographs in Fig. 6-2 and Fig. 6-3, and in the diagram in Fig. 6-4.

When a muscle contracts, these filaments of actin and myosin slide over each other so that each fiber shortens. In preparations made from very strongly contracted muscles, the ends of the filaments have been found to be slightly enlarged as a result of their pressing together very firmly as in the bottom diagram of Fig. 6-4.

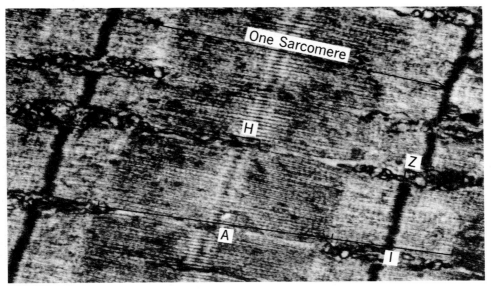

Fig. 6-2 Fine structural detail of mammalian striated muscle myofibrils in longitudinal section as shown by an electron microscope. Note the distribution of the thick myosin and thinner actin filaments. (Drs. K. A. Siegesmund and C. A. Fox, Marquette University School of Medicine)

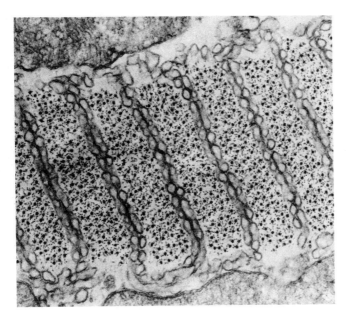

Fig. 6-3 Electron micrograph showing myofibrils of insect striated muscle in cross section. Each thick myosin filament is surrounded by several thinner actin filaments. Note that insect muscle structure is similar, but not identical, to that of mammals. (Drs. K. A. Siegesmund and C. A. Fox, Marquette University School of Medicine)

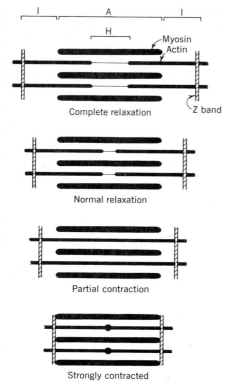

Fig. 6-4 *Diagram showing how the filaments of actin and myosin move during the contraction of a sarcomere.*

The neuromuscular junction.
The fine branches of nerves (Fig. 6-5) which spread over the muscle cells, make contact at specialized nerve endings called *neuromuscular junctions.* A chemical, *acetylcholine,* produced by the nerve end, transmits the stimulus to the muscle fiber where it triggers off a series of electro-chemical reactions that result in contraction. Since all fibers have these nerve endings over their surfaces, and since the impulse reaches all simultaneously, the entire muscle contracts at the same instant.

It would remain in this state of contraction if it were not for another substance, *cholinesterase,* formed by the nerve to neutralize acetylcholine. These two processes alternate very rapidly so that continued muscular activity can be maintained over a long period of time.

Sometimes a normal function can be well illustrated by a condition in which that function goes wrong. *Myasthenia gravis* (meaning "muscle weakness, severe") is a disease in which acetylcholine appears to be blocked at the neuromuscular junction. It is thought to be due to an abnormal accumulation of another chemical, possibly a waste product of muscle fatigue. Thus there is abnormally rapid exhaustion of voluntary muscles.

The eye muscles are most frequently affected, causing drooping eyelids and double vision. There may also be difficulty in smiling, chewing, speaking, swallowing, reaching, walking, or even breathing, in severe cases. All of these symptoms are worse with exertion and are somewhat relieved by rest.

Dr. Mary B. Walker, an English physician, noted the similarity of these symptoms to those of persons affected by the South American arrow poison, curare. An antidote for curare was known—neostigmine, which raises the level of acetylcholine by suppressing the activity of cholinesterase. Dr. Walker reasoned that if a patient with myasthenia were given this drug, improvement might result. Trial showed the drug to be fast acting and dramatically effective.

Patients may take neostigmine and similar drugs by mouth in tablet form, according to their needs. If an overdose is taken, unpleasant muscle twitching, stomach cramps, and diarrhea occur, due to the presence of excess acetylcholine.

Chemistry of Muscle Contraction

Striated muscle tissue consists of approximately 75 percent water, 20 percent protein, and 5 percent minerals and various organic substances. The most abundant minerals in muscle tissue are potassium and sodium, the former being found within the cell itself and the latter in the intercellular fluid. Calcium, phosphorus, and magnesium are also present in significant amounts, and all take the form of chlorides, carbonates, and phosphates. Some of these salts help regulate the osmotic pressure of the cells, and others take an active part in muscle contraction. Most important of the organic compounds are creatine, creatine phosphate, adenosine triphosphate, and glycogen.

Muscle contraction requires large amounts of instantaneous energy. The immediate source of this is the same as for other cells and has already been discussed in chapter 3. It is the high energy phosphate bonds of *adenosine triphosphate* (ATP). The breakdown of this compound takes place in the presence of the enzyme ATP-ase which catalyzes the splitting off of a phosphate group. This releases four times as much energy per gram molecule as

Fig. 6-5 Distribution of nerve endings over striated muscle fibers. The neuromuscular junctions are the small bodies at the ends of the nerve branches. (General Biological Supply House)

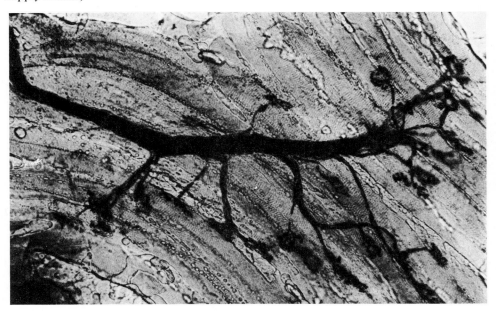

glucose, and all the energy is instantly available. The energy thus made available causes the filaments of actin to move and the muscle shortens.

$$ATP \rightarrow ADP + P + Energy$$

Reaction of ADP with *creatine phosphate* (CP), another high energy phosphate compound, results in its immediate reconversion to ATP and thus ensures a constant supply of ATP for contraction. Creatine phosphate is found only in muscle where it forms a reservoir of high energy phosphate bonds.

$$CP + ATP \rightarrow C + ATP$$

Creatine phosphate is later resynthesized by the reaction of creatine with ATP which has been formed as a result of the breakdown of glucose.

$$C + ATP \rightarrow CP + ADP$$
$$ADP + P + Energy \rightarrow ATP$$
from glucose

All these reactions can be summarized as follows:

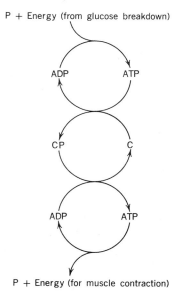

P + Energy (from glucose breakdown)

ADP   ATP

CP   C

ADP   ATP

P + Energy (for muscle contraction)

Glucose, the final source of energy for all the reactions resulting in muscle contraction, is stored in muscle in the form of glycogen, and sufficient is always present for normal activity. Its complete breakdown requires oxygen and makes all the energy stored in its chemical bonds available.

$$C_6H_{12}O_6 + 6O_2 \rightarrow 6CO_2 + 6H_2O + Energy$$
glucose   oxygen   carbon   water
dioxide

But enough oxygen cannot always be supplied fast enough by the blood and an incomplete breakdown to *lactic acid* with release of some energy occurs in muscle in the absence of oxygen. This accounts for the fact that a runner can cover several yards without taking a breath. However, the muscle must recover from its activity and the waste products of contraction must be removed. The accumulated lactic acid produced is transported by the blood to the liver where it is reconverted to glucose and then to liver glycogen. Glucose in the blood is taken up by muscle cells and used to replenish the store of muscle glycogen. With this last step the chemical processes of muscle contraction have come full circle. The final steps can be summarized as follows:

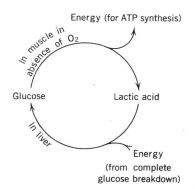

Energy (for ATP synthesis)

In muscle in absence of O₂

Glucose   Lactic acid

In liver

Energy
(from complete
glucose breakdown)

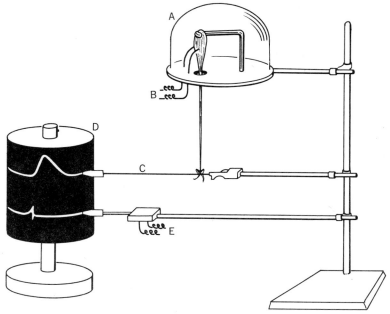

Fig. 6-6   *Method of recording a muscle twitch. The muscle is placed in a moist chamber* A, *with proper electrical connections,* B, *for its stimulation. The contraction of the muscle is recorded by the lever,* C, *on the smoked paper that covers the drum of the kymograph,* D. *In the base of the kymograph is a spring-driven motor that will make the drum revolve at a definite rate of speed. The exact instant of stimulation of the muscle is recorded by the signal lever,* E, *that has been wired into the same circuit as the stimulating electrodes. The writing points scrape the soot off the smoked paper and leave a white record on a black background.*

The Muscle Twitch

The classic method for studying the contraction of a muscle is that of the so-called *nerve-muscle preparation.* The calf muscle with nerve intact, from a fresh-killed frog, is usually used with an apparatus like that in Fig. 6-6. The upper of the two lines on the drum is a record of a single muscle twitch which results from stimulating the nerve by means of a small electric current.

From the record made on the kymograph, it can be seen that a single muscle twitch is divided into three distinct periods (Fig. 6-7). The electrical stimulus was applied at the point marked "S," and for 0.01 second thereafter no movement was recorded. This is the *latent period,* during which presumably there is a series of chemical and electrical changes within the muscle fibers. The *contraction period* commences with the breakdown of the complex

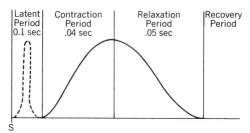

| Latent Period 0.1 sec | Contraction Period .04 sec | Relaxation Period .05 sec | Recovery Period |

Fig. 6-7 An analysis of the record of a muscle twitch shows that it is divided into three principal parts followed by a period of recovery. Each phase of this process is characterized by definite chemical processes occurring in the muscle.

chemicals within the fibers. This supplies energy for contraction. The contraction period lasts about 0.04 seconds during a single twitch. When contraction reaches its maximum peak, there follows a 0.05 second *relaxation period*. During the time the muscle is relaxing, the chemicals are reformed. Besides the energy produced in this reaction, further energy is being supplied by the breakdown of glycogen. These three periods can occur in the absence of oxygen. There follows a *recovery period* during which the muscle is supplied with oxygen for the oxidation of waste products that accumulate during the contraction.

Since the frog is a cold-blooded animal, these measurements are usually made at room temperature (20° C– 22°C). The latent, contraction, and relaxation periods occur in about 0.1 second, and under the conditions existing in making a kymograph record of a single muscle twitch, the recovery period is approximately one minute in length. The speed of reaction increases as the temperature is raised. In the human body, the effect of temperature is of minor importance because we maintain a fairly constant temperature at which the chemical processes occur with the greatest efficiency. In the human, the total time elapsing during a twitch of a muscle is about 0.001 second.

The strength of contraction in a twitch depends on the strength of the stimulus. The maximum contraction is obtained when a stimulus is strong enough to excite all the fibers of a muscle. Each individual fiber gives an "all-or-nothing" response.

If a muscle is stimulated repeatedly and with such rapidity that no complete recovery periods are allowed between stimulations, the condition of *fatigue* appears. The responses of the muscle to the stimulus become weaker and weaker until they fail entirely (Fig. 6-8). This condition of fatigue will last until the oxidation and removal of wastes has been completed.

A type of prolonged muscular contraction, which differs from the muscle twitch, is the state of *tetanus*. It is obvious that a single twitch of very short duration such as we have considered would not be effective in performing

Fig. 6-8 Fatigue. As a muscle is repeatedly stimulated, its responses become weaker and the recorded curves shorter.

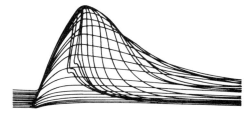

any work nor would it result in coordinated movements. However, if impulses are sent into the muscle at a very rapid rate (20–30 per second) so that each impulse occurs during the contraction period of the impulse preceeding it, the individual contractions add to one another and one much bigger contraction results. The muscle will remain contracted until the impulses cease. An example of this type of action is found in the motion of closing your hand slowly. Such motion involves a large number of muscle bundles acting together in a coordinated action for a considerable time interval.

Muscle Tone

A normal characteristic of all muscle tissue is that it remains in a state of partial contraction at all times. This condition is the result of a constant flow of weak nervous impulses to the muscle, and is considered a slight state of tetanic contraction. Because of this, we are able to stand erect without undue fatigue, since the muscles of the neck, back, and legs are in a state of partial contraction. A good example of how muscle tone affects muscles is seen in the fingers. If you allow your hands to hang naturally by your sides, you will see that the fingers are slightly bent. If the muscles were not constantly receiving a stream of impulses, the hands would be in a flat position with the fingers very loose and relaxed. An unconscious person shows the loss of muscle tone, which is one reason why it is so difficult to carry him in this condition. Normally, when a person is asleep, the tone of his muscles is at the lowest point short of total unconsciousness.

THE EFFECTS OF MUSCULAR EXERCISE

Increase in Muscle Tissue

The number of striated muscle cells present at birth is essentially the same as that found in the adult. In other words, muscle cells are unable to reproduce themselves, except for a very short period of time immediately following birth. Thus there must be some other explanation for the increase in muscle tissue. It is true that the nuclei increase in number by a process of simple nuclear division, and that some very few fibers may split, but there is a very small increase in number of cells. Nor do muscle cells have the power to replace themselves once they have been lost, except in those cases where a sufficient amount of their substance remains to permit regeneration.

As everyone realizes, continued muscular activity is accompanied by an increase in the size of the muscles that are being used. This enlargement is the result of an increase in the amount of cytoplasm within the individual cells. When any organ increases in size, it is said to *hypertrophy,* and the opposite, *atrophy,* occurs when the organ becomes smaller. Atrophy is seen in those muscles which have not been regularly used, either from a lack of exercise or because of injury to the nerves which control their activity. Muscles affected by paralyzing poliomyelitis frequently grow smaller. Muscle cells are not actually lost, but there is, rather, a destruction of the nerves in the spinal cord that control their activities. This loss of ability to move brings about a shrinking of the individual fibers in the unused muscles.

During exercise, not only do muscles themselves increase in total volume, but the products they produce cause changes in other organ systems. Respiratory, circulatory, excretory, and nervous systems all respond to the extra demands of the body during moderate or strenuous exercise.

Increase in Respiration

We have all experienced the increase in the rate of breathing during exercise. This is the body's attempt to supply oxygen used up in muscle activity. As we learned, some of the chemical reactions that occur in muscle contraction do not require oxygen. But the recovery phase of contraction does need oxygen in order to restore energy-producing substances. Deeper, more rapid breathing helps pay this *oxygen debt* so that continued work is possible. For example, it is possible for a runner to run 100 meters at top speed without taking more than one or two breaths or, perhaps, none. At the end of the race he may fall to the ground completely exhausted. This simply means that he has used up his oxygen supply in the effort of the race and does not have a sufficient amount for the processes of recovery. Prolonged, severe exercise does not have this same effect; the runner in a long race "gets his second wind" by an adjustment of the body to the demands that are made on it.

During severe muscular exercise, the rate of the necessary increase in the rapidity and depth of respiration depends on the amount of exercise taken. Consider what happens when a runner sprints for 110 meters. The following data will show his oxygen requirements at varying rates of speed.

Length of time to finish race	Oxygen requirements	
	Liters per 110 meters	Liters per minute
22.6 seconds	1.83	5.06
16.4 seconds	3.08	11.17
14.3 seconds	4.33	18.17
13.0 seconds	7.38	33.49

We have seen that more oxygen is needed during exercise to oxidize the increased amount of lactic acid formed in the cells during muscular contraction. Lactic acid may appear in the blood in quantities twenty times as great as that found when the muscles are normally resting. When this lactic acid is oxidized, the carbon dioxide formed becomes carbonic acid when absorbed into the blood stream. The change in the acidity of the blood has a profound effect on the nervous system, one result being an increase in the rate of breathing.

Changes in Circulation

Since it is extremely difficult to make direct chemical analyses of the changes in muscles during exercise, it is necessary to examine the body fluids. Both blood and urine furnish clues to some of the changes that go on in the muscles during exercise. During exercise, there is an increase in the number of red blood corpuscles, the rise being directly proportional to the severity of the exercise. Since the function of these cells is to transport oxygen around the body, the needs of the muscles can thus be met more easily. Two factors probably play a role in this increase in the number of red cells: first, some cells may be forced into the blood stream from the reservoirs such as the spleen, and second, there is a loss of water by sweating and the passage of some of the blood fluid into the tissues, which results in the concentration of the blood solids.

Following severe exercise, such as running a quarter-mile race, we find that the hydrogen ion concentration of the blood drops from a normal pH value of 7.35 to 6.95, indicating a marked shift to the acid side. Ten minutes after the race has been finished, the pH rises to 7.04, showing that some of the substances responsible for the shift have been removed. Four minutes after the race it is found that the lactate—a slightly acid salt of lactic acid—in the blood rises from a normal 5.6 milligrams per cent to 163, but at the end of ten minutes drops to 150 milligrams. The lactate is removed from the blood by the kidneys until a more normal level is reached.

Muscles are richly supplied with capillaries, the smallest types of blood vessels, to insure an adequate supply of blood to these active tissues. This supply is so great that in the gastrocnemius muscle of a cat, for example, there are 2341 capillaries per square millimeter of cross section of the muscle. This will give a surface area of 44.1 square millimeters of capillary surface for each cubic millimeter of muscle—a truly amazing amount of surface. In man, these values are approximately the same. When a muscle is at rest the blood flows through the capillaries at approximately 850 milliliters per minute, but when it is active, the rate is increased to about 6,920 milliliters per minute. The arterioles dilate or contract to give a greater or lesser flow to the capillaries. Since the capillaries themselves cannot contract, the mus-

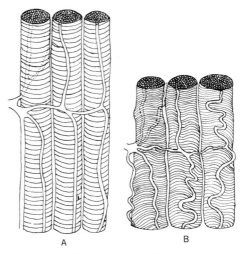

Fig. 6-9 *Three muscle fibers. In* A *the fibers are relaxed; in* B *the fibers have contracted and the capillaries are shown as folded.*

cle fibers over the surface of which they are spread tend to make the vessels fold on themselves. This is shown in Fig. 6-9.

One special effect of exercise is to assist in the return of the blood and lymph to the heart. The contraction of the muscles of the legs presses on the walls of the veins and lymphatics and drives their contents upward toward the heart. The presence of valves in these vessels prevents the backward flow of the fluids. Many of us have noticed that on long auto rides, or on standing without much movement for long periods, our shoes feel tight and our ankles are swollen. When walking is resumed, this swelling usually disappears as tissue fluid returns to veins and lymphatics and is forced upward through the vessels by the squeezing action of the large leg muscles.

Increase in Heart Rate

Studies indicate that the rise in the heart rate starts almost at the same moment that the exercise begins. In fact, the first beat after the work has started is shorter than the preceding one. The rate then rises very sharply but soon has a tendency to level off and remain more or less constant during the exercise periods. In trained athletes, the pulse rate following strenuous exercise varies between 126 and 152 beats per minute. After the exercise has stopped, the rate falls back to normal. At first rapidly; then slowly. If the work has been severe, the postwork rate may remain above the prework rate for as long as half an hour.

Elevation of the Blood Pressure

During severe exertion, a person's blood pressure may rise as much as 60 points above his resting rate. Recent work indicates that of the several factors involved, one may be the fact that the drop in the blood pH has a stimulating effect on the nervous system which results in the constriction of some of the larger blood vessels, thus increasing the tension on their walls.

Rise in Body Temperature

The heat produced during muscle activity is the principal source of heat in the body. Sixty percent of the energy liberated during muscle contraction is generated in the form of heat. During strenuous exercise, the rectal temperature (normal 37.5°C) rises swiftly to about 38°C and then drops rapidly back to normal at the end of the work period. If very fine needle thermocouples are introduced into the thigh muscles of men pedaling bicycles, the temperature recorded will be about 39°C.

The body has, however, a very complex system of liberating this excess heat over the surface of the body and by way of body wastes, so that the temperature is kept under control (see Chapter 27). There is a particularly interesting safeguard against too great a decrease in the body temperature. If a person is outside on a cold day and is not exercising, he may start to shiver. This phenomenon is simply a series of slight involuntary muscular contractions which raise the temperature by producing more heat in the muscles. The excess heat produced is then distributed through the body by the blood.

Change in the Urine

In the first place, the amount of urine is reduced and does not return to its previous prework level for about an hour and one half to two hours following the stoppage of work. It had been thought that this was due to the control of the nervous system over the flow of blood to the kidneys, reducing their activities. Investigation has shown that this may not be the case and that, instead, the production of urine during exercise may be under some other control. It may be that the increased absorption of fluids from the blood and the loss of water in perspiration reduces the amount excreted through the kidneys. A second feature is an increase in the amount of acid present in the urine. Because of the amount of lactic and other acids formed during exercise, the acidity of the urine increases sharply following the completion of the work and then drops back to normal quite rapidly as the body relaxes.

VOCABULARY REVIEW

Match the phrase in the left column with the correct word in the right column. *Do not write in this book.*

1	An enzyme produced at the neuromuscular junction.	**a** actin
2	A high-energy bond is split in this material to yield energy.	**b** acetylcholine
		c ADP
3	The smallest structural element of a muscle showing cross-striations.	**d** ATP
		e atrophy
4	The smaller of two important protein filaments responsible for muscular contraction.	**f** cholinesterase
		g creatin
5	Decrease in muscle size due to disuse.	**h** fibril
6	Is made up of the A and H bands.	**i** glycogen
7	A waste product of muscular contraction.	**j** lactic acid
8	The continued contraction of a muscle resulting from a continued flow of nervous impulses to it.	**k** muscle fiber
		l myosin
9	Is surrounded by the sarcolemma.	**m** striation
10	The chemical product of a broken high-energy bond.	**n** tetanus

TEST YOUR KNOWLEDGE

Group A

Select the correct statement. *Do not write in this book.*

1 A neuromuscular junction is the place of contact between a nerve and a (a) muscle fiber (b) nerve fiber (c) bone (d) sense organ.

2 Voluntary muscle fibers contain the structural protein (a) pectin (b) adenosine (c) actin (d) creatine.

3 The chemical that transmits a nerve stimulus to a muscle fiber is (a) cholinesterase (b) cholesterol (c) acetylcholine (d) adenosine triphosphate.

4 The most abundant substance in muscle tissue is (a) water (b) protein (c) mineral salts (d) lactic acid.

5 A state of prolonged muscle contraction is (a) fatigue (b) tetanus (c) tibia (d) myasthenia gravis.

6 A chemical substance closely involved in muscle contraction is (a) DNA (b) RNA (c) ATP (d) DDT.

7 Increase in the size of an organ is called (a) atropine (b) atrophy (c) hypertrophy (d) hypertension.

8 Muscle activity results in (a) decrease in body temperature (b) decrease in heart rate (c) elevation of blood pressure (d) atrophy of the muscles.

9   Of the following, the term that is *not directly related* to muscle activity is
(a) cholinesterase (b) acetylcholine (c) lactic acid (d) amino acid.

10   From the following terms, select *one* that is *not related* to the structure of
muscles (a) myofibrils (b) sarcolemma (c) striation (d) glycogen.

Group B

1   What role does each of the following play during the contraction of a muscle:
ATP, glycogen, lactic acid, oxygen, ADP?

2   Explain what is meant by the statement, "Muscles never push, they always
pull."

3   Briefly discuss the conversion of energy in a muscle fiber.

4   Briefly describe the structure of a muscle fiber.

5   What is a nerve-muscle preparation, and how can it be used in the study of
the contraction of a muscle?

6   What changes occur in the blood supply of a muscle during contraction?

Chapter 7 **Muscle Action**

We have seen that muscle action is the result of chemical processes within the individual muscle cells. A more general consideration of the anatomical location and action of some of these muscles will show how this muscular energy is transformed into mechanical energy in the form of motion.

Muscles are able to move bones because of the location of the muscle attachments. These points of attachment are called the *origin* and the *insertion*. The attachment to the bone that serves as a relatively fixed basis of movement is the origin. The insertion is the point of attachment to the bone which is moved. Most muscles are attached to the periosteum of a bone by means of a tendon; however, some make direct contact with the periosteum, while others are attached by a sheet of heavy connective tissue. The *belly* of the muscle contains the body of the muscle tissue itself. To understand the movement made possible by the action of a specific muscle, it is necessary to know its origin and insertion.

Consider, for example, the action of the *biceps brachii,* commonly called the biceps. This is a large muscle on the front of the upper arm that can be felt when the arm is bent at the elbow. Because of the placement of its origins, it not only brings about a movement of the forearm in a vertical plane, but also gives it a rotary motion, such as that used with a screwdriver. In Fig. 7-1, we can see that both the biceps and the *triceps brachii* or triceps (on the back of the arm) have more than one origin each; the triceps has three points of origin and derives its name from that fact (Latin, *tres,* three + *caput,* head), and the biceps has two. It will be noted that these muscles have several points of insertion, and because of this, complex action is possible.

The action of the biceps and the triceps when the arm is flexed or extended at the elbow illustrates an important characteristic of muscles: most skeletal muscles work in pairs. For each muscular contraction that brings about motion in one direction, there is possible a motion in the opposite direction which depends upon the contraction of the other member of the muscle pair. For example, when you flex your arm, the action is largely the result of the contraction of the biceps muscle. But when you extend your arm, the triceps muscle contracts. These opposing actions

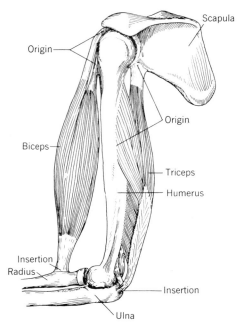

Scapula

Origin

Origin

Biceps

Triceps

Humerus

Insertion
Radius

Insertion

Ulna

Fig. 7-1   The biceps and triceps muscles of the arm, showing their origins and insertions. (From Miller and Haub, General Zoology, Holt, 1956)

are brought about by the so-called *antagonistic muscles.*

The contraction of pairs of antagonistic muscles is controlled by the central nervous system. Although the exact mechanism is not clearly understood, it appears that the contraction of a muscle stimulates nerve cells that lie within it. These nerve cells, known as stretch receptors, send a volley of nervous impulses to the central nervous system— to either the brain or the spinal cord. A

nervous impulse prevents the muscle from contracting rather than causing it to relax. The resulting smooth action of antagonistic muscles is due to this condition known as *reciprocal innervation.*

Superficial Muscles

The activities of many of the skeletal muscles pass unnoticed because they lie below the surface of the body. These are known as *deep muscles* to distinguish them from the *superficial muscles* which are responsible for the body's contour and are especially apparent during physical exertion. We will deal here only with those superficial muscles whose action we can readily see in our daily activities (Figs. 7-2 and 7-3).

Your facial expressions are produced by a series of interesting muscles that permit you to frown, smile, raise your eyebrows, squint, and perform a variety of conscious and unconscious facial movements. This is why the facial muscles are sometimes referred to as the "muscles of expression."

The muscle bundles that comprise these are generally not as distinct and separate as those found elsewhere in the body, due to their method of formation and attachment. Around the eyes their arrangement is circular. The *orbicularis oculi* acts as a *sphincter* muscle to draw the surface of the face around the eye into a circular form and close the lids, as in squinting. Similarly,

Fig. 7-2   The act of throwing an object (opposite page) calls into action the muscles of trunk and legs, especially. Some of the more obvious ones have been indicated in this drawing of a shot-putter.

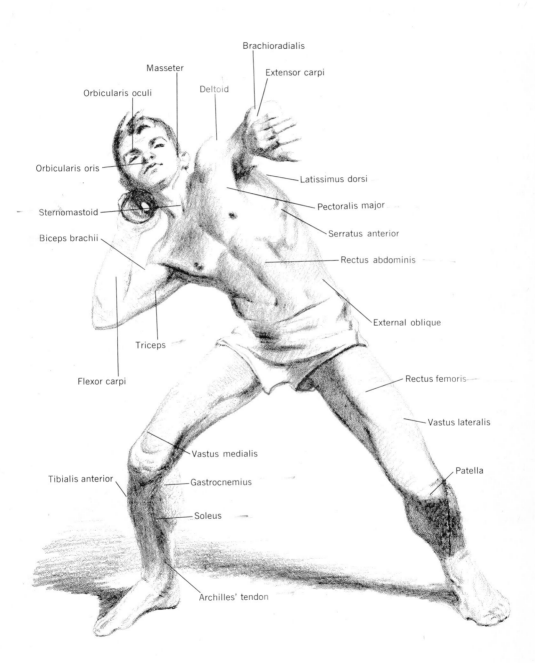

the *orbicularis oris* allows the lips to be pursed when whistling. Sphincter muscles such as these act in the same way as a drawstring around the opening of a bag or tube.

A powerful muscle on the side of the face, the *masseter,* is responsible for many of the motions of chewing. Its action may also be noticed during those moments of concentration when unconscious movements of the jaw occur. The masseter functions with the *pterygoideus internus* (the action of which is not visible from the surface of the body) in forming a slinglike structure that holds the lower jaw in place and permits it to move up and down with a slightly rotary motion.

A muscle of the neck, quite obvious when the head is thrown back, is the *platysma.* This muscle has its origin on the upper edge of the *pectoralis major* (described below) and spreads as a broad sheet upward toward the lower jaw. Its action is to draw the outer edges of the lower lip downward, especially when the head is thrown backwards. A runner gasping for air at the end of a hard race will show this muscle clearly.

The most conspicuous muscle on the front and side of the neck is a stout, often prominent band that arises from the inner third of the clavicle and the upper end of the sternum and passes obliquely over the side of the neck to a point behind the ear. This is the *sternocleidomastoideus* muscle; it is usually called the sternomastoid. The long name given to the muscle simply indicates that it has two points of origin, the sternum and the clavicle, and that its insertion is on the mastoid process behind the ear. Its action is to bend the cervical region of the spine to one side. At the same time the muscle draws the head toward the shoulder and rotates the chin upward. During this motion, the muscle on the opposite side of the neck from which the flexion is occurring is stretched and stands out like a thick cord.

The front upper half of the chest is covered superficially by the *pectoralis major* muscles, a large fan-shaped pair primarily concerned with the forward motion of the arms. These have their origins on each side of the sternum, on the clavicles, and on the cartilages of the true ribs near the sternum. From these points, the muscle fibers pass fanwise until they converge to form a strong tendon that inserts on the head of the humerus. As a result of the manner and extent of the origins of these muscles, their contracting action tends to bring the arms and shoulders forward and rotate the arms toward the middle line of the body. The most highly developed pectoral muscles are found in the flying birds and serve as the principal sources of power for flight; these muscles are the "breast meat" of fowl. In a shot-putter the pectoralis major gives the initial shove to the projectile; a swimmer depends on this muscle for the strength of his downward stroke.

Fig. 7-3   The extension of parts of the body (opposite page) can only be accomplished by the contraction of muscles. Those responsible for these acts show their position as clearly as those that cause flexion.

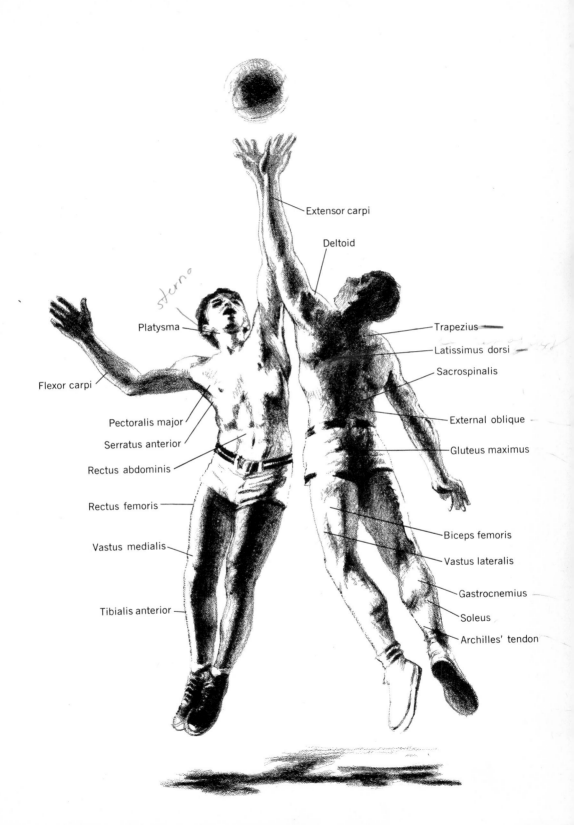

Extensor carpi

Deltoid

sterno

Platysma

Flexor carpi

Pectoralis major

Serratus anterior

Rectus abdominis

Rectus femoris

Vastus medialis

Tibialis anterior

Trapezius

Latissimus dorsi

Sacrospinalis

External oblique

Gluteus maximus

Biceps femoris

Vastus lateralis

Gastrocnemius

Soleus

Archilles' tendon

Another muscle that greatly affects the action of the arm as a whole is the *trapezius*. This is located on the back of the trunk and can be seen as a broad sheet extending from the base of the neck toward the top of the shoulder. Contractile action of this muscle will raise the scapula and draw it backward and downward, thereby giving added leverage to the shoulder for the powerful forward thrust demanded of the shot-putter or swimmer. The trapezius is also quite obvious when the shoulder is strongly extended in a downward position such as that shown by a person jumping for a basketball. The opposite (antagonistic) action on the scapula is brought about by the contraction of the *serratus anterior*. This muscle has its origins on the first eight or nine ribs and its insertion on the scapula. Its action is obvious in a person who is required to move a heavy object: the muscle appears as a series of ridges passing obliquely toward the shoulder from the region of the ribs as it draws the scapula forward.

Two other superficial muscles connected with the action of the arm are the *deltoid* and the *latissimus dorsi*. The deltoid forms the triangular pad of muscle which covers the shoulder joint, with the apex of the triangle pointing downward. The contraction of the major part of the muscle will draw the arm upward, but some of its fibers will draw the shoulder backward and outward. This complex action is possible because its origin is partly on the clavicle and partly on the spine of the scapula. The principal antagonistic muscle of the deltoid is the latissimus dorsi. As its name implies, it is the broadest (latissimus) muscle of the

back (dorsi). Its origin extends from the sixth thoracic vertebra downward until its lowest fibers are attached to the upper edge of the ilium. This broad triangular muscle has its insertion along the front of the humerus close to its head. Whereas the deltoid raises the arm and draws it forward, the latissimus dorsi pulls it downward and backward. Since the raising and lowering of the arm must be accomplished with considerable force at times, both of these muscles can become highly developed as very conspicuous features of the upper part of the body.

The muscles comprising the front of the abdominal wall serve not only to support the internal organs, but also to flex the trunk. Of the superficial muscles concerned in these activities, two are of special interest. In a basketball player or a wrestler, the *rectus abdominis* muscle may be quite conspicuous. It is a straplike muscle originating on the pubis and extending up the front of the abdomen to the fifth, sixth, and seventh ribs. The *linea alba* (midline) of the abdomen separates it into two parts. The linea alba is a thickened region of the sheath that covers the muscle. The action of the recti muscles, in common with that of other abdominal muscles, is to compress the internal organs and thereby aid in expelling feces from the rectum or urine from the bladder. They also flex the vertebral column by drawing the thorax downward. By their action, the pelvis may also be drawn up on the vertebral column, permitting motions such as climbing. Acting as antagonists of the back muscles, they are important in maintaining good posture. If the members of this pair of muscles act singly, they can bend the trunk to one side or the other and at the same time rotate it.

The *external oblique* muscles are a second pair of muscles in the abdominal region, concerned with bending the trunk and compressing the internal organs. These arise on each side from the borders of the eight lower ribs and then pass as a broad sheath of muscle to the crest of the ilium on that side. Contraction of these muscles results in the compression of the internal organs, and their action can be felt when one coughs or sneezes. When both members of this pair of muscles act together, they flex the vertebral column by drawing the

Fig. 7-4   Distribution and relation to each other of the superficial muscles of the human body.

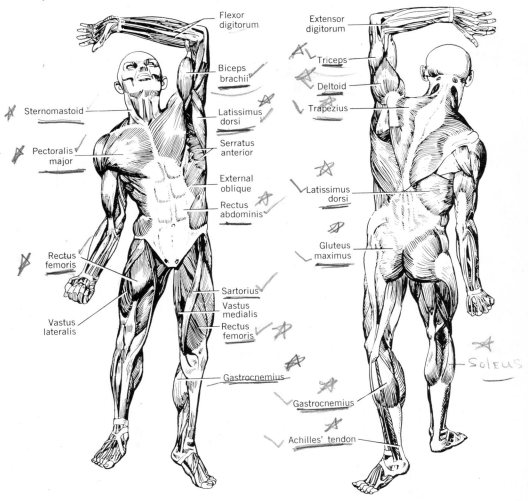

pelvis toward the sternum. When only one of the pair contracts, it bends the vertebral column to the side and rotates it, bringing the shoulder of the same side forward.

In the upper limb, or arm, there are many muscles that play a role in its flexion or extension. We have already spoken of the action of the biceps brachii and the triceps and pictured their relationships in Fig. 7-1.

In the forearm are the muscles responsible for the major movements of the fingers, and some that aid in flexing the arm at the elbow. One of the more obvious muscles of this region is the *brachioradialis* which appears as a band along the upper part of the forearm on the radial (thumb) side. It extends from the lower end of the humerus, where it has its origin, to the lower end of the radius. Its action is to work with the biceps in flexing the arm. When the arm is being forcibly flexed against pressure on the hand, this muscle stands out.

On the front and back surfaces of the forearm can be seen some of the muscles whose action is responsible for the motion of the fingers. Tendons from these extend over the wrist, where their action can be clearly seen when the fingers are moved. These groups of superficial muscles are the *flexor* and *extensor carpi,* some of which have their origins on the lower end of the humerus. At the wrist the tendons of these muscles become indistinct because of the presence of broad bands of ligaments which hold them snugly in place and prevent them from forming an arch between the forearm and the fingers.

The muscles of the lower limbs present an intricate pattern of the interaction of opposing muscle bundles that brings about an integrated motion of the limb. The large muscle at the top of the thigh and covering the hip joint in back is the *gluteus maximus.* This, together with two smaller muscles in the same region, has the action of extending the thigh and rotating it outward.

The muscles of the thigh itself are responsible for many of the more forceful movements required of the body as a whole. You need only watch a person walking to realize some of the roles these muscles play. On the back of the thigh is a large muscle, the *biceps femoris,* which is one of a group of three that are sometimes spoken of as the "hamstring muscles." The biceps of the leg has two points of origin: one on the ischium and the other on the femur. Its principal point of insertion is on the fibula with a small portion passing to the tibia. Its connection with the fibula can be felt as a heavy tendon on the back of the knee toward the outside of the leg. The action of this muscle is twofold: it flexes the leg at the knee and pulls it slightly to the side, and extends the thigh backward and rotates it.

On the front of the thigh are two prominent muscles. One is the *vastus lateralis,* which passes laterally from just above the kneecap across the outer side of the thigh to the pelvis. The second is the *vastus medialis,* which lies slightly inside the middle line of the front of the thigh. Both of these exert a powerful extensor effect on the thigh, and both have their origins on the femur. They have a common tendon which is inserted on the sides and top of the patella. The action of the individual parts of this tendon can be felt if you swing your leg back and forth over the edge of a chair.

In the leg (below the knee) are two muscles that form the calf of the leg, the more prominent being the *gastrocnemius*. It has its origins on the base of the femur, with its insertion in the heel bone (calcaneus) by means of the heavy *Achilles' tendon*. Closely associated with the gastrocnemius is the *soleus*. This originates on the upper surfaces of the tibia and fibula and its tendon blends with that of the gastrocnemius. The action of both of these muscles is to point the foot downward, and their general form and course can be noted with each step as the heel is raised from the floor in walking.

The front of the leg is characterized by a complex group of muscles, some of which are quite conspicuous in their action, while others show merely as rippling motions under the skin. There is one muscle of especial interest because of its ability to flex the foot strongly at the ankle. This is the *tibialis anterior* which lies just outside of (lateral to) the shin bone. The origin of this muscle is on the front of the tibia and its insertion is on one of the tarsal bones. As a result of its contraction, the foot is drawn upward, allowing the weight of the body to fall on the heel. The tibialis anterior is frequently affected by the type of poliomyelitis that injures the nerves controlling the muscles which are situated in the front of the legs.

VOCABULARY REVIEW

Match the phrase in the left column with the correct word in the right column. *Do not write in this book.*

1 The process involved in bending the arm at the elbow.

2 The end of the muscle that is attached to a relatively immovable part of the body.

3 Jumping for a basketball brings about this state of the parts of the body.

4 A muscle that closes a circular opening upon contracting.

5 One muscle of a pair relaxes when the other contracts as a result of this.

6 The end of the biceps femoris that is attached to the radius.

7 The opposing action of pairs of closely associated muscles.

a antagonism
b extension
c flexion
d insertion
e origin
f reciprocal innervation
g sphincter
h superficial muscles

TEST YOUR KNOWLEDGE

Group A

Select the correct statement. *Do not write in this book.*

1 The sternomastoid muscle (a) moves the sternum (b) is concerned with chewing (c) is easily seen when the head is bent to the side (d) moves the shoulder.

2 The muscle of the calf of the leg that is most prominent when taking a step is the (a) gastrocnemius (b) deltoid (c) platysma (d) vastus medialis.

3 Chewing motions result from the action of the (a) masseter (b) soleus (c) orbicularis oris (d) orbicularis oculi.

4 Muscles which clench the hand are (a) in the palm of the hand (b) on the back of the upper arm (c) on the front of the forearm (d) in the wrist.

5 Superficial muscles (a) are responsible for the body's contours (b) lack insertions (c) lack tendons (d) are only found in the arms and legs.

6 A point of attachment of a muscle is the (a) belly (b) termination (c) origin (d) bursa.

7 The muscle around the eye is the (a) orbicularis oris (b) masseter (c) orbicularis oculi (d) pterygoideus.

8 Of the following, select *one* muscle that is *not associated with arm movement* (a) pectoralis major (b) deltoid (c) trapezius (d) rectus femoris.

9 A pair of antagonistic muscles is (a) masseter and platysma (b) biceps and triceps brachii (c) gluteus maximus and gastrocnemius (d) vastus lateralis and soleus.

10 From the following terms, select *one* that is *not a muscle* (a) flexor carpi (b) external oblique (c) linea alba (d) latissimus dorsid.

Group B

1 What is the advantage of having muscles that flex the fingers located in the forearm?

2 Briefly describe the motions involved in some sport, such as serving in tennis, throwing a ball, or swimming. Name the principal groups of superficial muscles that are involved in the activity.

3 The scientific names of muscles are sometimes quite long. (a) Explain the basis for these names, and cite specific muscles to illustrate your answer. (b) What is the advantage of having names of this type?

Chapter 8 Nervous Coordination

THE NERVOUS SYSTEM

Man has attempted to duplicate certain aspects of the human brain by means of the computer. There are about 14 billion nerve cells in the outer layer of the brain alone, and the entire brain weighs only about three pounds. One can see why a computer that could even come near partial duplication of the human brain would fill a building covering several blocks and would probably require the entire electrical output of Niagara Falls to operate it.

All of the coordinated activities of a many-celled animal are made possible by nervous tissue. As we ascend the scale of animal development from lower to higher forms, the arrangement of this tissue into nervous systems becomes more and more complex. The human nervous system enables us to be aware of our environment, both internal and external. This awareness or perception is dependent upon *receptors,* the nerve cells in the skin and the specialized nervous tissue like the inner ear or the retina of the eye. The receptors are sensitive to specific changes in our surroundings. In the eye and ear, they transform light or sound waves into *impulses,* which are carried, as an electro-chemical charge, along a nerve cell *(neuron)* to one or more parts of the *central nervous system.*

With the awareness of these impulses by the central nervous system, we are able to adjust to our environment, physically as well as mentally. In addition, this nervous system has the marvelous ability to retain and to organize information for immediate or future use. When called upon to do so, the nervous system responds to the incoming impulses, analyzes information in the light of past experience, and makes an adjustment in a coordinated manner. Thus, the physiological basis for the nervous system involves three specialized characteristics: reception, also known as the ability to receive stimuli; response (irritability), the ability to respond to stimuli; and conduction, the process of transmitting impulses to and from the brain and spinal cord.

The Central Nervous System

As was previously mentioned, the nervous system is a coordinating mechanism. That part of the nervous system directly responsible for coordinating the activities and functions of the human body is the *central nervous system,* composed of the *brain* and *spinal cord.*

In man, no function of the body can be performed independently of some type of control by the central nervous system; even a simple response to a stimulus involves at least the spinal cord.

A more detailed discussion of the structure and function of the central nervous system will be presented in Chapter 9.

The Autonomic Nervous System

The activities of the stomach and intestines, of glandular secretions, as well as many other involuntary body functions, are under the control of the *autonomic nervous system,* which will be discussed in detail in Chapter 9.

The Peripheral Nervous System

A stimulus is any change in environment, outside or within the body. The central nervous system reacts to stimuli from the external or internal environment. It analyzes the information and initiates body movements in response to it. But this organizational success is dependent upon cooperation with that portion of the nervous system which lies outside of the central part and is called the *peripheral nervous system.*

Basically, the peripheral nervous system is a system of nerves, and ganglia which are groups of nerve cell bodies. A nerve consists of numerous individually insulated nerve cell fibers. All the fibers of a nerve may be responsible for conveying sensory impulses *toward* the central nervous system. These are *afferent* fibers. Such a group of fibers is called a *sensory nerve.* Again, all the fibers of a nerve may carry motor impulses from the central nervous system to particular muscle fibers or glands. These are *efferent* fibers and make up a *motor nerve.* Finally, a nerve may consist of both sensory and motor fibers. To this type has been given the name *mixed nerve.* (See Fig. 8-1.)

The large nerves directly connected to the brain and spinal cord are grouped according to their point of origin.

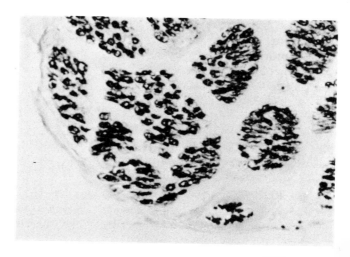

Fig. 8-1 Cross section through some of the bundles of nerve cell fibers composing a nerve. In this photomicrograph the insulation surrounding the fibers is black. The structure, lower right, is a blood vessel containing blood cells.

Twelve pairs of nerves originate from the brain and are called *cranial nerves*. They are motor, sensory, and mixed nerves which emerge or enter through openings at the base of the skull and proceed to or from the point of motor activity or sensory reception. Thirty-one pairs of *spinal nerves* originate from the spinal cord. All of the spinal nerves are mixed nerves having sensory and motor nerve cell fiber components. The cranial and spinal nerves will be discussed in Chapter 9 because of their close relation to the central nervous system.

The branching nerves of the peripheral nervous system terminate in the special sensory and motor end organs; the sensory end organs of taste, smell, sight, and sound, and those of the general senses of pressure, touch, pain, heat and cold, and the motor end organs in muscles and glands. The latter group of sensory end organs are often referred to as the cutaneous senses, since they are associated primarily with the skin (Fig. 8-2).

Receptors. The special sensory end organs are the receptors which make us aware of changes in our external environment.

Receptors are located on the peripheral ends of sensory neurons, or in the various sense organs, such as the eye and ear. They may be naked, free nerve endings (for pain and certain chemical senses); they may contain a light-sensitive chemical which initiates impulses for vision; or they may be specialized nerve endings sensitive to certain stimuli. Receptors vary widely in structure according to the sense they serve and also to their location in the various tissues of the body. They receive stim-

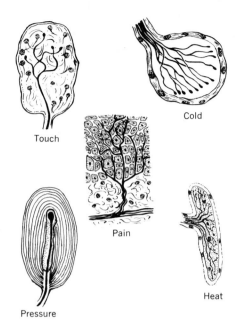

Touch

Cold

Pain

Pressure

Heat

Fig. 8-2 Five types of receptor found in the skin.

uli and translate each one into an electrical "code."

The sense organs responsible for taste are called taste buds. They are constructed of modified epithelial cells, and they predominate on the surface of the tongue, but may also be found on the walls of the soft palate, the pharynx and adjacent parts (see Chapter 15). Sense receptors for smell are seen in each nasal passage as bare nerve fibers hanging from the roof of the nasal canal (Chapter 18).

As you have probably observed, the sense of taste and the sense of smell are both chemical senses. The receptor is stimulated by a portion of the substance itself in molecular or ionic form. It is equally true that those substances that are volatile (for the sense of smell) or soluble (for the sense of taste) will cause greater sensory stimulation than nonvolatile or insoluble substances, such as metals. Organic compounds provide the greatest variety of tastes and smells. However, strong solutions of inorganic substances may also have pronounced stimulatory effect.

Recent work on the sense of smell suggests that the receptors in the nose are sensitive to the three-dimensional shape of molecules. There appear to be seven different primary molecular configurations corresponding to seven distinct smells; other smells result from predictable combinations of the primary seven.

The sense organ responsible for vision is the eye. It is able to receive light waves which cause impulses to be conveyed to the brain. There the impulses are recorded, analyzed, and interpreted as vision. The details of the structure of the human eye and the physics and physiology of vision will be covered in Chapters 10 and 11.

Reception of sound, *audition,* is dependent upon the ear and its transmission of impulses to the auditory center in the brain. You will see in Chapter 12 that hearing is one of our most important senses. As a means of communication, spatial understanding, and esthetic satisfaction, the ear has few competitors. To sit before an orchestra and be able to pick out individual instruments or to identify individual voices in a chorus is truly marvelous. It is said that in a completely soundproof room, such as the one on the campus at the University of California at Los Angeles, a person can hear the molecules of gases in the air striking his eardrum.

Receptors are generally classified in three main groups which are as follows: *exteroceptors, proprioceptors,* and *interoceptors.* Exteroceptors are on the external body surface, and receive stimuli from outside. They include touch, skin pain, temperature, light pressure, sight, hearing and smell. Some, dependent on direct contact, are called contact receptors, while others which are stimulated from a distance are called *teloreceptors.* The latter include smell, sight and hearing, and receptors for distant heat.

Proprioceptors receive their stimuli from deeper parts, especially muscles, joints and tendons, and thus make possible recognition of movement and position. They also receive sensations of deep pain and pressure.

Interoceptors receive impressions of internal body organ activity such as digestion, circulation, excretion, hunger, nausea, thirst, sexual feeling and feelings of sickness or well-being.

It has been found that receptors do not relay information to different nerve fibers as a discharge of only a single impulse, but that a *repetitive* discharge occurs which is a "code" of the external stimulus. The frequency of the repetitive discharge, the intensity of the stimulus, and the type of receptor stimulated determine the sensory quality "felt" by the central nervous system.

Receptors can adapt to stimuli. That is, they may become accustomed to certain stimuli and send out less frequent discharges. For example, you pay little attention to the weight and touch of your clothes most of the time.

Effectors. Effectors are the muscles and glands put into action by motor nerves which are activated in response to impulses received by sensory nerves from receptors.

THE NEURON

The neuron, or nerve cell, is the anatomical unit of the nervous system. Each neuron is derived from a single embryonic cell. The neuron is also the functional unit of the nervous system and is the only part that conducts nerve impulses. All neural circuits are composed of chains of neurons. The nervous system is made up of billions of these highly specialized cells, bound together by connective tissue, the *neuroglia,* found only in the central nervous system.

Nearly all of the body's nervous tissue arises from cells set aside early in the embryonic life of the individual.

Fig. 8-3 A typical motor neuron.

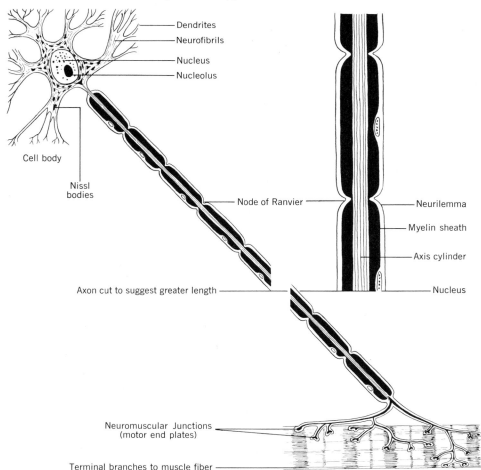

Dendrites

Neurofibrils

Nucleus

Nucleolus

Cell body

Nissl bodies

Node of Ranvier

Neurilemma

Myelin sheath

Axis cylinder

Axon cut to suggest greater length

Nucleus

Neuromuscular Junctions (motor end plates)

Terminal branches to muscle fiber

Why these neurons grow and elongate to connect the various sense receptors with the central nervous system and the central nervous system with the effectors of the body (muscles and glands) is not clearly understood. There is evidence to indicate that the direction of growth is determined by certain biochemical factors appearing along predetermined pathways, but more confirmation is needed to clarify the point.

In the structural make-up of neurons in man and higher animals, we find three distinct parts: the *cell body,* one or more *dendrites,* and a single *axon.* The latter two parts, which extend beyond the cell body, are collectively called nerve cell processes (Fig. 8-3).

The cell body consists of a mass of *neuroplasm* surrounding a nucleus in which is found a nucleolus. Within the nervous system, the shape of the cell bodies varies a great deal: some are round, as in various sensory ganglia; others are diamond or pyramidal in shape, as in the cerebral cortex; still others may be star-shaped, such as those in various motor neurons (see Fig. 8-4 which shows neurons of various types).

The processes of a neuron have a gel-like internal consistency. With the aid of an electron microscope, a thin outer membrane can be seen covering the central core (axis cylinder) of the fiber. Surrounding and protecting the fiber may be found a myelin sheath and a neurilemma.

In the neuroplasm of the cell body are granular structures that absorb basic dyes such as methylene blue. They are named *Nissl bodies* after their discoverer Franz Nissl, a German neurologist (1860–1919). The Nissl bodies apparently have a metabolic function in the cell for they contain ribonucleic acid. The amount varies according to cellular activity. A resting cell shows a relatively greater amount than a cell in a state of fatigue. In addition, these Nissl bodies show losses of stainability with certain fevers, lack of respiration (asphyxia) or when a nerve process has been crushed or cut.

The nerve cell processes, like the cell body, vary greatly in form (see Fig. 8-4). The dendrite is a branching structure (Greek, *dendron,* tree). Its function is to convey impulses toward the cell body, and it is thus referred to as the afferent process (Latin, *ad + fero,* carry toward). The axon is typically long and less branching than the dendrite. It carries impulses away from the cell body and is called the efferent process (Latin, *ex + fero,* carry away from).

The number of processes which extend from the body of the cell provide one means of classifying neurons. A neuron may be *unipolar,* consisting of a cell body and a single process. Unipolar neurons are found in lower animals, but are not present in man. In the human body we find some *bipolar* neurons, made up of a cell body, and only one dendrite, and one axon. Most frequently, however, we find *multipolar* neurons, with a single cell body and numerous dendrites plus a single axon (see Fig. 8-4).

Microscopic Structure

As we have seen, the structure called a nerve is actually many nerve fibers (axons, dendrites, or both) bound together. The fine structure of these fibers varies, depending on their function.

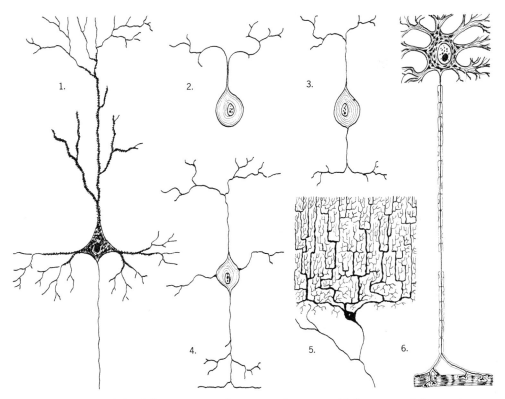

Fig. 8-4 A few of the different types of neurons: 1, a pyramidal neuron of the cerebrum; 2, a unipolar neuron; 3 a neuron with a short axon; 4, a neuron from the cerebral cortex; 5, a neuron from the cerebellum (Purkinje cell); and 6, a motor neuron from the spinal cord.

The central part of the nerve fiber is called the *axis cylinder*. In peripheral neurons, this cylinder is protected by an insulating covering, the *myelin sheath*. This is composed of protein and fatty or lipoid substances which give the process a pearly white appearance. Myelinated fibers conduct more rapidly than those without myelin, thus myelin may furnish a source of energy. Those neurons concerned with conduction within the central nervous system are also covered with a myelin sheath. With the peripheral neurons, they make up the *white matter* of the nervous system. (See Fig. 8-5.) Neurons lacking a myelin sheath, when together, appear rather dull and darker than myelinated fibers. Unmyelinated fibers are found in the brain and spinal cord. With the nerve cells and ganglia, they form the *gray matter*.

On all peripheral neurons the myelin sheath is surrounded by an outer covering—the *neurilemma*. It is indented at various points, forming segments much like a chain of sausages. Each indentation is a *node of Ranvier* (Fig. 8-6).

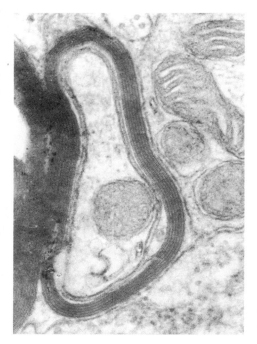

Fig. 8-5   Myelinated axon from the cerebellum of an opossum in cross section. This electron micrograph shows the axis cylinder (light central area) surrounded by closely packed layers of myelin (black). (Drs. K. A. Siegesmund and C. A. Fox, Marquette University School of Medicine)

In each segment is a nucleus. The neurilemma is important in the repair of any damage to the process. If a peripheral nerve is cut, the section farthest away from the cell body *degenerates* (Fig. 8-7). However, in time it can *regenerate* by growth of fibers through the neurilemma sheath.

Within the central nervous system, neurons lack a neurilemma and thus are unable to regenerate (repair) if injured. The cell body cannot reproduce, so that damage to this portion of a neuron is always permanent.

The Nerve Impulse

Many of you probably drive. You may at times have been confronted with an emergency situation, such as another car suddenly pulling away from the curb directly in front of you. Then you quickly put your foot down hard on the brake pedal. But there was a lapse of time between seeing the danger and your muscular reaction to avoid what might have been a disastrous situation. That was your *reaction time.*

How was your brain informed of this sudden problem, and why did your muscles react? For many years, this type of situation perplexed neurologists. Recent knowledge gained from experiments on the neuron indicates that there are both electrical and chemical changes involved whenever you respond to a stimulus.

The passage of the impulse over the neuron was once thought to resemble the flow of an electrical current through wires. However, two important differences were observed between the nerve impulse and the current. First, the speed of a nerve impulse (from 1 to 120 meters per second) is very slow when compared to that of an electrical current, which flows at a rate approaching the speed of light (3×10^5 kilometers per second). Second, there seems to be an electrochemical change in the membrane surface of the axis cylinder of the neuron process which, once started, is self-propagating. The neuron itself supplies energy for the transmission of the impulse, whereas the wire is a passive conductor and is dependent on the "push" of electrons.

Knowledge of the reaction of the neuron to stimulation has been largely

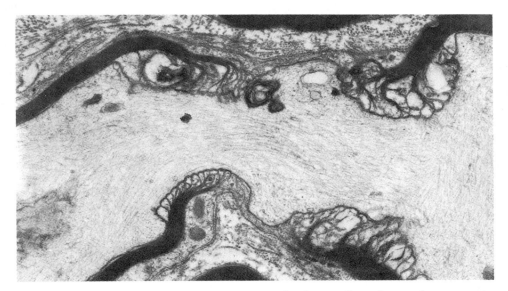

Fig. 8-6   Node of Ranvier. Electron micrograph of a myelinated axon from a rat in longitudinal section, same magnification as previous micrograph. Note how the myelin sheath is interrupted at the node (center) while the neurilemma continues across it. (Drs. K. A. Siegesmund and C. A. Fox, Marquette University School of Medicine)

Fig. 8-7   Regeneration of a nerve fiber. 1, the nerve fiber is cut; 2, the myelin sheath breaks up into droplets of fat; 3, the myelin sheath and neurilemma begin to form; 4, neurofibrils begin to sprout from the cut end of the original fiber.

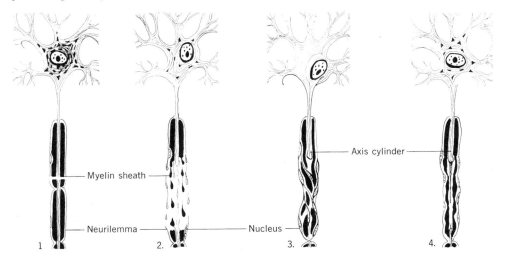

gained through the evidence of *electrical activity*. A great deal of experimentation has been conducted with highly sensitive galvanometers, instruments which measure tiny electrical currents. By employing a cathode-ray oscilloscope, which uses a form of tube, like that in your TV receiver, the electrical effects of a nerve impulse can be observed and recorded as small variations in potential (voltage).

Nerve tissue is highly developed to respond to stimuli and to conduct impulses. When a receptor responds to a stimulus, such as a mild electrical shock, an impulse is initiated and conducted along the neuron associated with the receptor. Under laboratory conditions, nerve fibers may be isolated for stimulation and the resulting impulse can be recorded either visually or on paper. Electrodes are placed on one end to stimulate the fibers by giving a brief shock with an electrical current. The nerve impulse so initiated

travels down the fibers to the recording electrodes in the form of a weak electrical disturbance and it must be amplified before it is observed by means of an oscilloscope. The characteristic pattern observed is called the *action potential* or the *spike potential*. This pattern occurs because the disturbance which passes over the recording electrode is electrically negative with respect to the rest of the tissue on the outside of the fiber (Fig. 8-8) and positive on the inside (Fig. 8-9).

Along with these changes in electrical activity occur others which affect the neuron and limit its ability to react to further stimulation. Its *excitability* is reduced to zero during the action (spike) potential, so that it is impossible to send another impulse along the cell during the *absolute refractory period*. This lasts 0.001 second (or 1 millisecond). Thus, the theoretical maximal rate at which impulses can be transmitted along a nerve is 1000 per

Fig. 8-8   Graph showing changes in electric potential on the outside of the membrane as an impulse travels through a nerve fiber.

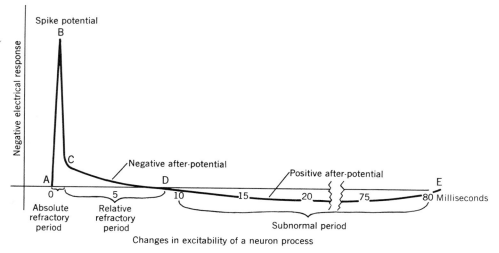

Changes in excitability of a neuron process

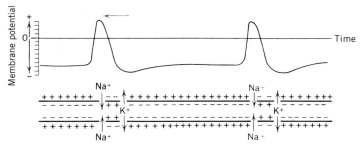

Fig. 8-9 Ionic movements across the nerve fiber membrane (below) coincide with changes in electric potential (above) measured on the inside of the membrane. As sodium ions move to the inside in front of the impulse, the membrane becomes positively charged and a spike potential results. Movement of potassium ions to the outside of the fiber follows and restores the original potential.

second. For about 5 milliseconds a much stronger stimulus is required during the *relative refractory period* for transmission of the impulse. Within a few milliseconds, the neuron recovers its ability to transmit an impulse. A *subnormal period* (positive after-potential) follows before the neuron completely recovers. During this time, however, it can readily transmit another impulse.

Electrical conduction in nerves does not depend on movement of electrons, as it does in conducting metals, but by the movement of ions. The main ones concerned with nerve conduction are sodium ions (Na^+) and potassium ions (K^+). Na^+ is ordinarily in higher concentration than K^+ outside the nerve cell membrane, while K^+ is higher than Na^+ inside. However during an action potential there is a movement of Na^+ into the cell and of K^+ from the cell. This ionic movement is the main generator of the action potential (Fig. 8-9).

Behavior of the nerve impulse. Certain conclusions can be drawn from observation of the way the nerve impulse is initiated and conducted. In some cases the reasons for the observed facts are not clearly understood. How strong must a stimulus be before an impulse is produced? This question brings in a new term—*threshold*. You might think of this in terms of turning on a light. No matter how slowly you throw the switch, you finally reach a point where the electrical contact is made. This point may be compared to the threshold for a nerve impulse. Any single stimulus below this critical point will not initiate an impulse.

If two or more stimuli of less than threshold intensity (individually not strong enough to initiate an impulse) are received within a very short period of time by a receptor, we may find that the second stimulus will reinforce the first and a third stimulus will add its effect to the second. This continues until the threshold level is reached, and an impulse starts. The effect of adding subthreshold stimuli is *summation*.

We have seen that the nerve impulse, unlike the electrical current, is self-propagating. That is, its strength remains constant because the nerve itself supplies energy for transmission.

The impulse weakens only when a section of the nerve is damaged or weakened. Thus, if a resistance is placed in an electrical circuit, the flow of electricity is reduced. However, if we expose a section of nerve to alcohol vapor (which acts like a resistance in an electrical circuit), we depress conduction of a nerve impulse through this area, but if it gets through at all, it is as strong beyond the affected sections as it was before.

Another aspect of this characteristic of the nerve impulse is that the strength of the nerve impulse carried over a single fiber does not depend on the strength of the stimulus. Once the threshold is reached, the fiber responds to its full capacity. We refer to this factor as the *all-or-none* response or law. The law applies also to the speed at which the impulse is conducted. Different fibers may have different rates of conduction, depending on their size and structure, but each fiber conducts at its own maximum rate, regardless of the strength of the impulse. For example, large fibers will conduct impulses at the rate of 5 to 120 meters per second. Thinner fibers will conduct impulses at a speed of 3 to 15 meters per second. Unmyelinated fibers are thinnest and will carry impulses at the slowest rate—from 1 to 2 meters per second.

We have seen that the changes in our environment are detected by highly specialized nerve endings. Each one is extremely sensitive to a particular type of stimulus and relatively insensitive to others. All receptors have a definite threshold, and must obey the all-or-none law. To that stimulus for which it is specialized—heat, pain, touch, light, taste, etc.—the receptor has a relatively low threshold. To the other stimuli, the particular receptor has a high threshold. It is possible that an extremely intense stimulus might trigger an impulse in a receptor normally insensitive to that stimulus. Thus, the boom of a cannon or a jet plane flying over your head may cause pain in your ears, and though a blow on the eye should be felt and not seen, it may result in the sensation of visible flashes or "stars," even though the eye received no light stimulus.

Let us now turn briefly to the evidence for the chemical activity underlying the nerve impulse. Less is known about this subject than about the electrical activity just described.

Oxygen is necessary to the life of all cells. We know that the rate at which oxygen is used can be correlated to a degree with the cell activity. This has been clearly shown for both nerve and muscle tissue. A neuron at rest consumes a small amount of oxygen at a constant rate in maintaining the concentrations of Na^+ and K^+ on each side of the cell membrane, against the concentration gradient. Upon activity there is a marked increase in this oxygen consumption. This higher oxygen utilization lasts for a short period of time after the impulse has passed and is correlated with the re-establishment of the concentration of sodium and potassium ions on either side of the axon membrane. Neurons found within the central nervous system cannot operate in the absence of oxygen. Their use of oxygen is greater than is that of peripheral nerves and the dependence of the central nervous system on oxygen is more immediate.

Neural Connections

Thus far, we have been discussing the passage of the nerve impulse along a single neuron. However, impulses must travel over pathways connecting many different neurons. When a receptor is stimulated, the impulse travels along the dendrite (afferent process), through the nerve cell, and along the axon (efferent process). In order to reach the central nervous system, the impulse must be transferred to other nerve fibers. The junction between the axon of one neuron and the dendrite or cell body of the next neuron is called a *synapse*. When a synaptic junction is observed with the aid of an electron microscope, it is seen that the cytoplasm of adjacent cells remains separated by distinct membranes. In fact, the intercellular distance between adjacent membranes is found to measure between 100 and 200 Ångstroms. Since there is no direct contact, a synapse should be considered a "functional" rather than a "structural or anatomical" connection.

Synapse never occurs between the dendrites of two different neurons, nor

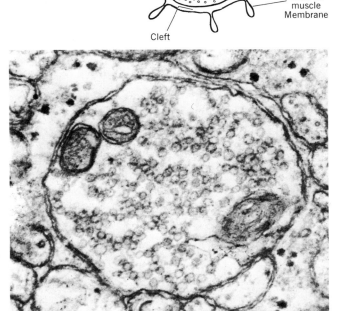

Fig. 8-10 Neuromuscular and synaptic junctions. Diagram of an axon in cross section terminating in muscle, to show the neuromuscular junction, and electron micrograph of an axon just before its termination at a synapse. Note in both fibers the absence of a myelin sheath, the presence of mitochondria and the presence of vesicles believed to contain the transmitter substance which is released into the junction gap upon arrival of the nerve impulse and excites the membrane of the muscle fiber or the post synaptic neuron. (Micrograph: Drs. K. A. Siegesmund and C. A. Fox, Marquette University School of Medicine)

does it occur between the two axons of different neurons. In some instances, a number of axons from different neurons will converge on the dendrites of a single neuron and, in like manner, the axon of one, by means of branches (collaterals), may diverge to make synaptic contact with a number of other neurons.

The neurons in the central nervous system excite other neurons at synapses or the junction of one fiber with a dendrite or cell body of another neuron. It is believed that transmission at these synapses is accomplished by the release of a small amount of a specific chemical transmitter substance. The chemical identification of this substance is not yet known and in fact, it is suspected that a number of such transmitter substances are present within the central nervous system. They probably have quite a bit of similarity in their action to acetylcholine at the neuromuscular junction (see Fig. 8-10). The action of acetylcholine and the enzyme cholinesterase has been described before, in Chapter 6, page 82. There it has been learned that acetylcholine is secreted by the end fibers of the motor axons. This substance when released at the nerve endings excites a special response within the end-plate region of the muscle fiber—*the end-plate potential*—and this end-plate potential in turn is responsible for exciting the propagated action potential in the muscle fiber. The enzyme cholinesterase terminates the acetylcholine action in a very brief time, of the order of 1 millisecond. It is believed that the unknown mediator, or mediators, within the central nervous system act at the synapses in like fashion. Acetylcholine is a transmitter substance in one certain class of synapses within the central nervous system, but it has been proved not to be responsible for transmission at a number of other synaptic junctions.

With respect to the enzymatic actions which acetylcholine and cholinesterase show in tissues, we are indebted to the work of Dr. D. Nachmansohn. He has shown that in the electric organs of the torpedo and the electric eel which can generate up to 800 volts, cholinesterase and acetylcholine are important materials in the production of these large voltages. These electric organs are believed to have come from motor end-plates that developed this specialized function in the process of evolution.

Acetylcholine is found in some places in the central nervous system, and it is possible that it has some function in cells of the central nervous system besides that involved in transmission of nerve impulses.

VOCABULARY REVIEW

Match the phrase in the left column with the correct word in the right column. *Do not write in this book.*

1 Conduct nervous impulses away from the central nervous system.

2 Groups of nerve cell bodies.

a afferent fibers
b cutaneous senses
c efferent fibers

3 A nerve which consists of both sensory and motor nerve fibers.

4 Conduct nervous impulses toward the central nervous system.

5 Special sensory end organs in the skin.

6 Insulating covering of a neuron.

7 Outer covering of a neuron.

d ganglia
e mixed nerve
f dendrite
g Nissl bodies
h neurilemma
i axon
j myelin

TEST YOUR KNOWLEDGE

Group A

Select the correct statement. *Do not write in this book.*

1 The outer layer of the human brain contains (a) 14 million cells (b) twelve hundred thousand cells (c) 14 billion cells.

2 Ganglia are (a) nerve fibers (b) groups of nerve cell bodies (c) receptors.

3 Receptors are (a) centers in the brain that receive nerve impulses (b) specialized nervous tissue sensitive to changes in our surroundings (c) nerve centers in the spinal cord.

4 Nerve fibers conducting impulses away from the central nervous system are (a) efferent fibers (b) receptor fibers (c) afferent fibers.

5 Conduction in nerves depends on (a) movement of electrons as when current passes through a wire (b) an electro-chemical process which is self-propagating (c) an entirely chemical process.

6 When a section of nerve is exposed to alcohol vapor an impulse going through it is depressed. If it gets through this section it is (a) stronger (b) weaker (c) the same.

7 A synapse is (a) the junction between the axon of one neuron and the dendrite or cell body of the next neuron (b) the junction between cranial and spinal nerves (c) the junction between the axons of two different neurons.

8 End-plate potential is produced by (a) an electrical discharge (b) acetylcholine (c) cholinesterase.

Group B

1 What is the difference between the nervous system of an amoeba and of a human being? What do they have in common?

2 What are the three characteristics of the nervous system?

3 Why is the neuron classified as the unit of structure of the nervous system?

4 What is the difference between a subthreshold and a threshold stimulus? What happens if a series of subthreshold stimuli encounter a receptor?

5 If a stimulus is strong enough to excite a neuron, an impulse is created which travels at maximum speed for the individual neuron. What is this activity called? What happens if the stimulus becomes stronger?

Chapter 9 The Central Nervous System

There are millions of miles of wire in the telephone systems of our country. This wire would be useless if it were not connected to central offices where calls could be connected with, and relayed to, their destinations. Similarly, our peripheral nervous systems, with their many miles of nerves and their synapses, would have no value if they were not connected with a central coordinating nervous system.

Since we have already discussed the general structure and functions of the neuron and the peripheral nervous system, we are now ready to study, in more detail, the structure and physiology of the central nervous system: the spinal cord and the brain.

THE SPINAL CORD

In the adult, the spinal cord occupies the upper two thirds of the *vertebral canal,* formed by the *neural foramina.* The foramina are the openings formed by the neural arches and the bodies of the vertebrae, see Fig. 5-13. You may ask why the spinal cord occupies only two thirds of the vertebral column. The answer lies in the fact that the spinal cord is composed of millions of neu-

rons, all of which lose their ability to divide at the time of birth or even before. The protective vertebrae grow longer and enlarge faster than the nervous tissue of the cord. In a newborn infant the spinal cord occupies nearly the entire length of the cavity formed by the vertebrae. Continued growth of the vertebrae during childhood results in a spinal column which is longer than the spinal cord.

Starting at the *foramen magnum,* the opening in the skull at the base of the brain, the spinal cord extends downward to the level of the disk between the first and second lumbar vertebrae. In the region of the coccyx and sacral vertebrae, the vertebral canal contains only the spinal nerves as they descend from their origin in the lower portion of the spinal cord. The cord thus has the appearance of being drawn up inside the vertebral canal.

Vital in its functions, and unable to replace itself, the spinal cord has great need of protection from injury. Any injury might permanently destroy the body's ability to coordinate sensory and motor activities. This vital structure is protected in several ways. First, the bones of the spinal column afford

118

the greatest protection. Because the diameter of the spinal cord (about 13 millimeters) is less than that of the canal in which it is housed, the spine may be moved freely without injuring the cord. Second, the three membranes, collectively called the *meninges,* cover the spinal cord. The outermost membrane, the *dura mater,* is a tough protective covering lining the vertebral canal. Within this we find the second membrane, the *vascular arachnoid,* whose blood vessels provide nourishment. On the surface of the spinal cord is the thin, delicate *pia mater.* As we will see later in this chapter, the meninges of the spinal cord are continuous with those covering the brain.

Between the arachnoid and pia mater is the *subarachnoid space.* Within this space we find the spinal portion of the *cerebrospinal fluid,* which protects the spinal cord from mechanical shocks. This fluid aids in the diagnosis of diseases of the meninges. A sample is taken by a spinal puncture and analyzed for signs of bacteria or viruses. In this way such infections as poliomyelitis and meningitis can be de-

tected. Spinal anesthetics are also injected into the subarachnoid space just below the end of the spinal cord, by means of a hollow needle thrust between the lumbar vertebrae.

A cross section of the spinal cord reveals that it consists of two types of nerve tissue. The central portion is darker; its shape is roughly the letter H. This portion is made up of gray matter, the cells within it act as an integrating center for incoming sensory impulses and outgoing motor impulses. Surrounding the gray matter and rounding out the oval shape of the spinal cord is white matter, composed of sheathed (myelinated) nerve fibers for transmitting impulses to and from the brain.

The four arms of the H formed by the gray matter are called horns. The two to the front are the *anterior horns,* and the two to the rear, the *posterior horns.* These horns give rise at intervals to the *roots* of the *spinal nerves,* the anterior roots leading efferent (mostly motor) fibers from the anterior horns, and the posterior roots bringing afferent (sensory) fibers to the posterior horns (Fig. 9-1).

Fig. 9-1 *Diagram of a cross section of the spinal cord showing the anterior and posterior roots.*

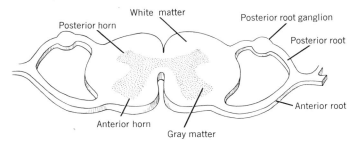

Posterior horn

White matter

Posterior root ganglion

Posterior root

Anterior root

Anterior horn

Gray matter

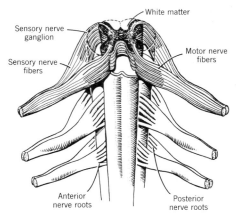

Fig. 9-2   Diagrammatic representation of a section of the spinal cord showing the sensory and motor branches of the spinal nerves.

The anterior and posterior roots join a short distance outside of the cord to form a trunk, the spinal nerve, which is a mixed nerve (Fig. 9-2).

The gray matter with its horns divides the white matter surrounding it into three areas or *columns,* anterior, lateral, and posterior, on each side of the cord. These columns are made up of bundles of longitudinal nerve fibers called *fasciculi,* or *tracts.* Ascending tracts carry impulses up to the brain, while descending tracts carry impulses down from the brain. These tracts are individually named according to their places of origin and their destinations.

The anterior columns contain descending fibers from the motor areas of the brain *(anterior corticospinal* or *direct pyramidal tracts),* carrying voluntary motor impulses to anterior horn cells, and ascending fibers carrying sensory impulses to the brain *(anterior spinothalamic tracts).*

The lateral columns contain the main descending motor tracts from the motor areas of the brain *(lateral corticospinal* or *crossed pyramidal tracts),* as well as ascending tracts carrying sensations of heat, cold, and pain to the brain *(lateral spinothalamic tracts).* Other ascending tracts carry sensations of body position from muscles and joints and receptors in the skin to the cerebellum *(spinocerebellar tracts).*

The posterior columns contain only ascending tracts which carry sensations of touch and "muscle and joint sense," which keeps us aware of the position of our limbs and body even when we are in the dark or have our eyes closed *(fasciculus gracilus* and *fasciculus cuneatus).*

Spinal Nerves

As previously mentioned, the 31 pairs of spinal nerves are mixed nerves, combining fibers from the anterior, or motor roots, and the posterior, or sensory roots, from the spinal cord. The uppermost eight pairs of the spinal nerves are cervical, the next twelve pairs are thoracic. Then come five pairs of lumbar nerves, followed by five pairs of sacral nerves. The last pair is called the coccygeal nerves. The spinal nerves, as you can see, are named according to the regions of the spine from which they emerge. On the posterior root of each nerve is a small swollen area, the *posterior root ganglion,* a group of nerve cell bodies supplying sensory fibers to the spinal nerve. Note that the fibers supplied by the anterior root come from motor neurons which are located in the gray matter of the spinal cord (Fig. 9-3).

A short distance from the cord, just outside the intervertebral foramina, the mixed nerve trunk splits, giving off two branches, the *posterior and anterior primary rami.*

The posterior primary rami go first to the muscles of the back of the skull and trunk, and their motor fibers supply these muscles. Their sensory fibers then pass through the muscles to the skin overlying the same areas.

The anterior primary rami of all except the thoracic area run forward to join in a series of complex nerve junctions called *plexuses,* where they are rearranged and combined to form the final nerves that supply the muscles and skin of the arms and legs. These plexuses are named, according to the areas they supply, *cervical, brachial, lumbar,* and *sacral.* Since the anterior primary rami of the thoracic area run along the lower borders of the ribs, they are called *intercostal* nerves. They supply the muscles between the ribs, the intercostal muscles, as well as the skin over the anterior thorax and abdomen. The general distribution of the spinal nerves is illustrated in Fig. 9-4, which shows the posterior (dorsal) view of the body.

Reflex Activity

The term *reflex* is derived from the word *reflect,* meaning "to turn back." When a receptor is stimulated, as for example, when the finger touches a hot stove, an impulse is carried to the cell body in the central nervous system. The axon of this *sensory neuron,* usually makes contact with a *connecting* (associative) *neuron* in the spinal cord. This contacts a *motor neuron,* which then sends an impulse down its axon to muscles which lift the burned finger off the stove. The impulse is thus turned back to a place near where it started. All of this happens in a split second, even before pain, or consciousness of the situation, occurs. Thus reflexes are involuntary, often unconscious.

Actually, when you touch a hot stove, more than one sensory neuron is stimulated and more than one motor neuron responds, stimulating many muscles to move the entire arm away from the offending heat.

Fig. 9-3   The relation of the various rami to the spinal cord and the sympathetic ganglion chain.

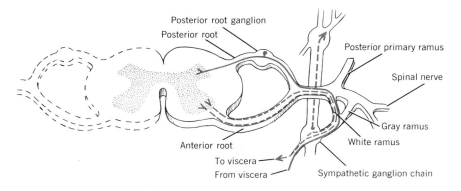

Posterior root ganglion
Posterior root
Posterior primary ramus
Spinal nerve
Gray ramus
White ramus
Anterior root
To viscera
From viscera
Sympathetic ganglion chain

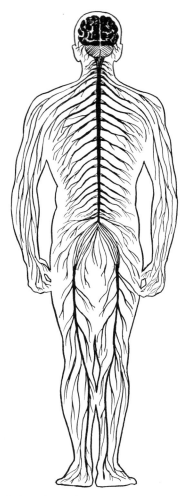

Fig. 9-4 The relationship of the brain, spinal cord, and spinal nerves.

The coordination of sensory and motor neurons involved in a reflex act is known as a *reflex arc* (Fig. 9-5).

Reflexes are of great protective value, guarding us from harm in many ways, for example, the corneal reflex of the eye, which makes us wink involuntarily when a bit of cinder touches the cornea; sneezing or coughing when an irritating substance gets into the nose or trachea. These acts are effective in promptly ridding the body of harmful foreign materials.

Reciprocal Innervation

When the arm is bent (flexed) at the elbow by the action of the biceps muscle, the triceps muscle, which normally straightens (extends) the arm at the elbow, must relax, or the two muscles would be pulling against each other and nothing would be accomplished. The reverse must be true when an attempt is made to extend the arm.

This is accomplished by a process called *reciprocal innervation,* in which cooperation is brought about between the nerves supplying any antagonistic pair of muscles. When one receives an impulse to contract, the other relaxes because it does not receive an impulse to cause contraction. It is therefore *inhibited* at the same instant that its antagonist contracts. Without this reciprocal innervation, coordinated muscular activity would be impossible.

THE BRAIN

The adult human brain weighs about three pounds. It occupies the cavity formed by the bones of the cranium and is thus one of the best protected organs in the body. (See Fig. 9-6 and Trans-Vision following page 212). Further protection is afforded by the continuation of the three meninges of the spinal cord. Adherent to the bones of the skull is the dura mater, attached to the gray matter of the brain itself is the pia mater, and in between is the archnoid. Between the pia and arachnoid, in the subarachnoid space, is the cerebral por-

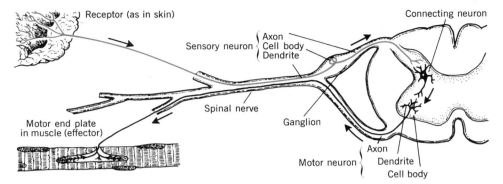

Fig. 9-5 *The nerve pathways involved in a simple reflex arc.* (*Redrawn from Otto and Towle,* Modern Biology, *Holt, Rinehart and Winston, 1965*)

tion of the cerebrospinal fluid, so very important in nutrition and protection.

Among invertebrate animals, the brain is relatively simple and quite small. In some it is simply a large ganglion that functions as the center of control for sensory activities. On the other hand, the vertebrates all have brains of much greater complexity to control body activities. Brains of various vertebrates, from fish through mammals, show great increases in complexity. In the higher animals, there is an increase in size and weight of the brain, and also in the brain weight considered in relation to the total body weight. The

Fig. 9-6 *A section through the human brain showing its major subdivisions.*

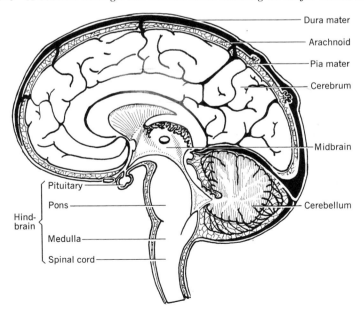

size of the brain, however, is not the primary factor by which to judge its effectiveness. There are several species of mammals that have a brain that is larger in proportion to their body weight than man's, but we do not consider them to be more intelligent. The complexity of the brain, in terms of the number of pathways and centers for the control of specialized functions, is the basis on which it should be judged. In man this complexity has reached its most developed state.

The human brain is derived from the forward end of the embryonic neural tube. This portion of the central nervous system contains the greatest concentration of nervous tissue. In the course of prenatal development, the brain, or encephalon, is first seen as three swellings in the neural tube. The smallest of the three is directly connected to the spinal cord and is called the *hindbrain,* or rhombencephalon, which later develops into the *medulla* and *pons.* Anterior to the hindbrain, is a larger swelling called the *midbrain,* or mesencephalon. The medulla, pons, and midbrain are called the *brainstem.* The *cerebellum* or "little brain" is anterior to the medulla and connected to the pons. The third and largest swelling in the neural tube forms the *forebrain,* or prosencephalon, which later divides into the *cerebrum,* or telencephalon, and the *interbrain* or diencephalon.

The Medulla

The lowest part of the brainstem is called the *medulla oblongata.* This is a vital brain center. In it, as in the spinal cord to which it is attached, the white matter is on the outside, and functions as a pathway for impulses from the higher parts of the brain and from other parts of the body. Also, as in the spinal cord, the gray matter is on the inside. The gray matter of the medulla acts as a coordinating center for incoming sensory impulses and outgoing involuntary motor impulses to the muscles and glands. Certain cells in the medulla's gray matter regulate many vital functions, such as the rate of breathing, heart rate, and circulation of the blood. Disease in this important area, as in injury or in bulbar poliomyelitis, can cause cessation of respiration.

The Pons

Just above the medulla, the brainstem enlarges because of an increase in bundles of fibers. Some of these come down from the cerebrum and cerebellum. There are also ascending fibers from the medulla, and bundles of transverse fibers partly surrounding the brain stem. This enlarged area is called the *pons.* Many cranial nerves originate or emerge from the brain in the region of the pons.

The Midbrain

The highest portion of the brainstem is called the midbrain, which is divided into ventral and dorsal areas (Fig. 9-7). The ventral area is made up of the *cerebral peduncles,* composed mainly of descending (efferent) fibers from the cerebrum. Just above the peduncles are crescent-shaped areas of dark-colored tissue called the *substantia nigra.* This tissue divides the peduncles from the *tegmentum,* which carries ascending (afferent or sensory) fibers to the brain. The tegmentum also contains areas of gray matter.

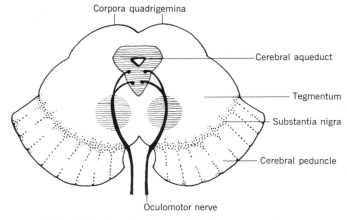

Corpora quadrigemina

Cerebral aqueduct

Tegmentum

Substantia nigra

Cerebral peduncle

Oculomotor nerve

Fig. 9-7 Section through the human midbrain.

Above the tegmentum are the *corpora quadrigemina* (two on each side), which are important reflex centers that regulate movements of the head and eyes in response to sound and light stimuli. Between, and forming part of these structures, runs the *cerebral aqueduct,* the walls of which are of gray matter containing the nuclei and fibers of the oculomotor nerve.

The brainstem is much larger and more complex than the spinal cord. It contains not only the important nuclei of the cranial nerves, but also mechanisms coordinating and integrating each higher center with lower ones.

The Cerebellum

The cerebellum is located above the medulla and is connected to the posterior surface of the pons. The internal structure of the cerebellum differs from that of the brain stem, in that its gray matter, *cortex,* is on its external surface, while white matter fills the space inside except for some masses of gray matter, the *cerebellar nuclei.* The largest of these is the *dentate nucleus.*

Numerous furrows are located transversely on the outer surface of the cerebellum, increasing the area of the gray matter. The cerebellum has two *lateral hemispheres,* divided by a worm-shaped mid-portion appropriately called the *vermis* (Latin, *vermis,* worm). (See Fig. 9-8.)

Cerebellar functions. The cerebellum is constantly receiving impulses concerned with balance, motion, and muscle tone from all parts of the body. When the cerebellum is damaged

Fig. 9-8 The cerebellar hemispheres: left, external appearance; right, sectioned.

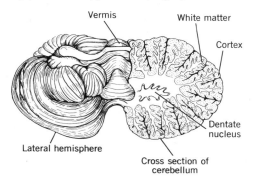

Vermis

White matter

Cortex

Dentate nucleus

Lateral hemisphere

Cross section of cerebellum

by disease or tumor, muscular movements become jerky, uncoordinated, and unpredictable. Fine movements become impossible. Muscles become weak, tremble severely, and lack tone.

Although motor impulses do not originate in the cerebellum, this structure is responsible for coordinating impulses originating in the highest centers of the brain. Thus muscular activities are smoothly and efficiently controlled. In addition, the cerebellum aids in maintaining the muscular tone and balance of the body. All of these activities are carried on below the level of consciousness.

The Interbrain

Between the midbrain and the cerebrum lie two important masses of gray matter, the *thalamus* and the *hypothalamus*. A bundle of nerve fibers connects the hypothalamus and the neural lobe of the *pituitary* gland (discussed in Chapter 30) and controls the liberation, by this part of the gland, of chemicals which affect several basic activities. The functions of the thalamus and hypothalamus have been largely deduced from the symptoms of animals and humans in whom these areas have been injured or diseased.

Injury to the thalamus brings about either increased sensibility to pain or complete unconsciousness. Since the thalamus is primarily a relay station for sensory impulses on their way to the cerebral cortex, injury to one side of it causes a reduction in the appreciation of these stimuli from the opposite half of the body, often associated with spontaneous pain on that side or a perversion of sensation. Thus a stimulus may be unduly painful or even pleasant.

Injury to the hypothalamus causes disorders of gastrointestinal functions, temperature control, blood vessel control, regulation of water balance, metabolism, sleep, and emotional response.

If the connections between the cerebrum and the hypothalamus are cut in an animal which is normally affectionate, it will become a wild, spitting, bristling creature. This condition is called *sham rage*. Brain territories around the medial surface and including basilar parts of the brain have been related with fundamental life functions; i.e., feeding and emotional control. Stimulation or injury within these areas cause remarkable changes in emotional display. Some stimulation sites give reactions similar to sham rage but directed to the source of provocation. Stimulation in one part of the hypothalamus may cause animals to stop eating even though they are hungry. Drinking centers are also revealed by stimulation or injury.

The Cerebrum

The *cerebrum* is the highest part of the brain, both in its location, and in the type and degree of its functions. It occupies two thirds of the cranial cavity and weighs about two pounds.

The cerebrum is responsible for the highest sensory and motor activities. It is divided by a deep groove, the *longitudinal fissure,* into two *hemispheres.* Covering the outer surface of these hemispheres is a layer of gray matter called the *cerebral cortex.* It has the appearance of a walnut, with many ridges and shallow grooves. Each groove is called a *sulcus* (plural, sulci), and the ridge between any two sulci is called a

Molecular	small cells; many fibers
Outer granular	tightly packed small cells; few fibers
Pyramidal	large pyramid-shaped cells; many fibers
Inner granular	smaller star-shaped cells; many fibers
Inner pyramidal	giant pyramid-shaped cells; more of them found in motor area of cortex
Fusiform	tightly packed small spindle-shaped cells

gyrus (plural, gyri), or, more commonly, a *convolution.* These convolutions result in a proportionately large amount of gray matter.

Each hemisphere is further divided into four lobes. The name of each lobe corresponds to the name of the bone of the skull adjacent to it. Thus we find the *frontal, parietal, temporal,* and *occipital lobes.* These lobes are divided from each other by two main *fissures,* the *fissure of Rolando,* dividing the frontal from the parietal lobes, and the *fissure of Sylvius,* dividing the frontal and parietal lobes from the temporal lobe. (See Fig. 9-9.)

Under the microscope, the cerebral cortex is found to be made up of six definite layers from the outside to the center, as indicated in the table (left).

The axons of the cortical cells are of three types, according to their destinations and functions: *projection fibers* which go to other areas of the nervous system, as from motor (pyramidal) cells to the spinal cord; *association fibers* which make connections with other neurons of the same hemisphere; and *commissural fibers* which connect to neurons in the opposite hemisphere.

Cerebral Function

The function of the cerebral cortex makes man the most intelligent of all animals. It provides for conscious thought, reasoning, judgment, will power, and memory. All of these activities are grouped as associative functions. So that we can become aware of our external and to some extent internal surroundings, the cerebrum interprets the sensory activity of the special senses of sight, hearing, taste, smell, the cutaneous or skin senses and the visceral sense organs. Finally, the

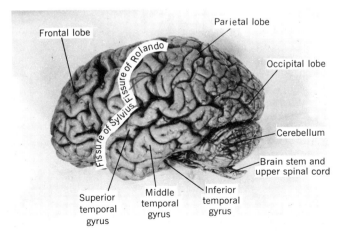

Frontal lobe

Parietal lobe

Fissure of Rolando

Occipital lobe

Fissure of Sylvius

Cerebellum

Brain stem and upper spinal cord

Superior temporal gyrus

Middle temporal gyrus

Inferior temporal gyrus

Fig. 9-9 A lateral view of the human brain showing its external features. (American Museum of Natural History)

cerebral cortex is responsible for the voluntary control of our skeletal muscles. As a result of this, it is possible for us to perform purposeful acts.

It is now recognized that no single area of the cerebrum is entirely responsible for total intellectual capacity. However, if an animal's entire cerebral cortex is removed, it will remain conscious, express rage or fear, eat, and respond to loud noises or visual stimuli. Yet it will show no real intelligence, and its emotional reactions (sham rage, for example) will be uninhibited, because lower centers, such as those in the thalamus and hypothalamus, are no longer under the controlling influence of higher brain levels.

Intelligence develops, in part, through storage, in various cortical areas, of impressions received by our special senses (sight, hearing, smell, taste, and touch). These stored impressions are linked by association fibers so that they can, in most cases, be recalled to the conscious level at will, or remembered.

Memories may exist for a lifetime which suggests that memory depends on some permanent change in the neurons. This is especially true of remote memory. Recent memory is less stable.

Motor activities of the cerebral cortex. It is possible to map accurately the *motor areas* of the cortex by experiments on anesthetized animals, and on humans during brain operations. This is done by applying electrical stimuli to the exposed cortex. It has been found that the motor area *(precentral gyrus)* occupies a long band of cortex just in front of the fissure of Rolando in the posterior part of the frontal lobe. Here the centers for various muscle groups are arranged in reverse order; those concerned with movements of the jaws and tongue are lowest on the lateral surface of the brain, while those sending impulses to the muscles bending the knees are at the top, and toe movements are on the medial surfaces of the hemisphere, in the longitudinal fissure. (See Fig. 9-10.)

Sensory activities of the cerebral cortex. Across the central fissure from the motor area, in the front of the parietal lobe, is the *sensory area,* also called the *somesthetic area,* of the

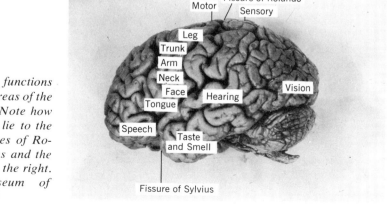

Fig. 9-10 The functions of the different areas of the cerebral cortex. Note how the motor areas lie to the left of the fissures of Rolando and Sylvius and the sensory areas to the right. (American Museum of Natural History)

Fissure of Rolando
Motor / Sensory
Leg
Trunk
Arm
Neck
Face
Tongue
Speech
Taste and Smell
Hearing
Vision
Fissure of Sylvius

cortex. This area receives impulses from receptors of sensations of movement in muscles and joints (kinesthetic sense), touch, heat, cold, and, at its lower end, taste.

Not all sensory impulses are received in the main sensory area. Those for hearing are received in the upper region of the temporal lobe, those for vision in the occipital lobe (inner or medial surface), and those for smell in the anterior part of the temporal lobe.

Motor pathways from the cerebral cortex. Impulses which govern voluntary muscular movements begin in the pyramidal cells of the motor area of the cortex. The axons of these neurons pass down the pyramidal tracts through a mass of fiber tracts called the *internal capsule,* and through the brain stem to the medulla, where a small number of them continue directly via the anterior columns of the spinal cord to its anterior horn cells *(direct pyramidal tracts).* However, a large number of pyramidal fibers cross in the medulla to the opposite sides and descend in the lateral columns of the spinal cord *(crossed pyramidal tracts),* to anterior horn cells.

The pyramidal tracts, which constitute the *upper motor neurons,* join,

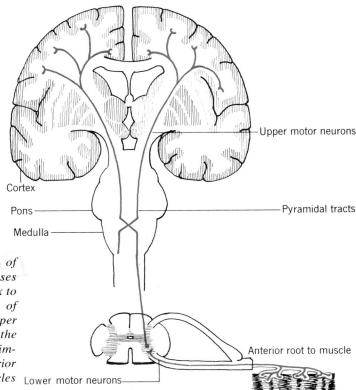

Upper motor neurons

Cortex

Pons

Pyramidal tracts

Medulla

Anterior root to muscle

Lower motor neurons

Fig. 9-11 The path of cerebral motor impulses from the cerebral cortex to the anterior horn cells of the spinal cord (upper motor neurons). From the anterior horn cells, impulses from the anterior root travel to the muscles (lower motor neurons).

through synapses, the anterior horn cells of the spinal cord, which then send their axons to their respective muscles. The anterior horn cells with their axons are *lower motor neurons* (Fig. 9-11).

Injury to the motor neurons of the cortex or pyramidal tract fibers, as by a cerebral hemorrhage or stroke, results in a spastic or "tight muscle" paralysis. Because of the crossing of the pyramidal tracts, injury to fibers on one side of the brain causes paralysis on the opposite side of the body, called hemiplegia. This is *upper motor neuron paralysis.* In this type of paralysis, the tendon reflexes are overactive, and wasting of the muscles, though it may occur, is not usually severe.

On the other hand, injury to anterior horn cells, as in poliomyelitis, or injury to motor nerves, causes a flaccid or "flail-like" paralysis. This is *lower motor neuron paralysis.* In this type of paralysis, tendon reflexes are reduced or absent, and great wasting or atrophy of the muscles usually occurs.

Sensory pathways to the cerebral cortex. Sensory nerve cell bodies, located in the posterior root ganglia of the spinal cord, receive impulses

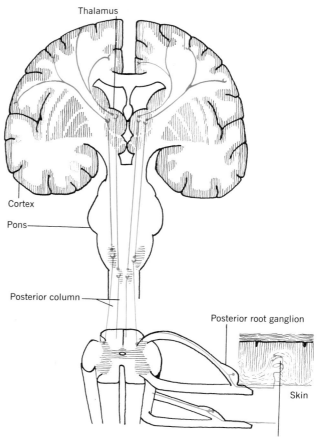

Thalamus

Cortex

Pons

Posterior column

Posterior root ganglion

Skin

Nerve ending for touch

Fig. 9-12 Sensory pathways from the skin to the sensory areas of the cerebral cortex.

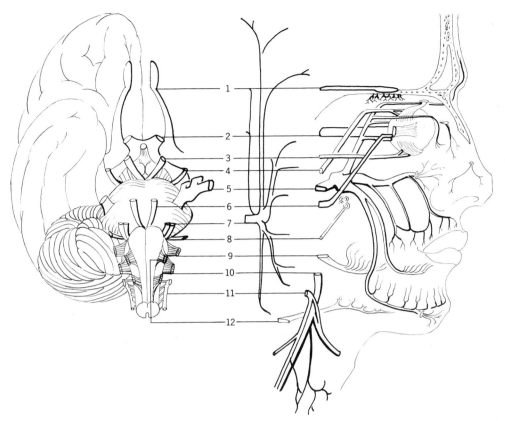

Fig. 9-13 Left: the base of the brain showing the cranial nerves. Right: the relationship of the cranial nerves to the organs of the head. Refer to the table on the next page for details of the nerves and their functions.

from receptors located all over the body. Axons of cells identified with position, joint, and muscle sense ascend directly in the posterior columns to the medulla. Here they form synapses with cells whose axons then cross the mid-line and ascend through the pons and midbrain to the thalamus. In the thalamus, synapses are formed with cells which then send axons to the appropriate sensory areas of the cortex. Axons of posterior root ganglion cells conveying pain, temperature, and touch sensations form synapses in the pos-

terior horn of the gray matter, cross the mid-line of the spinal cord and ascend the lateral and anterior columns of the white matter to the thalamus, where they again form synapses and project to the parietal cortex. (Fig. 9-12.)

The Cranial Nerves

Nerves arising from the brain are called cranial nerves. These nerves are numbered according to their point of action, from above downward (Fig. 9-13). With the exception of the 10th, or *vagus nerve,* which supplies the

THE CRANIAL NERVES

Name and Number	Action
1 Olfactory (Sensory)	Sense of smell (Chap. 18)
2 Optic (Sensory)	Sense of vision (Chap. 10)
3 Oculomotor (Motor)	Movement of extrinsic muscles of the eye except superior oblique and external rectus; through parasympathetic fibers, control of iris constrictor muscles (Chap. 10)
4 Trochlear (Motor)	Movement of superior oblique muscle of the eye (Chap. 10)
5 Trigeminal (Sensory and motor)	Sensation in parts of the face, eyes, nose, teeth, gums, and tongue (Chap. 15) Movement of muscles of jaw in chewing (Chap. 15)
6 Abducens (Motor)	Movement of external rectus muscle of the eye (Chap. 10)
7 Facial (Sensory and motor)	Sense of taste in anterior two thirds of the tongue (Chap. 15) Movement of muscles of face, scalp, outer ear, and neck Stimulation of secretion, through parasympathetic fibers, in the sublingual and submaxillary glands (Chap. 15)
8 Auditory (Sensory)	Sense of hearing; balance (Chap. 12)
9 Glossopharyngeal (Sensory and motor)	Sense of taste for posterior third of tongue; sense of touch and temperature in palate, tonsils, and pharynx (Chap. 15) Movement of stylopharyngeus muscle in pharynx Stimulation of secretion, through parasympathetic fibers, in parotid gland (Chap. 15)
10 Vagus (Sensory and motor)	Movement of muscles in larynx and pharynx Movement (autonomic) of muscles of heart (inhibition of heart rate), bronchi, esophagus, stomach, pancreas, gall bladder, small intestine, first third of colon Stimulation of gastric and pancreatic secretions (Chaps. 10, 20, and 25)
11 Accessory (Motor)	Movement of trapezius and sternomastoid muscles and, in part, of muscles of larynx (Chap. 7—the muscles involved are mentioned only)
12 Hypoglossal (Motor)	Movement of muscles of the tongue (Chap. 15—the muscles involved are mentioned only)

heart, lungs, and abdominal organs as well, these nerves function in the head and neck areas. Some are entirely sensory, some entirely motor; others contain both sensory and motor fibers and still others carry *autonomic fibers* to innervate involuntary (smooth) muscles, glands, and blood vessels. The table on page 132 lists the cranial nerves and their functions.

Electroencephalography

As early as 1874, Caton, an Englishman, observed electrical fluctuations from the cortex of animals and believed them to be related to the functional activity of the brain. Not until 1929, however, in the laboratory of Hans Berger, was the electrical activity of the human brain successfully recorded. When electrodes are affixed to the scalp, a record (an electroencephalogram or EEG) of changes in the electrical potential of the brain can be obtained. These electrical changes appear to be composed of waves of varying frequencies (see Fig. 9-14). The alpha

Fig. 9-14 A reproduction of the alpha, beta, and delta waves from a typical electroencephalogram.

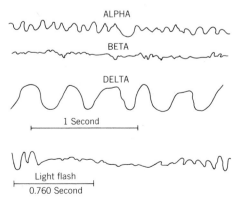

waves (8–14 cps) are the predominant wave forms in the relaxed, normal adult. Also present may be beta waves (15–30 cps) which may be exaggerated by certain diseased states and the administration of certain drugs. The slower rhythms, such as delta (0.5–3 cps) are seen normally in infants, children, and adolescents, and during sleep, and are also fairly characteristic of diseased states of the cerebral hemispheres, such as brain tumors. Other more specific wave forms, such as spikes and spike wave complexes, are seen in association with epilepsy.

The electroencephalogram has thus become a valuable tool in many of the diseased states which affect the intracranial contents of the nervous system. Electrospinograms and electromyograms, using similar electronic principles, are useful techniques for evaluating normal or diseased states of the spinal cord, peripheral nerves, and muscle.

Conditioned Reflexes

Earlier in the chapter we considered *unconditioned reflexes*. These reflexes are present when we are born, and most of them occur even in the lowest animals. They are not altered by, nor do they depend on, past experience. These include such responses as are required for taking food, removing waste, and simple defense from obvious danger.

Any reflex which is modified as a result of experience or training is known as a *conditioned reflex*. The Russian physiologist Pavlov developed methods whereby conditioned reflexes could be studied in animals. For example, he found that food placed in the mouth of a very young puppy which had not been fed before would bring secretion of saliva. This is, of course, an unconditioned reflex. If, however, a bell was rung every time the animal was offered food, eventually saliva would flow every time he heard the bell, even in the absence of food. The puppy was now conditioned to the sound of the bell, that is, he remembered the association between the bell and food. Conditioned reflexes obviously require the participation of the cerebral cortex of the brain although some very simple types may be learned by subcortical structures.

Another type of conditioning is called *instrumental* or *operant* conditioning illustrated by the following example. An animal is placed in a cage in which there is a small lever. As he moves about he accidentally strikes the lever which releases a small pellet of food. Eventually he associates striking the lever with release of food and thereafter he does so with a purpose, as he *remembers* or has learned the association of lever and food. The lever might then be fixed so that moving it only in a certain direction would release the food. If the animal, by *trial and error,* learned to push the lever in the right direction each time he would have performed on a higher plane than mere conditioned reflex. Many types of human behavior result from operant conditioning.

Sleep

The true cause of the phenomenon we know of as sleep remains unknown in spite of much intensive research. There are many theories, none of which is immune to challenge. They range

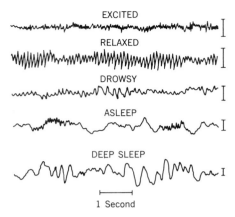

Fig. 9-15 *Electroencephalograms showing various states of alertness.*

from reduced blood supply to the brain to Pavlov's theory of conditioned inhibition of brain activity.

Experiments with the EEG have shown that lower brainstem centers control sleeping-waking states, but the pathway from these centers to the cerebral cortex is yet to be found. Other brain centers, such as the thalamus, are also related to sleep.

When the EEG was used experimentally on sleeping individuals, eyeball movements occurred 3 or 4 times during a 7 hour period. Subjects awakened at such times reported they had been dreaming. During movement of the eyeball, the EEG record looked like that from an alert person. This indicates that the cerebrum is involved in the phenomenon called dreaming.

We do know that sleep is necessary for all mammals and that lack of it may lead to serious mental disturbances and possibly death. Man cannot go sleepless much longer than 6 or 7 days without serious and possibly permanent impairment of function.

During sleep, most of the body's functions are carried out on a greatly reduced level as compared with the waking state. The body temperature may drop as much as one or two degrees, the heart rate may be decreased as much as 40 beats a minute, blood pressure falls, muscle tone and reflexes are reduced or abolished, and the volume of urine production drops. On the other hand, secretion of both perspiration and stomach acid may be increased. The EEG shows definite changes in sleep as compared with the waking state (see Fig. 9-15).

The need for sleep varies for different people, but in general children require more than adults.

THE AUTONOMIC NERVOUS SYSTEM

We now come to another part of the nervous system, one often called the autonomic or involuntary nervous system. It is so called because the individual has very little conscious control over the many activities performed by it.

Frequently the autonomic system is considered a part of the peripheral nervous system because of its extensive distribution of nerves away from the brain and spinal cord. However, since there are important direct connections to a part of the forebrain (the hypothalamus) and parts of the midbrain and medulla, as well as the spinal cord, it seems best not to classify the autonomic nervous system directly with any other division of the nervous system, but to consider it a part whose functions are essential to all other parts.

In general, the autonomic nervous system is responsible for the control of our internal environment; the rate of the heartbeat, the peristaltic movements of the stomach and intestinal tract, the response the body makes to temperature changes, and the contraction of the urinary bladder. In addition, we have all experienced the dry mouth, the pounding heart, and sweating palms in response to an emergency, as well as the red blush of embarrassment or anger. These are also responses that are under the control of the autonomic nervous system.

Organization

The cell bodies of the efferent nerves of the autonomic nervous system are located in areas ranging from the midbrain to the sacral division of the spinal cord. The axons of cells within the central nervous system are called *preganglionic fibers,* while the axons of cells lying in ganglia outside of the spinal cord or in or near the muscles or glands innervated are called *postganglionic fibers.*

The autonomic nervous system is divided into two parts, on both anatomical and physiological grounds. They are the *parasympathetic* and *sympathetic* divisions, also called the *craniosacral* and *thoracolumbar* divisions, respectively (Fig. 9-16). The latter terms refer to the location of the cell bodies of each portion in the gray matter of the spinal cord.

The two divisions are largely antagonistic to each other, although they have independent functions as well. If you drive a car with one foot on the accelerator and the other on the brake pedal (as some racing drivers do), you are in a position either to speed up or slow down very quickly. Similar quick changes are achieved by the autonomic nervous system. If you observe the iris of someone's eye you will note frequent changes in the diameter of the pupil. The nerve control of this reaction can be traced to the autonomic nervous system. The craniosacral division is responsible for the contraction while the thoracolumbar dilates the pupil.

Parasympathetic division. The cranial part of the parasympathetic division takes its origin from cells in the

Fig. 9-16 *Schematic drawing of the autonomic nervous system and the organs it supplies. The sympathetic division is in red; the parasympathetic division is in black. The heavy black line represents the vagus nerve.*

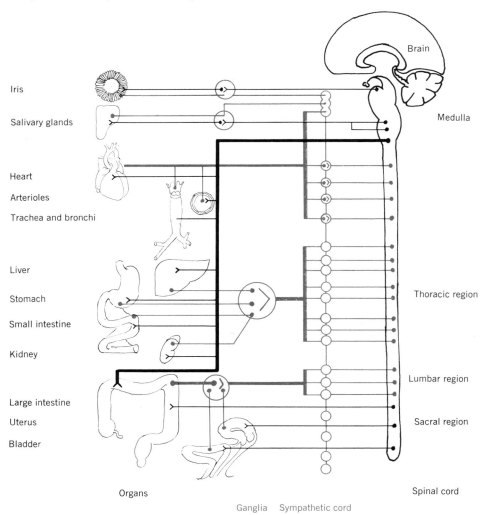

Iris

Salivary glands

Heart

Arterioles

Trachea and bronchi

Liver

Stomach

Small intestine

Kidney

Large intestine

Uterus

Bladder

Brain

Medulla

Thoracic region

Lumbar region

Sacral region

Spinal cord

Organs

Ganglia   Sympathetic cord

midbrain, pons, and medulla. Four cranial nerves carry autonomic preganglionic fibers. These include the oculomotor (to constrict the pupil), the facial and glossopharyngeal (secretory to salivary glands), and the vagus, which is mainly a parasympathetic nerve (motor and/or secretory to the heart, bronchi, esophagus, stomach, small intestine, first third of the colon, liver, gall bladder, bile ducts, and pancreas).

The sacral part of the parasympathetic division comes from cells in the second, third, and fourth segments of the sacral spinal cord. They leave the cord with the anterior roots, then leave the roots and come together to form the *pelvic nerve,* which supplies motor fibers to the involuntary (smooth) muscle of the lower two thirds of the colon and urinary bladder. The pelvic nerve also sends inhibitory fibers to the bladder and anal sphincters.

Sympathetic division. The nerve cell bodies of the sympathetic or thoracolumbar division are in the gray matter of the spinal cord, from its first thoracic to third lumbar segments. Preganglionic sympathetic fibers run from these cells down the anterior roots to their junctions with the posterior roots. Here some of them leave, as slender strands called *white rami communicantes,* to join the sympathetic ganglia. The ganglia are arranged as two beaded *cords* for each side of the body, extending from the base of the skull to the coccyx. Those situated along the sides of the vertebrae, the *vertebral cords,* have smaller ganglia than those running down the front of the vertebrae along with the aorta, the *prevertebral* cords. Each ganglion contains cells with which some of the white rami form synapses, although many white rami send fibers along the cords to higher or lower ganglia. (See Fig. 9-3.)

Axons of cells in the ganglia leave as *gray rami communicantes* to join the spinal nerves, which distribute them.

The sympathetic division is of great importance in preparing the body for

AUTONOMIC FUNCTIONS

Structure Involved	Parasympathetic	Sympathetic
Iris of eye	Contraction of pupil	Dilation of pupil
Salivary glands	Increased salivation	Decreased salivation
Heart	Depressed activity	Accelerated activity
Arterioles	Dilation of vessels	Constriction of vessels
Trachea and bronchi	Constriction	Dilation
Liver	Conversion of glucose to glycogen	Conversion of glycogen to glucose
Stomach	Increased motility	Depressed motility
Intestinal wall	Increased peristalsis	Decreased peristalsis
Adrenal medulla	Decreased production of adrenaline	Increased production of adrenaline
Urinary bladder	Contraction	Inhibition

action at times of stress. It sends accelerator nerves to the heart, and vaso-constrictor nerves to superficial blood vessels. Thus, this system speeds the heart's action, dilates the blood vessels in the muscles (including the coronary vessels of the heart muscle), and constricts the vessels in the skin. This diverts a greater blood supply to muscles and other vital organs. The sympathetic system also dilates the pupils of the eyes for maximum peripheral vision; quiets the gastrointestinal tract, which is not needed in the emergency; and stimulates the production of adrenaline (epinephrine), from the adrenal glands. This hormone liberates sugar from the liver for immediate use in energy production.

Autonomic Chemicals

Until recently, the effects of autonomic nerves on the muscles and glands they stimulate have been thought to be due to the nerve impulses themselves.

Now the parasympathetic nerve endings throughout the body have been found to liberate acetylcholine. Sympathetic nerve endings have been found to liberate adrenaline or substances resembling adrenaline. This has led to use of the terms *cholinergic* for those fibers liberating acetylcholine, and *adrenergic* for those liberating adrenaline substances.

It has been known for some time that the liberation of acetylcholine at the myoneural junction brings about initiation of voluntary muscle contraction (see Chapter 6). It has also been discovered that acetylcholine plays some part in the transmission of impulses across both sympathetic and parasympathetic synapses, as well as performing its known function in synaptic and nerve impulse transmission in the rest of the nervous system.

VOCABULARY REVIEW

Match the phrase in the left column with the correct word in the right column. *Do not write in this book.*

1 Convey sensations of touch and "muscle and joint sense" which keep us aware of position.
2 A series of complex nerve junctions.
3 Large opening in skull at base of brain.
4 Cooperation between the nerves supplying any antagonistic pair of muscles.
5 Area of brainstem from which many cranial nerves originate or emerge from the brain.
6 Supplies sensation in parts of face, eyes, nose, teeth, gums and tongue, and movement of jaw muscles.
7 Lowest part of brainstem.

a foramen magnum
b arachnoid
c posterior columns
d plexus
e reciprocal innervation
f medulla
g pons
h sulcus
i gyrus
j trigeminal nerve

TEST YOUR KNOWLEDGE

Group A

Select the correct statement. *Do not write in this book.*

1 Neural foramina are (a) nerve junctions (b) openings in the vertebral column through which spinal nerves pass (c) nerve cells in the spinal cord.
2 The dura mater is (a) a thin delicate covering of the surface of the spinal cord (b) a tough protective covering lining the vertebral canal and skull (c) a membrane containing the blood vessels that nourish the spinal cord and brain.
3 The anterior roots of the spinal cord (a) bring efferent fibers to the cord (b) bring motor fibers to the cord (c) bring (afferent) fibers to the cord.
4 The connecting neurons within the spinal cord (a) connect the tracts in the cord with one another (b) connect the receptor neurons with the effector neurons (c) connect the brain with the spinal cord.
5 The spinal nerves are all (a) efferent nerves (b) afferent nerves (c) mixed nerves.
6 If the spinal cord is cut, it will (a) regenerate fully (b) will regenerate partially in time (c) will never regenerate.
7 A simple reflex arc represents (a) a learned response (b) an unlearned response (c) a partially learned response.
8 The cerebellum is concerned with (a) balance, coordination of motion and muscle tone (b) coordination of impulses between the cortex of the brain and the thalamus (c) sleep and emotion.
9 The sensory (somesthetic) area of the brain is located in (a) the frontal lobe (b) the parietal lobe (c) the occipital lobe.
10 The autonomic nervous system is responsible for (a) involuntary control of our internal environment (b) voluntary control of respiration (c) coordination of muscular activity.

Group B

1 (a) What are the meninges? (b) Name the three membranes of the meninges. (c) What is the function of each?
2 What is the function of the spinal fluid? Where is it found?
3 Describe the activities of the brainstem by considering the following: (a) the structure and function of the medulla, pons, and midbrain, (b) location, appearance and function of the cerebellum, (c) activities performed by the thalamus and hypothalamus.
4 What part of the human brain is responsible for the highest motor and sensory activity?
5 How is the surface area of the cerebrum increased without an increase in total volume?
6 How does a conditioned reflex differ from an unconditioned reflex? Give examples of each type.

Chapter 10 **The Eye**

The human eye is one of the most remarkable organs in the body. Your eyes convert light waves traveling at a speed of 3×10^5 kilometers per second into nerve impulses which are transmitted by way of the optic nerves to the brain. It is in the cerebral cortex of the occipital lobe that these impulses, resulting from the light stimuli, are interpreted as vision.

As just mentioned above, the center for vision is found in the occipital lobe of the cerebral cortex. The eye is simply a highly specialized receptor, sensitive to light stimuli. We cannot see until impulses from the eye reach the brain. Even though the eye may receive a light stimulus and respond by initiating impulses which travel along the optic nerve, we are totally blind if the impulses are stopped before they reach the visual center in the brain. On the other hand, if during brain surgery, the surface of the occipital lobe of the cortex is stimulated by a weak electric current, a visual sensation is frequently reported by the patient.

Visual function is a very complicated process. We have vision in dim light as well as in bright daylight or artificial light; we have black and white vision as well as color vision. We can see objects at a distance as well as close by; we have acute vision of what we look at directly, and hazy vision of the surrounding area. Some of these aspects of vision will be discussed in Chapter 11. For the present, let us consider the structure of the eye and see how certain aspects of physics apply to its function.

STRUCTURE OF THE EYE

The Orbital Cavity

Each eye lies within a bony socket, the orbital cavity. This is a funnel-shaped depression, made up of seven different bones. If you look at a human skull, you will see a rather prominent ridge around the anterior surface of each orbital cavity. This protective ridge is made up of three bones: on the upper surface and continuing into the eye socket is the *frontal bone;* the bone forming the medial side and about one half of the lower ridge is the *maxilla;* the lateral portion of this ridge and the outer lower portion is made up of the *zygomatic* bone. Forming the interior walls of the orbital cavity are the *sphenoid, lacrimal, ethmoid,* and a small portion of the *palatine.* These seven bones form a protective, unyielding housing for each eye (Fig. 10-1).

Two openings are found in the bones of the orbital cavity. One of these, the *optic foramen,* the smaller of the two, allows both the *optic nerve* and the *ophthalmic artery* to enter the eye. The second opening, the *superior orbital fissure,* provides an entrance for other arteries and veins and the nerves which carry impulses to the muscles that move each eye.

The orbital cavities house, in addition to the eyeball, numerous other structures: the extrinsic eye muscles; the lacrimal apparatus, which is responsible for producing tears; blood vessels; nerves; various tissues called *fascia,* which hold these structures in place; plus a rather heavy padding of fat. This fat tissue cushions the eye itself. It absorbs most of the shock of blows on the external surface of the eye. During illness, as fat is used up by the body for energy, the eyes may appear to sink into their sockets and give a person a ghostly appearance.

Muscles of the Eye

Two groups of muscles operate in the process of vision. One group, the *intrinsic muscles,* is found within the eye. Among these are the muscle of the ciliary body. This structure is mainly responsible for the ability of the eye to focus on near or distant objects. The *iris* contains a second group of intrinsic muscles which control the amount of light entering the eye. We will discuss these muscles later. For the present, let us turn to those muscles responsible for eye movement—the *extrinsic eye muscles.*

Three pairs of striated muscles are responsible for the movement of each eye. The insertions of these muscles lie along the mid-line or equator of the eyeball; and their origins except for one, the inferior oblique muscle, are near the apex of each orbital cavity. Two pairs of muscles in each eye are attached to the eye at right angles to each other and each is called a *rectus* muscle (Latin, *rectus,* straight). Depending upon their position, they are: *superior rectus, inferior rectus, medial rectus,* and *lateral* or *external rectus.*

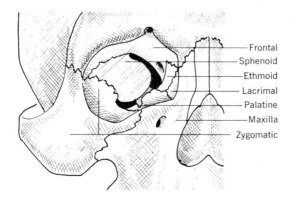

Frontal
Sphenoid
Ethmoid
Lacrimal
Palatine
Maxilla
Zygomatic

Fig. 10-1 Bones of the orbital cavity.

When the superior rectus muscle contracts, the eye turns upward; opposing this muscle is the inferior rectus, which turns the eye downward. Attached to the inner surface of the eye is the internal rectus, which turns the eye inward. The lateral or external rectus is attached to the outer surface of the eye and turns it away from the midline of the body. (See Fig. 10-2.)

The remaining two muscles of each eye are set at an angle to the other four. The *superior oblique* is attached to the upper part of the eye and passes through a small fibrous ring, called the *trochlea,* located on the upper medial surface of the orbital cavity. After passing through the trochlea, the muscle attaches near the apex of the socket

Fig. 10-2 Eye movements caused by the extrinsic eye muscles. The eye at the top illustrates the normal forward looking position. (Redrawn from Best and Taylor, The Human Body, *Holt, Rinehart and Winston, 1963)*

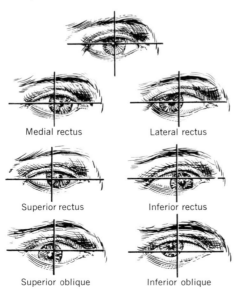

Medial rectus Lateral rectus

Superior rectus Inferior rectus

Superior oblique Inferior oblique

with the four recti muscles. Opposing this muscle is the *inferior oblique.* It arises from the orbital wall of the maxilla and inserts on the under portion of the eyeball. These two muscles act to rotate the eye, probably helped by the recti muscles.

The coordination of the eye muscles is complex. If you look at the eyes of a newborn child, you will probably see that each eye moves independently. As the child grows older, the ability to follow moving objects becomes coordinated in both eyes.

Three cranial nerves supply the connection between the central nervous system and the controlled movements of the eyes. The *oculomotor nerve* (third cranial nerve) sends a number of branches to the superior, inferior, and medial rectus, and the inferior oblique. The *abducens nerve* (sixth cranial nerve) supplies the lateral or external rectus. The *trochlear nerve* (fourth cranial nerve) supplies the superior oblique muscle.

As has been mentioned, a nerve consists of many individual neuron processes. That fact is important, considering that the same cranial nerve controls both opposing muscles used in moving the eye upward and downward (the superior rectus and the inferior rectus). In like manner, when the right eye turns outward, the left eye must turn inward and vice versa, in order to keep both eyes properly focused on an object. Higher centers in the brain are responsible for the complexity of coordinated eye movement.

Protection of the Eye

Except for the extrinsic eye muscles, the accessory organs of the eye are concerned primarily with protection.

The *eyebrows* protect the eye from small particles falling from above the eye. This skin projection, containing rather short stiff hairs, also shades the eye from bright illumination.

The two *eyelids* are folds of skin which protect the eye by closing over its surface. Sphincter-type muscles, the *orbicularis oculi* close the eyelids while the *levator palpebralis superior* aids in opening the eyelids. Although they are under voluntary control, their conscious action is not usually fast enough to prevent injury to the eye. The eye is therefore protected by an automatic reflex action which closes the lids whenever the eye is liable to be damaged by a blow or by some foreign object. The visual receptors within the eye must detect the approaching object and send impulses to the brain. The brain responds unconsciously, and returns impulses to the muscles which close the eye—all in less than a second.

Attached to the anterior margin of both lids is a row of short curved hairs, the *eyelashes*. Near these protective hairs are oil secreting sebaceous glands which lubricate the hairs. If one of these glands becomes infected, the result may be a sty.

Attached to the inner surface of the eyelids and continuous over the surface of the eye is a thin transparent mucous membrane, the *conjunctiva*. When the eyelids close, the conjunctiva attached to the inner surface of the lids comes in contact with that portion of the conjunctiva which is folded back over the surface of the eye. Any inflammation of this membrane is called *conjunctivitis*. A good example of this is "pink eye," a highly contagious infection.

The last protective accessory structure we will consider is the *lacrimal apparatus*. The production of tears is the responsibility of the *lacrimal gland,* located in the orbital cavity in a slight depression of the frontal bone along the lateral superior wall. The tears produced by the lacrimal gland are slightly germicidal and composed mostly of water with a little salt and mucin. The secretion, which is under the control of the autonomic nervous system, flows from the gland through several ducts to the surface of the eye, where it performs its lubricating function. Most of these tears evaporate from the eye's surface, but those which do not are drained into the nasal cavity by way of two *lacrimal ducts,* located at the inner margin of the eye. These ducts connect to the *nasolacrimal duct* which leads to the nasal cavity. When the production of tears increases, due to pain or some emotional state, the lacrimal ducts are unable to carry away the added quantity, and tears overflow the lower lid.

Fine Structure of the Eyeball

The eyeball (Fig. 10-3) is nearly spherical, measuring approximately 24 millimeters in diameter, although the distance from front to back is slightly longer than from side to side. This difference is due to the slight bulge in the front of the eyeball. This region of the eye, the *cornea* is transparent, while the rest of the eye wall is opaque. Structurally, the wall of the eye is composed of three layers.

The outer layer is called the *sclera* or *sclerotic coat*. It is a grayish-white, tough, fibrous membrane which helps to maintain the spherical shape of the eye. The six extrinsic eye muscles insert onto this layer. The cornea is part of the sclerotic coat. This transparent "window" has no blood vessels to interfere with the transmission of light waves. The cells of the cornea are fed by the movement of tissue fluid through the intercellular spaces. The cornea is supplied with receptors for touch and pain, and is extremely sensitive to any object coming in contact with its surface.

At each margin of the cornea we see the sclera as the white of the eye. This layer has few blood vessels. It may appear inflamed, but this is usually due to

Fig. 10-3   Structure of the eyeball. Above: external view. Below: sagittal section to show internal structures.

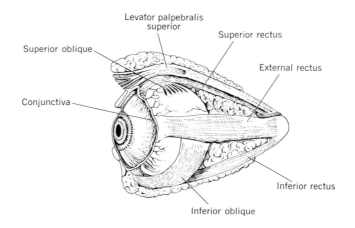

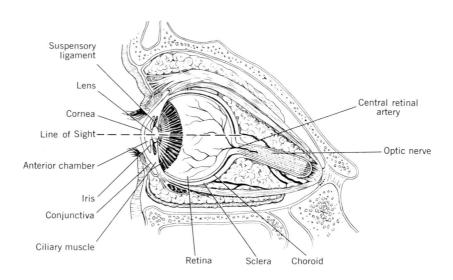

the dilation of the small blood vessels within the conjunctiva which covers this portion of the eye, rather than to enlargement of the vessels in the sclera.

The middle layer of the wall of the eye is known as the *choroid* layer. This contains many blood vessels which bring food and oxygen to it as well as to the other coats. It is colored a deep reddish-purple by the presence of pigment granules. This prevents reflection of light within the eye in much the same manner as the black paint within a camera eliminates stray rays of light that would fog the film.

The anterior part of the choroid layer is known as the *iris*. It may be of a blue, brown, gray, green, or black color depending on the amount of pigment that is present in it. Structurally, the iris is composed of two distinct layers. The layer toward the back of the eye contains a pigment (melanin) in the form of large granules; the layer in front of this may lack pigment entirely or may contain some pigment in either small or large granules. If there is no pigment present, the light rays striking the iris are dispersed and, the iris appears blue. (By a similar circumstance, certain dark-colored birds seem blue. Actually, such birds are brown, but fine lines on the feathers break up the light and reflect it in the form of blue wave lengths.) If there is a small amount of pigment in the outer layer, the eye appears to be green or gray. As the amount of melanin increases, the depth of color approaches black. The distribution of pigment is controlled by the genes (Chapter 33) and is, therefore, inherited.

Within the iris of the eye are two sets of smooth muscles. The inner set is circular (sphincter) and has the ability to constrict the inner edge of the iris. The second set is perpendicular to the circular muscles. When this set contracts, the inner edge of the iris is drawn open.

The hole in the center of the iris is the *pupil*. The pupil is not a structure, but an opening in the iris which varies continually as the two sets of muscles vary in contraction. Thus the amount of light which can enter the eye is determined by the size of the pupil. These changes in the pupil are evident as the eye adjusts to variations in illumination. If a bright light is flashed into the eye, the sphincter muscle contracts and the size of the pupil is reduced very rapidly. In dim light the pupil becomes larger. (See Fig. 10-4.)

Fig. 10-4   Reactions of the pupil to moderate light conditions (above) and dim light conditions (below).

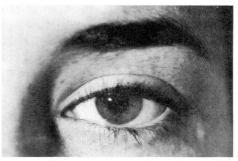

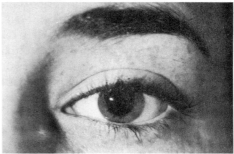

The muscles of the iris are under the control of the autonomic nervous system through the oculomotor (third cranial) nerve.

The innermost layer of the wall of the eye is called the *retina*. It completely lines the inner surface of the eye up to the ciliary body and is composed mainly of different kinds of neurons. The blood supply to this important receptor layer is from two sources: the choroid capillaries which supply a small amount of the nutritional requirements; also, and of greater importance, the *central retinal artery* which spreads throughout the retina. Waste materials are removed by the retinal veins which parallel the arteries.

The retina and vision. Within the retina are the neurons responsible for vision. If the cellular structure of the retina is analyzed, it is seen to consist of many layers. Three of these layers contain neurons which are of special importance. The layer closest to the choroid coat is responsible for converting visual light into impulses. These receptors are called *rods* and *cones* because of their appearance. The axons of the rods and cones make synaptic connections with the bipolar neurons in the second important layer, found in the middle of the retina. Finally, the axons of the bipolar neurons form synapses with the neurons in the third important layer, the multipolar *ganglion cells*. The axons of these neurons all pass across the inner surface of the retina and collect in one area at the posterior part of the eye. These axons collectively make up the *optic nerve* which leads to the brain (Fig. 10-5).

From this description of the retina, it should be noted that light must first pass through the ganglion layer of neurons, then the other layers, including the bipolar cells, before they come in contact with cells directly stimulated by the light: the rods and cones. Once these receptors are stimulated, nerve impulses must again retrace the path of the incoming light waves to the ganglionic axons before they enter the optic nerve. Such an arrangement of cells is called an *inverted retina,* and is typical of all vertebrates.

Rods and cones. These elongated neurons, the rods and cones, are both photosensitive receptors. The rods are more cylindrical than the cones, and measure between 40 and 60 microns in length and about two microns in width. Each rod is divided into an inner and outer segment. In the outer segment we find the chemical responsible for this cell's photosensitivity—*rhodopsin* also called *visual purple*. It is estimated that there are as many as 125 million rods in each retina.

The cones are conical or flask-shaped, with the pointed end toward the choroid layer. Like the rods, these neurons are made up of an inner and an outer segment. There are three kinds of cones each containing a different photochemical substance within the outer segment. The nature of these pigments is not clearly understood. The cones measure about 25 microns in length and up to six microns in width. There are only about 7 million cones in each retina, but they are concentrated in the posterior portion of the eye.

The rods have a low threshold of light sensitivity and are thus responsible for vision in dim light. It is also believed that they are insensitive to

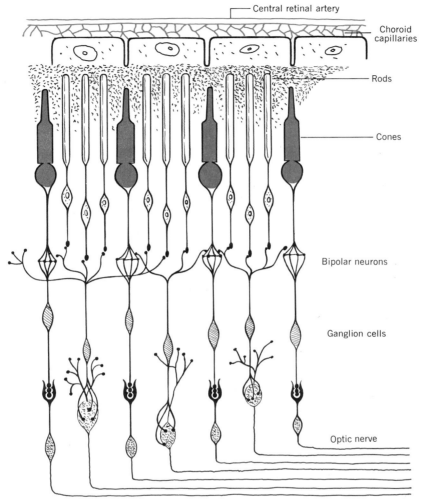

Fig. 10-5 Simplified diagram of the important layers of cells in the retina of the eye.

color. The cones, on the other hand, have a relatively high threshold of sensitivity and are responsible for our perception of color, or *chromatic vision.* This topic will be covered in more detail in Chapter 11.

Other structures in the retina. With the aid of an ophthalmoscope, an instrument which shines a light on the retina and reflects it into the operator's eye, you can see the retinal surface. The most obvious feature on the posterior wall is a yellow disk, known as the *macula lutea.* In the center of this disk is a small area called the *fovea centralis,* the center for distinct and color vision, containing only the cones. The area around the fovea centralis, called the *extrafoveal* or *peripheral region* (Fig. 10-6), contains mainly rods for dim or peripheral vision.

Slightly toward the nose from the fovea centralis is a pale area in which blood vessels can be seen entering and leaving the eye. It is in this area, the *optic disk,* that all of the ganglion axons converge. There are no rods and cones in the optic disk. Thus, we have no visual reception in this area, so it is called the *blind spot.*

There are many ways of explaining why we are not conscious of the blind spot, but probably the explanation easiest to understand is that there is an overlapping of the retinal images of the two eyes, and what one cannot see, the other eye can.

The lens and associated structures. Just behind the pupil is a transparent cellular body, the *crystalline*

lens. This structure is disk-shaped and somewhat elastic. The two surfaces are convex, forming a biconvex lens; but the posterior surface has a greater curvature than the anterior surface. The crystalline lens is held in place by *suspensory ligaments* which are attached to the ciliary body of the choroid layer.

Between the cornea and the iris we find the *anterior chamber.* This chamber is filled with a watery fluid called *aqueous humor,* which helps to maintain the shape of the cornea and also has a nutritive function. The aqueous humor is transparent and is constantly replenished by the blood vessels behind the iris.

Behind the lens and suspensory ligaments and within the space lined by the retina is the vitreous body. It is a transparent jellylike substance which helps to maintain the spherical shape of the eyeball.

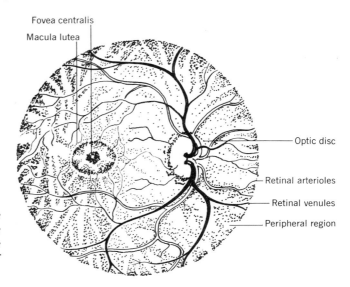

Fovea centralis

Macula lutea

Optic disc

Retinal arterioles

Retinal venules

Peripheral region

Fig. 10-6 View of the posterior surface of the retina. This part of the eye can be examined by use of an opthalmoscope.

Fig. 10-7 Refraction of light. A striped plastic drinking straw appears to bend at a sharp angle where it enters the water.

If the amount of fluid in the eye increases, the eyeball becomes very hard because of the increased pressure caused by the fluid. The optic disk is depressed, and there is restriction of the field of vision. This condition, called *glaucoma,* necessitates immediate medical attention. Blindness may result if the internal pressure is not relieved.

The Physics of Vision

Light is the form of radiant energy which can stimulate the receptors in the retina of the eye. Only through the presence of light are we able to respond visually to our surroundings. The wave lengths of radiant energy to which the eye is sensitive vary from 3.9 ten thousandths of a millimeter to 7.6 ten thousandths of a millimeter. This is often written as 3900 to 7600 Ångstrom units. These wave lengths are very short and travel at the tremendous speed of 3×10^5 kilometers per second.

Light possesses properties of both waves and matter. It is composed of electric and magnetic packets of energy known as *photons.*

For all practical purposes, light may be considered to travel in straight lines through a transparent medium of constant density. However, light rays are bent or *refracted* when they pass from one medium to another of different density. This refraction of light is easily seen if a straw or a pencil is put into a bowl of water so that it leans against the side of the bowl (Fig. 10-7). The object seems to be bent at the surface of the water. This refraction explains why light rays bend as they pass through the cornea and the lens of the eye, and form a clear image on the retina. Of the two structures in the human eye, the cornea has the greater responsibility for refraction.

Light rays traveling from a less dense medium and entering a more dense medium are refracted (bent) toward the perpendicular to the surface of the denser medium. However, if the rays strike perpendicularly to the surface, they are not refracted, regardless of density of the media.

When light passes through a transparent object, the rays are refracted both on entering the object and upon leaving it. Just how much the light will bend depends not only on the curvature of the object but also on its density. If a transparent object, such as a lens, varies in thickness, the light rays will bend toward the thicker portion. A lens that has a uniformly curved surface and is thicker at the edge than in the middle, is called *concave*. Parallel rays of light passing through this type of lens will diverge or spread outward. If on the other hand, parallel light rays pass through a lens that is thicker in the center than at the edge, a *convex* lens, the rays will *converge* and be brought to a single focal point or focus (Fig. 10-8). The distance from the center of the lens to the focal point is called the *focal length* of the lens. If both sides of the lens have a similar curvature, the lens is called *biconcave* or *biconvex* depending upon the location of the thickest part of the lens. Thus the crystalline lens of the human eye is biconvex, even though the posterior surface of this lens has greater convexity than the anterior surface.

Convex lenses invert the images of objects that are further from the lens than its focal length. The image is also reversed from side to side. This can be seen easily by looking at the frosted plate on the back of a portrait camera. If the camera is directed toward an individual, his image appears upside down and reversed in the view plate.

The inversion of the image produced by a convex lens can also be demonstrated by holding such a lens between the eye and an object at some distance from the eye. If you move the lens backward and forward, you will see at one point a clear inverted image of the object. The way in which the visual center of the cerebrum interprets the inversion and reversal of the image produced by the convex lens of the eye will be discussed in Chapter 11.

If a normal eye observes an object at a distance of six meters or more, the light rays entering the eye are almost parallel and the resulting image is in focus on the retina. However, if the object is less than six meters away, the light rays entering the eye are not parallel. If there were no changes in the eye, the image would be blurred since it would not be focused on the retina. In order to form clear images of objects nearer than six meters, two corrections must be made by the eye to sharpen the image on the retina. First, each eye must turn slightly toward the object. This is called *convergence*. To accomplish this, the extrinsic muscles draw the eyes inward so as to point them directly at the object in view. Second, and of greater importance for sharp focus, the eye must *accommodate*. Many theories have been proposed to explain accommodation, but the one most widely accepted involves contraction of the ciliary muscle, and a resulting change in shape of the lens. The action of the muscles is transmitted to the lens by the sling-like suspensory ligament. This tissue surrounds the lens and, as the muscles contract or relax, allows the lens to change shape and permits accommodation. As was mentioned earlier, the crystalline lens is an elastic body. If unsupported, it would assume more of a spherical shape than it has when held in position by the sus-pensory ligament. When the ciliary muscles are relaxed, the suspensory ligament exerts tension on the lens, thus decreasing the lens's curvature.

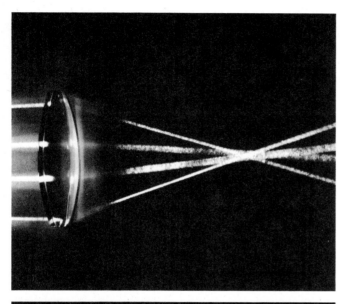

Fig. 10-8 The effects of lenses on the light rays that pass through them. Above: parallel rays pass-ing through a biconvex lens are bent to a focus. Below: parallel rays pass-ing through a biconcave lens diverge.

This is what happens when we look at distant objects. Since the ciliary muscles are relaxed there is little or no fatigue.

When we must focus on an object less than six meters away, the ciliary muscles contract. This in turn decreases the supensory ligament tension. The lens assumes a greater curvature, thus decreasing its focal length. This brings the image into focus on the retina. If you read a book, or do other close work, you may begin to feel the fatigue of the contracted ciliary muscles after several hours.

The entire process of accommodation is automatic. As the image becomes blurred, when the object changes position within the six meter range, sensory impulses are carried to the brain by way of the optic nerves. The brain in turn sends motor impulses by way of autonomic fibers in the oculomotor nerve to the ciliary muscles, which adjust their contraction to bring the image into clear focus again.

VOCABULARY REVIEW

Match the phrase in the left column with the correct word in the right column. Do not write in this book.

1　Chemical responsible for photosensitivity of the rods.
2　Produces tears.
3　Bony socket in which the eye is situated.
4　Middle layer of wall of the eye.
5　Increase in amount of fluid in eye with increased pressure.
6　Electric and magnetic packets of energy composing light.
7　Thin transparent membrane, continuous over the surface of the eye.

a　orbital cavity
b　iris
c　conjunctiva
d　lacrimal gland
e　choroid
f　rhodopsin
g　optic disk
h　photons
i　glaucoma
j　vitreous body

TEST YOUR KNOWLEDGE

Group A

Select the correct statement. *Do not write in this book.*

1　Color vision is due to the activity of (a) rods (b) cones (c) rhodopsin.
2　Peripheral and dim vision is due to the activity of (a) rhodopsin in the rods (b) rhodopsin in the cones (c) opsin in the cones.

3 The blind spot is due to (a) the fovea centralis (b) deficient action of the extrinsic eye muscles (c) lack of receptors (rods and cones) in the optic disk.

4 The anterior chamber of the eye is filled with (a) vitreous humor (b) mucus (c) aqueous humor.

5 Light rays traveling from a thin medium and entering a more dense medium are (a) not bent (b) bent toward (c) bent away from the perpendicular to the surface of the denser medium.

6 When parallel light rays pass through a concave lens they (a) converge (b) diverge (c) go straight.

7 Accommodation is thought to be accomplished by (a) contraction of the ciliary muscle resulting in a decrease in lens curvature (b) relaxation of the ciliary muscle resulting in decrease in lens curvature (c) contraction of the ciliary muscle resulting in a decrease in suspensory ligament tension and an increase in lens curvature.

Group B

1 How is the eye protected from injury?
2 What are the actions of the intrinsic eye muscles?
3 Discuss the actions of the extrinsic eye muscles.
4 What cranial nerves supply the extrinsic eye muscles?
5 What are the three layers of the wall of the eye? What is the function of each?
6 What is the inverted retina?
7 How do the eyes accomplish convergence?

Chapter 11 **Vision**

Probably none of our special sense organs has received as much experimental study as the eye. As we have seen in Chapter 10, the anatomy of the visual organs can be fully described to show how the structures within each eyeball form a mechanism which converts light stimuli into sensory impulses for vision. The accessory structures of the eye, except the extrinsic eye muscles that move the eye, are primarily concerned with protection. The optic components of the eye operate on physical principles to refract the light waves to those cells responsible for visual perception. These parts adjust to changes in distance and intensity of illumination. Finally, the photochemical receptors—the rods and cones—convert the radiant energy of light into nerve impulses. Just how the impulses are initiated and how they result in mental interpretation is still not completely understood.

In order to gain a better understanding of the human eye, let us now see how these structures operate.

General Characteristics

Field of vision. We little realize the full extent of our field of vision. When you drive an automobile down the highway, your visual attention is directed straight ahead of you on the roadway. Yet you are still conscious of movement at each side. This *peripheral vision* does not form a very sharp image, nor is it usually in color, because the associated areas of the retina do not contain cones. Peripheral vision is very important since it enables you to be somewhat aware of what is going on around you without the necessity of constantly turning your head.

You can determine the extent of your own peripheral vision by a simple experiment. Hold both hands together about a foot in front of your nose. Now, while looking straight ahead, move your hands slowly apart and around your head. Continue moving them until you can no longer see them. Most people can see through an arc of about 190°. This area is called your *field of vision*. Your peripheral visual field is highly important in driving a car, in various sports, and in many other activities. A test of a patient's field of vision is helpful in diagnosing certain types of brain tumors.

The retinal image. As we have learned, the image projected onto the retina is inverted and reversed. The visual centers in the brain are responsible for interpreting these changes in position. Experiments have been conducted in which special glasses are worn by a subject so that he sees everything reversed and inverted. After a period of time, the brain once again makes the correct adjustment and the subject sees objects in their true position even though he still wears the glasses. When the glasses are removed, everything seems turned upside down and the brain once again has to readjust. This experiment shows the importance of mental interpretation of retinal impulses in visual perception.

Binocular vision. When we close one eye, the brain receives visual impulses from the other eye only. This is monocular vision. Normally, however, the brain receives impulses from both eyes simultaneously. We know that although our eyes are set at a slight distance apart, we do not "see" two separate images, but one single image in which the two are fused. This phenomenon is called binocular vision.

A single impression is recorded in the brain because the two images fall on corresponding points in the retina. To understand this, we must consider the route the retinal impulses take as they move from the eye through the optic nerves to the visual centers in the brain. Impulses follow the *central visual pathways.* As shown in Fig. 11-1, each retina is divided into a nasal or medial portion, and a temporal or lateral portion. Due to the reversal of the retinal image, the visual field from a point straight in front of the eye to where the bridge of the nose blocks the view is recorded on the temporal

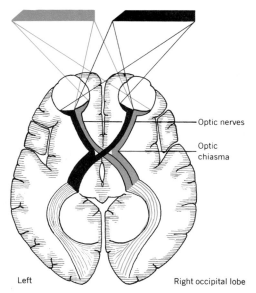

Fig. 11-1 *The visual pathways followed by light stimulated impulses from an object on the right (black) and on the left (red).*

(lateral) half of the retina. The area from straight ahead to the margin of our peripheral view (approximately 95°) is recorded on the medial or nasal half of the retina. If you follow the route of the optic nerves as shown in Fig. 11-1, you will see that the temporal fibers do not cross at the *optic chiasma,* but they join the optic tract on the same side and continue to the visual center in that side of the occipital lobe. The fibers from the nasal half of each retina cross at the optic chiasma and enter the optic tract with the temporal fibers from the opposite eye. The terminations of all these fibers form a field of interpretation in one area of the visual center.

You will note from Fig. 11-1 that the image of an object falls on one half of each retina. Impulses from corresponding points on the retinas are carried to the same part of the brain. Thus both images of an object on the right of the viewer are received in the left occipital lobe, and create a single visual impression.

The fact that the eyes can converge on a single point is an important factor in binocular vision (Fig. 11-2). As an object approaches the eyes, they must turn inward toward each other or the images would not strike corresponding points on the retinas.

One of the remarkable facts about the visual process is that we see things in three dimensions. Not only are we aware of the surface area of objects—the vertical and horizontal planes—but we are also conscious of their volume. We are thus equipped for *depth perception* or *stereoscopic vision*. Since the eyes are set slightly apart, each eye views a single object from a slightly different angle. The left eye sees a little more of the left side of the object than the right eye and vice versa. The images on the retina are not quite identical. However, the visual centers in

the brain intepret these images as one image that has depth and solidity.

Judgment of spatial relationships. Our judgment of distance is dependent upon many different clues which are interpreted by the brain. An infant will reach for the larger of two colored toys even though the smaller toy is held closer to him. Both seem equally near to his judgment. As we gain knowledge of the size of familiar objects, we use their size as a means of judging their distance. For example, we soon recognize the difference in height between a telephone pole and a mailbox. If we see a telephone pole apparently the height of a mailbox, we know that it is a considerable distance from us. We learn that of two objects moving at the same speed, the object closer to the observer appears to move at a greater speed and to cover a greater distance. We also know that those objects nearest us prevent us from seeing a portion (or all) of those objects which are farther away. The degree of accommodation and convergence, as communicated to the brain from the eye muscles, may also indicate the relative position of objects, the nearer object requiring greater effort in focal adjustment. Finally, gradations in the color of near and distant objects and the shadows on their surfaces are other clues used in our attainment of the judgment of depth.

After we have set up guideposts in our visual memory for spatial relationship of objects, we can then confuse the brain by artificially changing certain references. An *optical illusion* is the resulting error in visual judgment. Figure 11-3 gives two simple examples of optical illusions.

Fig. 11-2   The action of eyes in convergence.

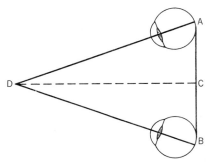

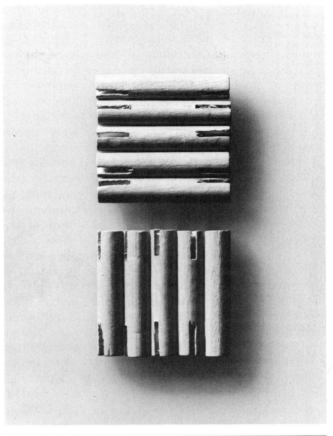

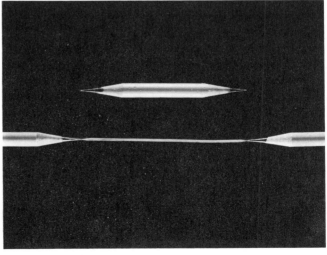

Fig. 11-3 Two illustrations of optical illusions. Judge the heights of the stacks of dowels, and compare the length of the single pencil with the distance between the two pencil points.

Night Vision

The retina of the human eye responds to varying intensities of light, or *achromatic* stimuli, and also to *chromatic,* or color, stimuli. In Chapter 10 we described the structure and location of the receptors responsible for this vision—the rods and cones. Let us now consider the physiological activity of these cells.

The actual process of converting nerve impulses into sight is still a mystery, but many of the chemical changes accompanying it are known. As stated in Chapter 10, the rods are responsible for vision in dim light, and contain the chemical *rhodopsin,* or *visual purple.* By extraction of rhodopsin from rod receptors of experimental animals, the molecular make-up of this chemical has been fairly well established. Its molecular weight is over 200,000, and it is composed of a protein combined with *retinene,* a pigment. When rhodopsin is present in the rods, rod vision is possible and the eyes show *dark adaptation.*

Rhodopsin is stable in dim light, but when exposed to bright light, it dissociates into a protein (opsin) and retinene. The speed at which this takes place depends upon the brightness of the light and the length of time the rhodopsin is exposed. The resulting opsin and retinene, called *visual yellow,* does not show dark adaptation. If the light intensity is then decreased, the two products recombine and after a short period dark adaptation returns. However, if the bright light continues, there is further breakdown of retinene to vitamin A. This protein is called *visual white* and is apparently insensi-

tive to light. The vitamin A is absorbed by the blood. If at this point the light intensity diminishes, a slower return of dark adaptation occurs as rhodopsin is reformed. Vitamin A from the blood supply to the retina is responsible for this return (Fig. 11-4).

Intense light bleaches rhodopsin very rapidly. Thus, bright light may dazzle the eye and cause pain, but fortunately this pain does not last long. If you have driven an automobile at night, you have had such a reaction to an oncoming car with its headlights on high beam. During the time it takes your eyes to readjust to dark adaptation, your level of visual sensitivity is below normal. It should be easily understood why this might prove to be a very dangerous situation, especially if you were traveling at high speed. Many states have passed laws which require you to change from high beam to low beam when approaching another car at night.

As was mentioned earlier, vitamin A is required in the formation of rhodopsin. If the supply of this vitamin is below normal or lacking in one's diet, dark adaptation occurs very slowly. At night, when light intensities are low,

Fig. 11-4 A schematic summary of the chemical changes in rhodopsin on exposure to light and dark conditions.

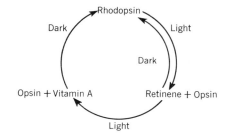

an individual lacking sufficient vitamin A would be nearly blind for a considerable time after a bright light had been shone into his eyes. This condition *is called night blindness*, or nyctalopia.

We have referred to dark adaptation as that adjustment the eye makes to a low intensity of illumination. Adjustment is made by the reconversion of rhodopsin. In order to allow the maximum available light to enter the eye, the pupil dilates. Vision in dim light has less detail and little or no color, and if movement is lacking, objects are seen as large masses with no clear outline. We can see things more clearly in dim light if we look out of the side of the eye, because rods are lacking in the fovea centralis where most of the incoming light is normally focused.

Color Vision

Light adaptation occurs when the eyes are exposed to continued bright illumination such as daylight. Under these conditions, the second type of receptor cell in the retina is used—the cones. When the eyes are light adapted, they see objects in fine detail and in color.

What is the kind of light that produces a rainbow? It is sunlight, which is classified as *white light*. If you hold a triangular glass prism in the sunlight, you will see a similar arrangement of colors (spectrum) on a nearby surface (Fig. 11-5). White light is a combination of all the wave lengths of this spectrum, and therefore black must be the absence of these wave lengths. The reason for this spectral effect is refraction. In passing through the prism, the various wave lengths of light are bent un-

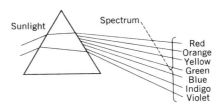

Fig. 11-5 *A prism separates white light into the colors of the spectrum.*

equally and thus separated. The rays of the sun striking droplets of water in the air are refracted and then reflected to produce a rainbow.

It has been stated before that the wave lengths responsible for the visible spectrum are from 3900 Å to 7600 Å. From the spectrum diagram shown in Fig. 11-6, it can be seen that any color is actually a name for light of a band of wave lengths. For example: the label "red" is given to the sensation stimulated by wave lengths starting at 7000 Å. The other colors range through orange, yellow, green, blue, and violet, terminating with a wave length of about 4000 Å. From this list, one should note that the colors in the red end of the spectrum have the longest visible wave length, while those with the shortest wave length fall in the violet end of the spectrum.

Those wave lengths just shorter than our visual spectrum are called *ultraviolet* rays. True ultraviolet rays cannot be seen by the normal human eye, but their presence can be detected when they shine on certain chemicals that have fluorescent qualities. Exposed to such rays, these chemicals appear to glow or emit their own light. Ultraviolet rays cause tanning or burning of the skin.

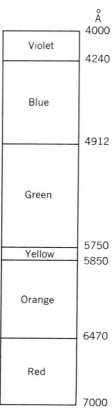

Fig. 11-6   The wavelengths (in Ångstrom units) of visible light and the colors associated with them.

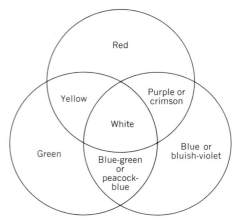

Fig. 11-7   When the primary colors of light are superimposed, various intermediate colors are obtained. White light results from superimposing all three primary colors.

Wave lengths longer than those within our visual range are felt as heat and are called *infrared*. Photographic film that is infrared-sensitive will take pictures in total darkness. The image on the film results from differences in the amount of heat given off, or reflected, by different parts of the photographed object.

White light is the combination of all the wave lengths of the visual spectrum. However, if only three of these wave lengths of light are superimposed upon a light surface, the result is white (Fig. 11-7). The colors used are red, green, and blue-violet. The actual wave lengths of these three colors have been assigned by an international standard: 7000 Å for red, 5460 Å for green, and 4358 Å for blue-violet. We call these three the *primary colors of vision,* or *simple colors.* To this list is often added yellow—5800 Å—since it is very difficult to produce this as a pure color from the combination of any of the above primary colors of light. All of the other colors represent certain combinations of these primary colors and are thus called *compound colors.* The primary colors of vision should not be confused with the primary colors of pigments. Primary pigment colors are red, blue, and yellow. From these three, nearly all colors can be produced except white and black.

Pigments, that is, the colors we see in objects around us, are dependent on the ability of the surfaces to reflect the various wave lengths of white light. If a surface reflects wave lengths of all colors, it appears white. If it absorbs all of the wave lengths and thus reflects none, it appears black. Between these extremes, a surface may absorb some wave lengths, and reflect one or more of them (Fig. 11-8). The waves that are reflected stimulate the cones of the retina and we say the surface is red, blue, green, etc.

Many theories have been proposed to explain color vision. One of the oldest but most widely accepted theories was proposed by Young and Helmholtz early in the nineteenth century. Part of this theory has recently received experimental support. According to the theory, there is a corresponding type of receptor in the retina for each of the three primary colors of light. The three kinds of cone receptors contain different photochemical substances, each chemical sensitive to one of the three primary wave lengths of color.

These three cone receptors and the three pigments, one primarily sensitive

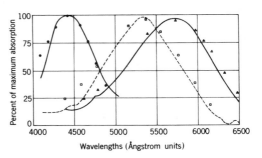

Fig. 11-9 The range of sensitivity of the three light sensitive pigments present in the retina. Left curve, blue; middle, yellow; right, red.

to blue light, one to green light, and one to red light, have now been shown experimentally to exist as predicted by Young and Helmholtz. Spectral sensitivity curves for each are shown in Fig. 11-9. It will be noticed that the red receptor pigment has its peak well into the yellow region of the spectrum, but it extends far enough into the red to sense red as well.

Recordings made with microelectrodes placed in different parts of the retina show that when illumination of the retina decomposes the light sensitive pigments, characteristic changes in

Fig. 11-8 Objects appear colored due to the colors they reflect to our eyes. Other colors of the spectrum are absorbed.

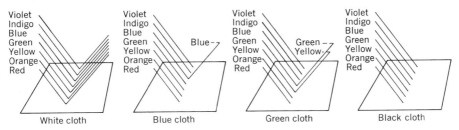

electric potential occur depending on the wave lengths of light involved. These changes are relayed to the ganglionic cells presumably by the bipolar cells. The ganglion cells respond to light by emitting bursts of impulses either at the beginning of illumination due to an excitatory stimulus ("on" signals), or at the end due to the cessation of an inhibitory stimulus ("off" signals), or at both due to the action of both excitatory and inhibitory stimuli. The firing characteristics of these ganglion cells are wave length dependent. For example, one cell may be excited primarily by a group of red cones and inhibited by a group of green cones. Another cell might be conversely affected, and still another might be equally excited and inhibited by these cones. All colors can be recorded and translated into this nervous code; not just the primary colors. In this way three-color information is processed and coded in the retina into two-element on/off signals originating in the ganglion cells. These signals are relayed to the visual cortex where final analysis takes place.

It is possible to determine the zones of primary color sensitivity on the surface of the retina by a fairly simple method. A person looks with one eye at a fixed point. Disks of each of the three primary colors are then slowly moved, one at a time, in a circle around his head. He reports when a disk loses color but is still visible as an object. By making measurements of his field of view for the three primary colors, it is possible to map his color receptors on the retina. Such a map is shown diagrammatically in Fig. 11-10. The zones of color sensitivity may be thought of

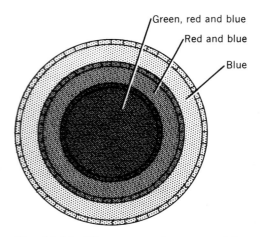

Fig. 11-10 The comparative areas of the retina sensitive to the three primary colors of light, shown diagrammatically.

as irregular disks superimposed. In the largest area we find cones receptive to blue wave lengths of light. In the area marked red, we find cones sensitive to this wave length as well as cones sensitive to blue. In the area recorded as green, we find an additional type of cone, as well as the other two. Beyond these three zones, there is an area composed predominantly of rod receptors, which will record the various intensities of white. The final peripheral area of the retina is insensitive to any intense light.

It should be noted at this point that as the intensity of white light gradually decreases, for example at dusk, there is a shift from cone reception to rod reception. This shift from light adaptation to dark adaptation is called the *Purkinje shift*. The shift also occurs when light increases, as at dawn, but in the opposite direction—from rod vision to cone or color vision.

Afterimages. Whenever we see a motion picture, we are actually seeing a series of still pictures flashed on a screen at a minimum rate of 16 pictures, or frames, per second. Because of an afterimage, we are conscious of a single picture in which the objects appear to move. The activity in the photochemical substance within the receptors cannot keep pace with changes in visual stimuli and thus there is a carry-over of visual impressions. If this carry-over is exactly the same as the original stimulus pattern, we say that it is a *positive afterimage.* When you glance at a very bright object, such as an electric light, and then close your eyelids, you are conscious of the same visual image even though the external stimulus has stopped. If, however, you stare at a colored object for a few minutes and then close your eyes or look at a blank wall of a neutral color, you are conscious of the image, but in the complementary color. Thus, a white figure on a dark background will appear as black on a very light background. A red image will appear to be green, since green is the complementary color of red. This *negative afterimage* is probably due to the fatigue of the originally stimulated receptors.

Characteristics of color vision. There are three qualities of color vision: *hue, saturation,* and *brightness.* Hue refers to the sensory response made by the eye to various wave lengths. If the eye is stimulated by light of a wave length of from 6750 Å to 7000 Å, we say that we are looking at red. If light of a wave length of from 5000 Å to 5500 Å enters the eye, we say we are looking at green, and so on. As each black and white key on a piano repre-

sents a different note on a musical scale, each wave length within the visible spectrum represents a different hue.

A hue of one single wave length is said to have complete saturation. Such a color is seldom encountered in nature, although it can be produced under controlled conditions in a laboratory. The light reflected from the objects around us contains varying amounts of white light. As a pure color is mixed with white light, it becomes less saturated and appears pale, or pastel. Thus pink is a red color of low saturation.

Brilliance or brightness is a quantitative factor determined by the intensity or energy of the light waves reflected by a surface. In terms of pigments, we say that brilliance depends on the amount of black added to the color. Navy blue is a color of low brilliance in contrast with royal blue which has high brilliance. Of course, the appearance of colored objects is affected greatly by the brightness of the light falling on them. Objects that seem brilliantly colored in bright light will appear darker and duller in color under dim illumination.

Color blindness. We have been speaking of color vision as if all individuals respond equally well to these visual stimuli. Actually, 6 to 8 percent of all males and about 0.4 to 0.6 percent of all females have some defect in color vision. In Chapter 33 you will learn about the heredity of certain defects in color vision.

Deficiency in color perception is called color blindness. Persons who have this hereditary condition may be classified in two general groups. One group is totally color-blind, apparently having either no active cone receptors or a defect in the visual center of the brain. This results in lack of ability to distinguish the various wave lengths of light responsible for color vision. Individuals in this group are called *monochromats*—they see objects only in terms of white, gray, and black. A second group, the *dichromats,* may be red-green or blue-yellow color-blind, the latter type being rather rare. In this second group, the defect may be due either to a defect in the receptors sensitive to red, green, or yellow; or to some neurological change in the complementary relationship of the red-green receptors.

Individuals are called *trichromats* if they are receptive to all three primaries: red, green, and blue-violet. This is normal color vision. There may be slight deficiency in determining closely related wave lengths of light, but this defect is negligible and certainly causes little hardship.

Visual acuity. Not only are the cones responsible for color vision, but as was mentioned earlier they are concerned with acuteness of vision. *Visual acuity* is the sharpness with which detail is seen. It is often measured in terms of the smallest distance that can be seen between two vertical black lines on a white background.

Many factors influence visual acuity: brightness or intensity of illumination, the size of the objects, the color of the objects, and the retinal area on which the image of the object falls. Certainly if the image falls outside of the fovea centralis, visual acuity is greatly reduced.

Doctors use a simplified visual acuity test when they ask you to read the letters on an eye chart. The *Snellen Eye Chart* contains series of letters, each series having a different standard size. If you can read at twenty feet the line of letters the average person can read at twenty feet, your vision is measured as 20/20 and is normal. However, if your visual acuity is poor, you may only be able to read at twenty feet what the average person can read at 100 feet. Your vision is then said to be 20/100. Many young people have better than 20/20 vision.

Visual Defects

We can classify abnormalities of vision into two general groups: defects characteristic of all convex lenses, and structural defects of the organs within the eye or of the eyeball itself.

Characteristics of all convex lenses. In all convex lenses, rays of parallel light that pass through the outer edge of the lens are refracted more than the rays passing through the center of the lens. This is called *spherical aberration*. It results in a series of focal points rather than the single focal point which is required for acute vision (Fig. 11-11). If the aberration is uncorrected, the central visual field may be in focus, while the perimeter of the image is fuzzy or out of focus. The eye corrects for spherical aberration automatically by shutting out the rays at the edge of the crystalline lens. This is accomplished by decreasing the size of the pupil and thus allowing only those rays of light near the center of the lens to pass through.

Usually an oculist puts drops into your eyes during an examination in order to dilate the pupils. If the eye is then exposed to intense illumination the effects of spherical aberration are

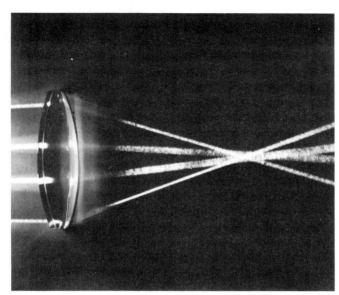

Fig. 11-11 Spherical aberration of a lens. Note how the rays that pass through the center of the lens focus further away than those that pass through the edges. The iris of the eye cuts off the peripheral rays to reduce the aberration. The diaphragm of a camera serves a similar purpose.

easily noted. The dazzling brilliance is almost painful and visual acuity is low due to the diffusion of focus.

A distortion very similar to spherical aberration occurs because white light passing through a convex lens breaks up into individual wave lengths. This is called *chromatic aberration.* When white light passes through a convex lens, the short wave lengths (violet) are refracted or bent more than the long wave lengths (red). In between these extremes, there is a progressive distribution of colors, like a rainbow. As a result, a halo effect of colors is seen surrounding each image (Fig. 11-12).

To correct this, the sphincter muscles of the iris contract and the pupil is constricted. This shuts out rays that pass through the perimeter of the lens where the greatest amount of aberration occurs. Once again we have sharp color vision without distortion.

Defects in eye structure. In high quality scientific instruments, like the microscope or telescope, all of the optic surfaces must be perfectly ground and polished. All of the light waves, either passing through or reflecting from these optics, must reach their predetermined focal point without distortion.

As you know, the optical parts of the human eye (the cornea and the lens) are composed of specialized living cells. A

Fig. 11-12 Chromatic aberration of a lens. Why does the lens act like a prism?

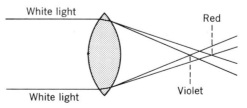

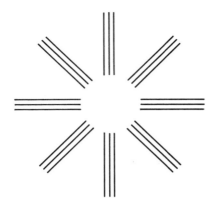

Fig. 11-13   This type of diagram is used to test the eyes for astigmatism.

slight irregularity of the cornea or lens often occurs. This results in a common eye defect called *astigmatism*. In this condition, light waves passing through the cornea (or lens) are not all refracted to a single point. There is a distorted image on the retina. Uneven curvature of the anterior surface of the cornea is most common. If the distortion is great enough, eyestrain and frequent headaches may result.

The eye specialist may use numerous devices to determine astigmatism. One such device is an *astigmatic dial* as seen in Fig. 11-13. It consists of radiating lines or spokes which all have the same width and darkness. A person with astigmatism will see some of these lines more clearly than others, because they are brought to a sharper focus on the retina. Astigmatism can be overcome by the use of eyeglasses with lenses ground to correct for any defects in curvature on the cornea or lens.

Another common eye defect is *myopia*. In myopia (near-sightedness), the eyeball is too long from front to back.

Parallel light waves come to a focus in front of the retina and distant vision is blurred or out of focus. Rays of light which diverge from near objects, with the aid of accommodation, are focused on the retina. Thus a person may be able to read a printed page with little difficulty, but will be unable to read a stop sign one block away. Myopia is common among young people because the eyeball may increase in length too rapidly during the early years, especially during puberty. The condition frequently adjusts to normal by the time the person becomes adult.

To correct for myopia, a concave lens which will diverge parallel waves of light is placed in front of the eye. This pushes the focal point back to the retina and results in a clear sharp image of distant objects. Correction for myopia is shown diagrammatically in Fig. 11-14.

If the length of the eyeball is too short, a person can see clearly at a distance but is not able to form a clear image of nearby objects because the image falls behind the retina. In this condition, called *hyperopia* (farsightedness), the lens cannot accommodate enough to produce a sharp image of a nearby object on the retina. However, images of distant objects are focused easily by slight accommodation of the lens.

We correct for hyperopia by the use of eyeglasses having convex lenses which aid the lenses of the eyes in bringing the images of nearby objects to a sharp focus (Fig. 11-14).

Nearly all muscles gradually lose some of their contractive power as they age. This is also true of the ciliary muscles in the eyes. In addition, the crystal-

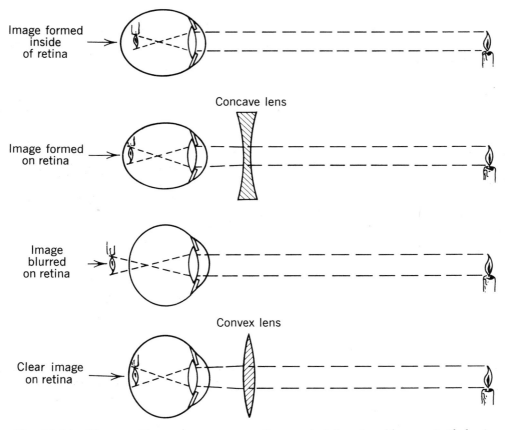

Fig. 11-14 *The use of lenses to compensate for myopia (above) and hyperopia (below).*

line lens loses some of its elasticity with age and is no longer able to assume the maximum convexity necessary for near vision. After about the age of 45 years, there is a decline in the accommodation that can occur in each eye. This condition is called *presbyopia*. An individual finds as he gets older that he must hold reading materials farther and farther from his eyes in order to have acute vision. From this time on, eyeglasses with convex lenses are necessary for reading without discomfort.

A person with myopia may find the onset of presbyopia no disadvantage, and he may never need corrective lenses for reading. Nearly all elderly individuals who can still read easily without glasses show histories of myopia in their younger years.

VOCABULARY REVIEW

Match the phrase in the left column with the correct word in the right column. *Do not write in this book.*

1   Decline in visual accommodation with age.
2   Visual impulses received by the brain from only one eye.
3   A visual deception.
4   Adjustment of the eye to low intensity of illumination.
5   Distortion of vision due to distortion of cornea or lens.
6   Nearsightedness.

a   monocular vision
b   stereoscopic vision
c   optical illusion
d   achromatic stimuli
e   scotopic vision
f   Purkinje shift
g   monochromats
h   astigmatism
i   myopia
j   presbyopia

TEST YOUR KNOWLEDGE

Group A

Select the correct answer. *Do not write in this book.*

1   The average field of vision is (a) 180° (b) 150° (c) 190°.
2   The image projected on the retina is (a) upright (b) inverted and laterally reversed (c) upright and laterally reversed.
3   Night blindness is due to (a) lack of vitamin D (b) lack of retinene (c) lack of vitamin A.
4   We can see things more clearly at night when we (a) look directly at them (b) look out of the side of the eye (c) look slightly above or below them.
5   The shortest wave length is that of (a) infrared (b) ultra violet (c) X rays.
6   Color vision is due to (a) a different kind of cone for each of the primary colors (b) the primary colors are seen differently by the rods, the cones, and the fovea centralis (c) the brain interprets the difference between colors.
7   Purkinje shift is (a) a shift from stereoscopic vision to monocular vision (b) a shift in an image as we look away and then back (c) a shift from photopic vision (cone vision) to scotopic vision (rod vision) at dusk.

Group B

1   How do two eyes, seeing separately, record a single impression on the brain?
2   How do we judge the relative distances of the objects we see?
3   What are the wave lengths responsible for our visual spectrum?
4   What percentage of both sexes are color blind?
5   What is chromatic aberration? How does the eye correct it?

Chapter 12 The Ear

Next to visual perception, hearing is probably our most important special sense. Through sound we have a marvelous means of communication. As children we learn to speak when we hear others speak. Later we construct complicated electrical and electronic devices to span great distances and reproduce sound. Even the micro-voice of satellites and radiations from distant stars can be converted to sounds. Sound enables us to judge the distance of objects and to locate them, even in the dark. We respond emotionally to various sounds: we receive pleasure from music, we are annoyed by a creaking hinge, and we may become frightened by unidentified sounds at night or by unexpected loud noises.

The human ear is responsible for converting sound waves into nerve impulses which result in our sense of hearing. It is also an important organ of equilibrium. This remarkable structure is divided into the *external ear,* the *middle ear,* and the *inner ear.* See Fig. 12-1.

The External Ear

As the name implies, part of the external ear extends beyond the lateral surface of the head. This is the *auricle,* or *pinna.* The pinna has a cartilaginous frame which is covered by skin. Below the area of cartilage there is a flap of skin filled with areolar and adipose connective tissue. This part of the auricle is called the *ear lobe,* or lobule, which may hang free or be attached along the medial edge.

The second part of the external ear is the *auditory canal* or *external acoustic meatus.* The auditory canal extends about one inch into the temporal bone. The canal, which is open at the auricle to allow sound waves to enter, is closed at the inner end where it meets the middle ear, by the resonating *tympanic membrane,* or eardrum. This membrane is not clearly visible without special instruments, since the canal is not straight but has an open S-shape.

The meatus is lined with a thin section of skin which adheres to the cartilage and bony framework, and covers the outer surface of the tympanic membrane. Just beneath the skin, in the outer portion of the canal, are numerous wax-producing, *ceruminous* glands. This wax, *cerumen,* is very sticky. In addition to these glands, we find a few short hairs which point outward. Both

169

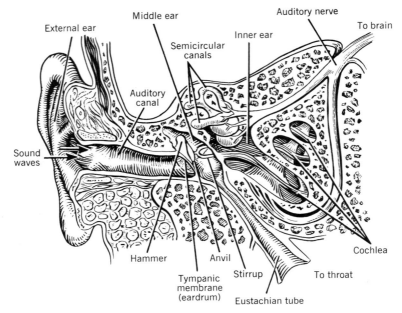

External ear

Middle ear

Semicircular
canals

Inner ear

Auditory nerve

To brain

Auditory
canal

Sound
waves

Hammer

Anvil

Stirrup

Tympanic
membrane
(eardrum)

Eustachian tube

To throat

Cochlea

Fig. 12-1   The structure of the human ear.

the hairs and cerumen help to prevent foreign particles from entering the meatus.

The Middle Ear

The second division of the auditory apparatus is the middle ear or *tympanic cavity*. This small air-filled cavity is housed within the spongy portion of the temporal bone near the mastoid process. The bony walls of the tympanic cavity are lined with a mucous membrane similar to that of the nasal cavity. This membrane also covers the inner surface of the ear drum. Between the ear drum and the inner wall of the middle ear we find a chain of three small but exceedingly important bones, the *ossicles* (Fig. 12-2).

Sound waves striking the ear drum are passed along to the inner ear by means of the ossicles, which form a system of levers.

The first bone responsible for transmitting the physical vibration of the tympanic membrane is the *malleus* or hammer. The handle of this mallet-shaped bone is attached to the inner surface of the eardrum so that the tip of the handle ends at the apex of the membrane. The head portion fits into a shallow socket at the base of the middle bone of the three—the *incus,* or anvil. There is little individual movement between these two bones because ligaments bind them rather firmly together as a single unit. The long process of the incus comes in contact with the head of the *stapes,* or stirrup, the third bone. This articulation is freely movable and results in a rocking action of the stirrup. The base of the stirrup oscillates against the membrane covering an opening, the *oval window,* in the inner wall of the middle ear. It is interesting to note that at birth the ossicles have already reached their adult size and from this time on do not change.

The *oval membrane* covering the oval window has a very much smaller surface area (3.2 square millimeters) than the tympanic membrane (64 square millimeters). Due to this, and to a small extent to the lever system formed by the ossicles, considerable vibration amplification takes place. In fact, the force of vibrations in the oval membrane is seventeen to twenty one times greater.

Two small striated muscles are responsible for controlling the extent of movement of the ossicles. Attached at one end to the handle of the malleus near the head and at the other end to the floor of the tympanic cavity, is found the *tensor tympani* muscle. When this contracts, the handle of the malleus is pulled inward, producing a greater tension of the eardrum. The second muscle is the *stapedius* muscle. It arises near the roof of the cavity and inserts on the posterior surface of the neck of the stirrup. It opposes the tensor tympani muscle.

Fig. 12-2   Ossicles of the ear showing their relation to the tympanic membrane and other structures within the middle ear cavity.

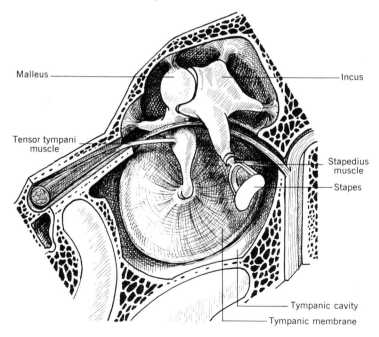

Malleus ——————— Incus

Tensor tympani muscle

Stapedius muscle

Stapes

Tympanic cavity

Tympanic membrane

As mentioned earlier, the middle ear is an air-filled cavity. The tympanic membrane lies between this cavity and the outer air. Equal pressure must be maintained on both sides of the membrane or it will bulge inward or outward. Under such tension, the membrane does not vibrate normally and hearing is impaired. The middle ear communicates with the outer air through the *Eustachian tube,* which extends between the middle ear and the nasopharynx. This tube is about 40 millimeters in length and, at the narrowest point, about 3 millimeters in diameter. The pharyngeal opening is oval with the inner edges normally touching. When swallowing or yawning, muscles pull the edges apart and air is allowed to enter or leave the middle ear, depending upon external air pressure. If the entrance to the Eustachian tube is inflamed or congested, as a result of a cold or a sore throat, the edges may not open. Inequalities of pressure are not adjusted and the eardrum cannot move freely. Slight temporary deafness results.

When an airplane descends rapidly, the pressure in the middle ear remains low as it was at the higher altitude, but the pressure in the airplane, unless it is well-controlled, increases rapidly. This pressure forces the eardrum inward, but may be equalized by swallowing rapidly to open the Eustachian tube.

The Eustachian tube is lined with the same type of mucous membrane that is found in the middle ear and in the nasal cavity. Infections in the throat often progress up through the Eustachian tube to the mucous lining of the middle ear. Such infection may then invade the mastoid process, a part of the temporal bone, and thus cause mastoiditis. In the past, this condition often required surgical treatment. Today antibiotics are used to control the infection.

The Inner Ear

Within the inner ear, the most important part of the auditory apparatus, we find the receptors that initiate nerve impulses which are interpreted by the brain as sound. Specialized receptors concerned with equilibrium are also found there. The inner ear is divided into three series of fluid-filled tubes or canals, collectively called the *bony labyrinth* (Greek, meaning maze). The three parts of the labyrinth are the *vestibule,* the *cochlea,* and the *semicircular canals* (Fig. 12-3).

Within the bony labyrinth is a tubular membrane which follows the same shape as the bony canal. This has been called the *membranous labyrinth.* It does not adhere to the bony wall, but is separated from it by a fluid called the *perilymph.* Within the tubular membranous labyrinth we find a second fluid, the *endolymph.*

The vestibule is connected to the middle ear by the *oval window.* It acts as an entrance to the other two divisions of the inner ear: the semicircular canals, which are concerned with equilibrium, and the cochlea, in which the sound receptors are found.

The cochlea resembles a snail's shell with its spiral canal making between $2\frac{1}{2}$ and $2\frac{3}{4}$ turns around a central pillar— the *modiolus.* Within this pillar we find a branch of the auditory nerve (eighth cranial nerve) that is called the *cochlear nerve,* concerned with conducting sensory impulses to the brain for our sense of hearing. Projecting from the central pillar is a small shelf of bone called the *spiral lamina,* which is continuous within the spiral canal and partially divides it in two. Attached to the outer border of the spiral lamina is a thin membrane—the *basilar membrane* —which stretches across to the cochleal wall and completely separates the canal into two passages, except at the apex. The upper passage is the *scala vestibuli,* and the passage below the lamina and attached basilar membrane is the *scala tympani* (Fig. 12-4). The oval window in the wall between the middle and inner ear communicates with the scala vestibuli. Below the oval window is the *round window,* the membrane-covered opening of the scala tympani.

From the upper edge of the osseous lamina across to the outer edge of the canal above the basilar membrane, extends an extremely thin membrane— the *vestibular membrane,* or the *membrane of Reissner.* This membrane divides the scala vestibuli unequally, forming a small triangular canal, the *scala media.* No direct function has been connected with the vestibular membrane and the resulting scala media.

The sense receptors for sound are found within the *organ of Corti,* which is located on the upper surface of the basilar membrane in the scala media and is just lateral to the bony spiral lamina. The cells of the organ of Corti include a series of columnar epithelial cells out of which small hairlike endings protrude (Fig. 12-4). These hair cells lie in two rows running the entire length of the canal. Covering these cells is a canopy attached to the lamina which arches over the hair-cell endings; this membrane is called the *tectorial membrane.* As vibrations pass through the noncompressible fluids of the cochlea, the basilar membrane and attached tectorial membrane also vibrate. It is believed that this movement causes changes in the length of the hair cells.

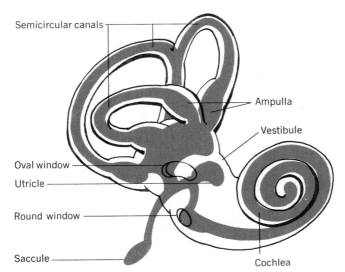

Semicircular canals

Ampulla

Vestibule

Oval window

Utricle

Round window

Saccule

Cochlea

Fig. 12-3 Structure of the inner ear.

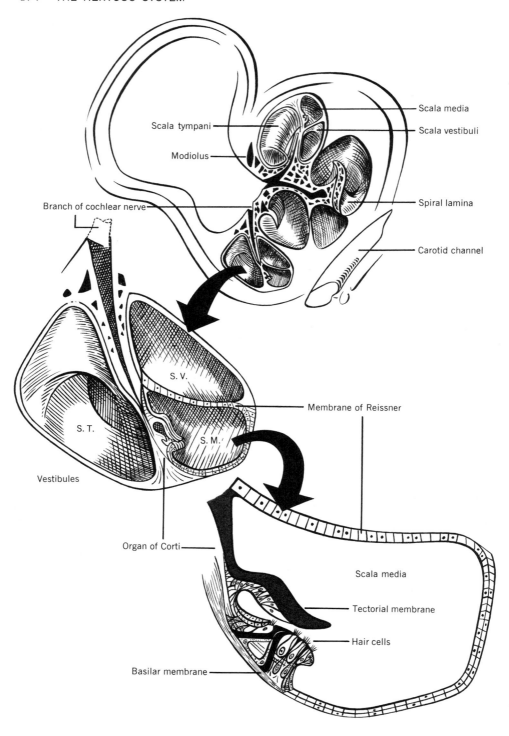

Scala media

Scala tympani

Scala vestibuli

Modiolus

Spiral lamina

Branch of cochlear nerve

Carotid channel

S. V.

Membrane of Reissner

S. T.

S. M.

Vestibules

Organ of Corti

Scala media

Tectorial membrane

Hair cells

Basilar membrane

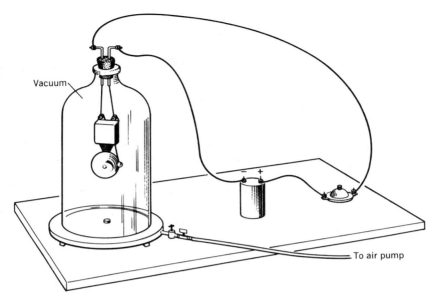

Fig. 12-5   The apparatus shown here can be used to demonstrate that sound waves are carried by the air.

These changes stimulate the terminal branches of the auditory neurons which surround the epithelial hair cells, initiating the auditory impulses.

The Nature of Sound

Sound can be produced only when an object vibrates, disturbing the molecules of the substance in which the object is found. If a ringing doorbell is suspended in a bell jar from which the air has been removed, it will not produce a sound. When the air re-enters the jar, the sound can be heard (Fig. 12-5).

When an object vibrates in air it produces regions in which the air mole-cules are squeezed together (compressions) alternating with regions in which the air molecules are farther apart (rarefactions). Thus, sound waves are alternate regions of compression and rarefaction in the air. They move outward in all directions like the waves produced by a rock thrown into a quiet pool of water. Sound waves may be reflected in the same way as light waves are reflected when they come in contact with a nonabsorbing surface. If the surface reflects the sound waves directly back to the source, the sound is heard at the source as an echo.

The speed at which sound waves travel through different substances

Fig. 12-4   Successive enlargements of a vertical section through the cochlea.

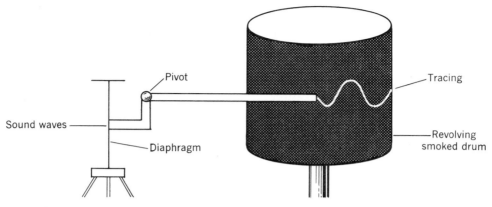

Fig. 12-6   Apparatus used to produce a graphic record of sound waves. (Redrawn from Wenger, Jones, and Jones, Physiological Psychology, *Holt, 1956)*

varies according to the elasticity and density of the substance. In air at sea level, sound travels about 332 meters per second, or about 760 miles per hour. In all cases, the speed is much slower than the speed of light. You have seen steam from a whistle before hearing its sound, and gun smoke before hearing the report, or the distant flash of lightning before hearing the thunder. The table below illustrates the speed of sound in various substances in meters per second.

Medium	Approximate Speed
Air	331.7 m/sec
Water	1441 m/sec
Wood (maple)	4110 m/sec
Iron	5130 m/sec
Glass	5000 m/sec
Marble	3810 m/sec

If we convert compressions and rarefactions into a line graph, as seen on an oscilloscope, we see certain auditory characteristics. The line moves up dur-

ing the compression phase and downward during the rarefaction phase (Fig. 12-6). The distance it moves up or down is called the amplitude and provides us with the characteristic of *loudness* or *volume.* The higher the curve, the louder the sound.

In determining loudness, we use a comparative scale with the starting point at the threshold of audible sound. The units in the scale are called *bels,* named after the great experimentalist, Alexander Graham Bell (1847–1922). The term *decibel* represents $\frac{1}{10}$ of a bel, since the decibel seems to be the smallest unit variance the human ear can distinguish. The following chart indicates the various levels of sound as expressed in decibels.

Type of Sound	Sound Level in Decibels
Threshold of hearing	0
Ordinary breathing	10
Whispering	15–20
Average suburban home	30–40

Automobile	40–50
Conversation	60
Average TV sound system	70
Heavy street traffic	70–80
Elevated trains	90–100
Thunder	110
Threshold of pain	120

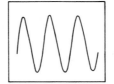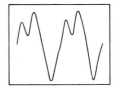

Fig. 12-7   Waveform pictures produced by the type of apparatus shown in Fig. 12-6. Left is a graph of a pure tone. Right is a graph of a tone from a French horn, showing the effect of an overtone (harmonic) on a pure tone.

The time interval between one complete compression curve plus the rarefaction curve and the next compression curve gives us the *cycle* of the sound. We call this *pitch,* or *frequency,* usually written as the number of cycles per second or *cps.* As the number of cycles increases, the pitch or frequency becomes higher. When you strike the keys of a piano, moving to the right on the keyboard, each succeeding string vibrates at a faster rate than those before, and thus each note is higher in pitch. As the eye has limits in perceiving light waves (we do not visually respond to ultraviolet or infrared), so the human ear has two thresholds of pitch. Our ears are unable to respond to sound waves fewer than 16 cycles per second—our low threshold—nor can most people detect sounds of above 15,000 to 20,000 cps—our high threshold. The human ear has its greatest sensitivity between 2000 and 4000 cps.

If several different musical instruments all play the same pitch, it is quite easy to distinguish one type of instrument from another. Most of us can tell the difference in sound between a piano, a clarinet, a trumpet, and a violin. The quality of tone gives these instruments their characteristic sound. In describing sound, we may say that a cello is "rich," a trumpet is "harsh," a celesta has a "tinkling" sound, and a bass drum "booms." All of these terms are attempts to describe the quality or

timbre of the instrument. Graphically, *quality* or *timbre* is represented by the smoothness of the sound wave as seen in Fig. 12-7. If the line produced is a simple sine wave, or curve, the tone is classified as a *pure tone.* Pure tones can be mechanically produced but they are never found in a musical instrument or in the human voice. If the sound wave produces a curve that contains many smaller deviations, the tone is composed of a number of tones sounding together. We call these *overtones* or harmonics. If these overtones are produced in a consistent pattern, we usually find the sound pleasing. If, however, the overtones are spasmodic and irregular, we usually classify the sound as a noise. The form of the sound wave determines the quality of the sound.

Hearing

What happens after sound waves enter the external acoustic meatus and set the tympanic membrane and the chain of middle ear bones in motion? It has already been explained that the base of the stirrup rocks back and forth in the oval window, which is separated from the fluid in the inner ear by a thin membrane. When any part of a confined liquid is subjected to pressure, the pressure is distributed equally throughout the liquid. So, when the base of the stirrup is forced inward, pressure is directed into the scala vestibuli and the smaller scala media. The fluid exerts a wavelike pressure on certain parts of the yielding basilar membrane, which curves downward, in turn exerting a pressure in the fluid-filled scala tympani. Pressure is directed against the membrane covering the round window, which bulges outward. As already mentioned, the round window is a membrane-covered opening between the middle ear and the inner ear just below the oval window and at the level of the scala tympani. As the stirrup rocks outward in the oval window, this series of movements is reversed.

The basilar membrane vibrates or curves in sympathy with the various characteristics of sound: amplitude, frequency, and quality of the sound waves. This delicate fibrous membrane is narrower at the vestibular end of the membranous labyrinth within the cochlea and becomes progressively wider up to the apex. It is believed that the higher frequencies result in curves of the basilar membrane near the shorter fibers at the bottom of the cochlea. Lower frequencies are associated with the longer fibers near the apex.

The activities just mentioned briefly describe the *Place Theory* of audition. This theory is based on the assumption that our discrimination of sound frequency and quality is based primarily upon the place where maximum movement of the basilar membrane occurs. Impulses seem to be initiated when the receptors of the organ of Corti are stimulated at that part of the basilar membrane corresponding to the pitch. It is estimated that by this mechanism we can distinguish over 10,000 different frequencies or tones. To accomplish this remarkable feat, we have an estimated range of from 15,000 to 30,000 hair cells and resulting fibers in the cochlear nerve for each ear. The resulting impulses travel to the auditory cortex of the brain in the temporal lobe, by way of the auditory nerve. Many of the fibers that make up this nerve cross to the opposite temporal lobe from the ear stimulated.

We have seen how visual judgment of distance is made possible through stereoscopic vision. Similarly, the judgment of direction is possible through binaural (two-ear) perception of sound. Sound vibrations reach one ear before they reach the other, except when we are facing the source of the sound. We are best able to determine the source of sound when the sound originates in front or to one side of us. When the sound is above or behind us, judgment is less accurate. The brain is frequently confused by sound reflections, so we often find ourselves facing a blank wall as we turn toward what we believe to be the source of the sound.

Auditory defects. As we have seen, hearing involves two basic activities: the receiving and conduction of sound waves and the nerve function by which impulses are set up and transmitted to the brain.

If any of the structures responsible for either of these activities fails to function normally, partial or total deafness results. Deafness may thus be caused by conduction failure or nerve failure. Partial conduction deafness may be due to an accumulation of wax in the external auditory canal. The common forms of partial deafness due to structural defects are thickening of the tympanic membrane, perforated tympanic membrane, adhesion of the ossicles, fusion of the stapes to the oval window due to over-growth of bone called otosclerosis, or loss of elasticity in the oval window membrane. Fortunately, nearly all forms of conduction deafness involving the tympanic membrane or the middle ear can be corrected with an electronic hearing aid. This device may either increase amplitude at the tympanum or give direct bone-conduction by vibrating against the mastoid bone which transfers the vibrations to the ossicles, completely by-passing the external ear.

An operation called stapes mobilization may restore partial hearing in 80% of people with otosclerosis. This operation is performed through a special microscope. The surgeon lays back the ear drum and loosens the stapes from the oval window.

Deafness due to nerve deterioration is more serious. Many people in heavy industry, such as boiler-makers, are subjected to continuous loud sounds. They frequently develop *tonal gaps* or islands which are insensitive to certain frequencies, while bordering areas have greater sensitivity. The explanation seems to be that there is usually a separation of the organ of Corti from the basilar membrane at some point.

As an individual ages, there is frequently a gradual decrease in sensitivity in the upper frequency range. In the same way that muscles gradually lose their firmness, the narrower end of the basilar membrane gradually loses elasticity where rapid vibratory activity occurs. There is some question as to the usefulness of hearing aids with high-frequency deafness.

Total deafness, without hope of recovery, may result from any of the following causes: destruction of the auditory nerve, destruction of any of the parts of the inner ear mechanism, and failure of the cochlea or its connected nerves to develop properly before birth. Some types of total deafness are hereditary.

Balance

We have described in some detail the structure and function of the auditory portion of the bony labyrinth. We must now consider those structures responsible for our sense of balance or equilibrium.

If your body is rapidly whirled around, you become dizzy—that is, you find it difficult to stand upright or walk in a straight line. Two different organs are involved in maintaining proper balance. One is a fluid-filled sac that is divided into two parts, the *utricle* and the *saccule*. The other is a combination of three fluid-filled tubes, the *semicircular canals*.

The utricle and saccule. These structures lie within the vestibular portion of the bony labyrinth. The utricle is the larger and is connected to the membranous ducts of the semicircular canals by a number of common openings. The floor of the utricle, called the *macula,* contains two types of modified columnar cells. One kind is made up of neurons which have short hairs extending into the endolymphatic fluid. Intermingled with these hair cells are other supporting cells. Surrounding and covering the hairs of the sensory receptors is a layer of gelatinous material in which are suspended many small solid particles of calcium carbonate known as *otoliths*. When the head is tilted, gravity causes these solid particles to come in contact with the hair cells. These send out impulses which normally enable us to determine the position of the head with respect to gravity.

The nerve impulses from the hair cells are carried through a branch of the eighth cranial nerve, called the *vestibular nerve*. They are conveyed to the medulla and then to the cerebellum where efferent responses result in changing the body position by muscular contraction.

Little is actually known about the function of the saccule. It is innervated by two groups of neuro-fibers from the vestibular nerve. As this structure is closely associated with the cochlea, there is some basis for believing that it may be concerned with hearing rather than balance. It may involve low frequencies of sound or even the sound sensation of one's own voice. We must not rule out the possibility that the saccule may also have minor functions like those of the utricle, since their internal structures are similar.

The semicircular canals. Three looped tubes, the semicircular canals, are embedded in the temporal bone anterior and superior to the vestibule. They are about 4 millimeters in diameter and vary in length from 12 to 22 millimeters. The semicircular canals are connected to the vestibule at five points. As seen in Fig. 12-2, two of the canals join to form a common canal which connects with the vestibule. The most obvious structural characteristic of the semicircular canals is the fact that the three loops are situated at right angles (90°) to each other in three different planes. Two of these structures, the anterior canal and posterior canal, are vertical and at right angles to each other. They are also situated at 45° angles to the mid-line of the head. The third canal is horizontal and at a right angle to the two vertical canals.

Within the semicircular canals we find a portion of the membranous labyrinth. As is true of the cochlea, this tubular membrane is filled with endolymph and is held in place by connective tissue and the surrounding perilymph. The diameter of the *semicircular ducts,* as these tubular membranes are called, is about one fourth the diameter of the bony canals which surround them. At the junction to the vestibule, the semicircular ducts join the utricle at the five openings mentioned before.

At one end of each canal, where it comes into contact with the vestibule, there is a swelling or bulging called the *ampulla.* In the ampulla, the semicircular ducts contain a crest-shaped group of hair cells called the *crista acoustica.* These cells can detect changes in acceleration and deceleration. Covering these hairs is a gelatinous substance called the *cupula* into which the hairs protrude. When there is a change in horizontal acceleration, as when we move forward, the endolymph within the ducts presses against the gelatinous cupula which in turn causes the hairs to bend in a direction opposite to the direction of motion. This stimulates the hair cells to produce sensory impulses. The branch of the vestibular nerve which innervates these receptor cells leads to the medulla of the brain. Here it forms synapses with other neurons, some of which carry impulses to the motor nerves that control eye movement. Others carry impulses directly to the cerebellum.

Equilibrium is not classified as one of our special senses, as are vision, hearing, taste, etc., since there is no known anatomical connection with the cerebral cortex. Through the vestibular and semicircular activities we respond to rapid rotation by becoming dizzy. This in turn may involve other side effects. If the rotary movement is accompanied by vertical motion, as is the case when we travel on the ocean, in the air, or in an automobile, we may show symptoms of motion sickness.

During and since World War II, many remedies have been developed to ease or prevent the helpless feeling of motion sickness. Since these drugs have no effect on the frequent changes in body position, it is believed that their action depresses impulses in the vestibular nerve. If these impulses are not controlled, there is a resulting confusion of the balance centers in trying to analyze the constantly changing waves of impulses.

Associated with the function of the semicircular canals and the vestibular apparatus in maintaining equilibrium are the kinesthetic and visual senses which provide clues to spatial and structural position. The resulting response to these sensations involves coordinated muscular activity under the control of the cerebellum which keeps the body in a vertical position.

VOCABULARY REVIEW

Match the phrase in the left column with the correct word in the right column. *Do not write in this book.*

1	Middle ear	a	auricle
2	Auditory canal	b	external acoustic meatus
3	Conducts sensory impulse to the brain for our sense of hearing	c	tympanic membrane
4	Extends between the middle ear and the nasopharynx	d	tympanic cavity
		e	oval window
5	External ear, or pinna	f	stapedius
6	Units in the scale of loudness of sound	g	binaural
		h	Eustachian tube
		i	cochlear nerve
		j	decibels

TEST YOUR KNOWLEDGE

Group A

Select the correct statement. *Do not write in this book.*

1 The Eustachian tube (a) drains fluid from the middle ear into the nasopharynx (b) keeps up pressure in the inner ear (c) keeps the pressure equal on both sides of the eardrum.

2 The semicircular canals are associated with (a) hearing (b) balance (c) draining fluid from the inner ear.

3 The tiny bone (ossicle) which oscillates against the oval window is the (a) incus (b) stapes (c) malleus.

4 The basilar membrane is within (a) the semicircular canals (b) the vestibule (c) the cochlea.

5 The human ear has its greatest sensitivity between (a) 15,000 to 20,000 cps (b) 2,000 to 4,000 cps (c) 1,000 to 3,000 cps.

6 Total deafness, without hope of recovery, may result from (a) otosclerosis (b) destruction of the auditory nerve (c) destruction of the eardrum.

7 The function of balance in the ear depends on the (a) auditory nerve (b) vestibular nerve (c) trigeminal nerve.

Group B

1 What structure divides the external ear from the middle ear?

2 How do the ossicles perform in the process of hearing?

3 How do changes in external air pressure affect hearing? How can these changes be controlled?

4 What membrane covered openings exist between the middle ear and inner ear? What is the function of each?

5 Where are the sense receptors for sound located? What theory seems to best explain the function of this structure?

6 Discuss the thresholds of hearing. How do these thresholds vary with a person's age?

Chapter 13 Foods

Foods have several characteristics which distinguish them from other materials which we take into the body. In the first place, a *food,* or *nutrient,* must be able to *supply energy* to the body. This must be done without injuring the body in any way. Alcohol, for example, will supply energy, but its side effects are frequently not beneficial. A second characteristic of foods is that they can be used to *increase the amount of protoplasm.* This will result in growth, or in maintenance of the body by replacing the protoplasm that is used up in the normal processes of living. A third requirement of food is that it can be stored and used as a *reserve supply* on which the body can draw in times of emergency. Thus, a food may serve as a source of immediate energy or may be temporarily stored for future energy and repair requirements.

Food Materials

Several groups of materials enter into the composition of foods. Three groups of organic compounds, proteins, fats and carbohydrates, provide the energy and building materials required by the human body. In addition, the body requires water, the universal solvent, mineral salts, and vitamins to enable it to utilize the energy and building materials stored in the three principal food groups. Only a combination of all these substances will meet the requirements for proper growth and development.

Water, the salts, and the vitamins cannot meet all of the requirements; neither can the organic nutrients supply all necessary substances needed by the body. This chapter will deal with the composition of foods and their uses by the body. The vitamins will be discussed in a later chapter.

Water. This simple chemical compound, H_2O, is essential for all living things; without it plants could not manufacture their food and animals, including man, would be unable to exist. The water intake of animals is largely dependent on the water contained in the food and also that produced by the processes of oxidation. In fact, some animals, like the kangaroo rat, drink no water. They depend on water produced by the oxidation of food materials such as glucose. When this particular material is oxidized, one molecule of glucose will produce six molecules of water.

$$C_6H_{12}O_6 + 6O_2 \rightarrow 6CO_2 + 6H_2O$$
glucose oxygen carbon water
dioxide

Water plays many roles. It will dissolve more different types of materials than any other solvent. Materials dissolved in water can pass into the blood stream quite easily and rapidly and food materials not in a state readily soluble in water are made more soluble by the process of digestion. These materials can then be carried in solution to all parts of the body where they will either be used or, if wastes, excreted.

A second function of water in the body is to maintain the osmotic pressure of cells, and supply them with the fluid necessary for chemical reactions. To insure an adequate amount of water for these purposes, certain areas of the body serve as reservoirs where water can be stored. The salivary glands are one of these depots, and this accounts for the dry sensation of the mouth that frequently accompanies loss of water.

Water is also largely responsible for the maintenance of body temperature. The fluid part of the blood, the plasma, is approximately 92 percent water, and this is warmed as it passes through active tissues. Later some of this fluid is secreted in the form of sweat, the evaporation of which is largely responsible for lowering body temperature.

Certain types of chemical reactions occurring in cells require water for their completion. In *hydrolysis*, for example, there is chemical addition of water to large molecules which results in the splitting of the large molecules into two or more smaller molecules.

Inorganic and organic compounds. All chemical compounds fall into one of two classes; they are either

Inorganic Compounds	Organic Compounds
Made up of combinations of any of the 92 elements.	Relatively few elements present. 9 principal elements with *carbon* always present.
Number of atoms in the molecules relatively few.	Molecules frequently very large and complex with carbon atoms linked to each other either as closed rings or complex open chains with side branches.
Usually quite soluble in water.	Usually insoluble in water; will form suspensions; soluble in organic solvents, such as alcohol, ether, chloroform.
Usually ionize in a water solution.	Ionization infrequent due to sharing of electrons (covalent bonding).
Chemical reactions usually rapid and simple.	Chemical reactions usually slow and complex; may require several steps for completion.

inorganic or *organic.* The original significance of these terms lay in the fact that inorganic matter was thought to be associated only with nonliving things, while the organic substances were believed to be limited entirely to the living world and its products. With the rapid advances made in the field of chemistry during the past century, this particular distinction between the two no longer exists. Today, one refers to those substances which contain *carbon,* usually in combination with hydrogen or with hydrogen and oxygen, as organic materials, while those lacking carbon are classified as inorganic substances. Carbon dioxide (CO_2) and its derivatives such as the carbonates ($-CO_3$), are generally considered as inorganic since they occur so frequently in compounds that have not been associated with life. The following table shows some of the principal differences between these two groups of compounds.

Mineral salts. The term mineral salt can be considered a synonym for inorganic salt because it refers to those substances that lack the element carbon, except in the form of carbonates. These substances are important in the regulation of cellular functions; each exerts its own osmotic action, and is able to take part in reactions with other compounds to form new materials. Thus the component parts of ordinary table salt, sodium chloride, may later appear as constituents of blood, cell proteins, or sweat.

The table on page 186 gives the functions of some of the more important elements in these salts.

The presence of inorganic salts in foods is determined either by chemical analysis of the ash obtained by burning the food, or by spectroscopic analysis. The former method is quantitative, while the latter is most often used to detect small traces of elements that cannot easily be measured by chemical analysis.

Organic nutrients. The organic nutrients serve as the principal source of materials for energy and protoplasm formation. Of the three classes of organic nutrients, the *carbohydrates* are the simplest. They contain only three chemical elements. The *fats* and *oils (lipids)* are more complex because their molecules may contain several different elements in addition to the three in the carbohydrates. The *proteins* are the most complex organic compounds. These compounds become constituents of protoplasm and give it those particular characteristics which differentiate it from all other known substances.

Carbohydrates

Carbohydrates are defined as chemical compounds, composed of carbon, hydrogen, and oxygen, in which the hydrogen and oxygen are present in the ratio of 2 to 1, as they are in water. This arrangement of the atoms gives the name carbohydrate to these substances. *Carbo-* refers to the element carbon, and *hydrate* is from the Greek word for water.

For convenience, the carbohydrates can be divided into *sugars* and polysaccharides. A few sugars are sweet to the taste; many are tasteless or even rather objectionable in flavor. The following table gives the relative sweetness of some of the common sugars compared with sucrose (table sugar):

Element	Biological Effects	Excellent Sources
Calcium	Building of bones and teeth; aids in clotting of blood; regulation of heart, nerve, and muscle activity; enzyme formation; milk production.	Asparagus, beans, cauliflower, cheese, cream, egg yolk, milk.
Chlorine	Regulation of osmotic pressure; enzyme activities; formation of hydrochloric acid in stomach.	Bread, buttermilk, cabbage, cheese, clams, eggs, ham (cured), sauerkraut, table salt.
Cobalt	Normal appetite and growth; prevention of a type of anemia; prevention of muscular atrophy.	Liver, sea foods, sweetbreads.
Copper	Formation of hemoglobin; aids in tissue respiration.	Bran, cocoa, liver, mushrooms, oysters, peas, pecans, shrimp.
Iodine	Formation of thyroxin; regulation of basal metabolism.	Broccoli, fish, iodized table salt, oysters, shrimp.
Iron	Formation of hemoglobin; oxygen transport; tissue respiration.	Almonds, beans, egg yolk, meat, heart, kidney, liver, soybeans, whole wheat.
Magnesium	Muscular activity; enzyme activity; nerve maintenance; bone structure.	Beans, bran, Brussels sprouts, chocolate, corn, peanuts, peas, spinach, prunes.
Phosphorus	Tooth and bone formation; buffer effects in the blood; essential constituent of all cells; muscle contraction.	Beans, cheese, cocoa, eggs, liver, milk, oatmeal, peas, whole wheat.
Potassium	Normal growth; muscle function; maintenance of osmotic pressure; buffer action; regulation of heart beat.	Beans, bran, molasses, olives, parsnips, potatoes, spinach, oranges.
Sodium	Regulation of osmotic pressure; buffer action; protection against excessive loss of water.	Beef, bread, cheese, oysters, spinach, table salt, wheat germ.
Sulfur	Formation of proteins	Beans, bran, cheese, cocoa, eggs, fish, lean meat, nuts, peas.
Zinc	Normal growth; tissue respiration.	Beans, cress, lentils, liver, peas, spinach.

Fructose	173.3
Sucrose	100.0
Glucose	74.3
Maltose	32.5
Galactose	32.1
Lactose	16.0

Polysaccharides may be of plant or animal origin: they differ from each other by the arrangement of the atoms within their molecules. Sugars are generally soluble in water, while polysaccharides are insoluble.

Structurally, we can consider a molecule with the general formula of $C_2H_4O_2$ (glycoaldehyde) as a carbohydrate, but it does not meet any of the requirements of a nutrient. In fact, no carbohydrate containing less than six atoms of carbon can be used in human cellular oxidations. Our body chemistry is of such a nature that our cells can get energy only from the six-carbon sugars *(hexoses)* and from the more complex carbohydrates which can be converted into hexoses. Nevertheless there are some important five-carbon sugars such as ribose and deoxyribose which are essential to the structure of RNA and DNA respectively.

The simplest carbohydrates that can meet the energy demands of the body are those having the molecular formula $C_6H_{12}O_6$. A *molecular formula* is one that shows the number of atoms of each element in a molecule of a compound but does not show the arrangement of the atoms within the molecule. A formula that does show the arrangement is of value to a chemist. It is called a *structural* formula. Examples of such formulas are shown in Fig. 2-6. The molecular formula $C_6H_{12}O_6$ represents the basic structure of the carbohydrate used in human nutrition.

This substance belongs to the group *monosaccharides* (hexoses) because it represents a single (mono-) sugar group (-saccharide). The three principal hexoses in our diet which form the basis of our sugar metabolism are *glucose* (grape sugar or dextrose), *fructose* (fruit sugar or levulose), and *galactose*.

The next more complex group of sugars are the *disaccharides*. These are formed by green plants from the hexoses that are the end products of photosynthesis. They consist of two molecules that have been chemically combined with the loss of a molecule of water as in Fig. 13-1.

In sugar cane and sugar beet plants molecules of glucose and fructose are joined together to form the principal sugar of commerce, the disaccharide *sucrose*. A similar process occurs in grains to form *maltose* (grain sugar) or in cows to produce *lactose* (milk sugar).

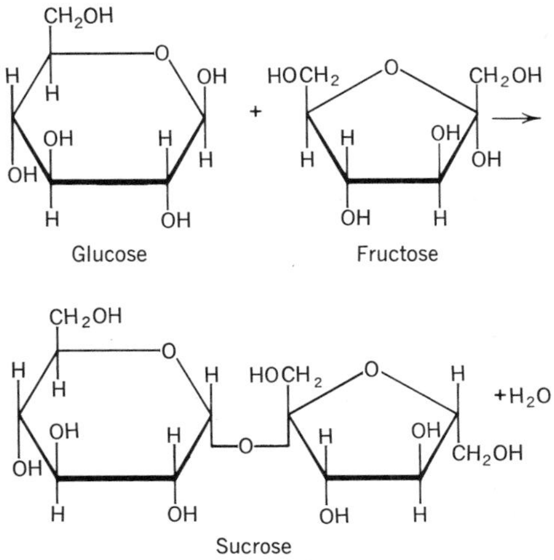

Fig. 13-1 The formation of a disaccharide from two monosaccharides.

When any of these sugars is present in food, the digestive system breaks them down into hexoses by hydrolysis, which reverses the chemical process by which they were formed. Hexoses can be further broken down in the cells to provide energy. In the three principal disaccharides, the process of hydrolysis results in the following chemical changes:

$$C_{12}H_{22}O_{11} + H_2O \rightarrow C_6H_{12}O_6 + C_6H_{12}O_6$$

Sucrose	Glucose + Fructose
Maltose	Glucose + Glucose
Lactose	Glucose + Galactose

Plants and animals are able to recombine the hexoses into higher carbohydrates of extremely complex structure. Such substances may be stored, or used in plants, to give added strength. These materials are called *polysaccharides (poly-,* many) and are represented mainly by *glycogen* (stored in animals), *starch* (stored in plants), and *cellulose* (an important plant structural material). Less common polysaccharides are dextrin, agar, inulin, chitin, and pectin.

The molecules of all these different substances are so complex (starch has almost 1000 hexose groups) that no attempt is made to write an exact chemical formula for them. The polysaccharides are therefore usually shown by the simplified formula $(C_6H_{10}O_5)_n$, the *n* standing for an unknown number. As in the case of the disaccharides, it is necessary to break a polysaccharide molecule down into its component monosaccharide parts before it can be used by the body. In some cases, starch for example, the structure is so complex that this breakdown cannot be accomplished in a single step. The course of the digestion of starch can be followed experimentally, and changes during the process noted. When a suspension of starch is mixed with a dilute solution of iodine, a dark blue suspension forms. If this suspension is mixed with saliva, the blue changes to red. Then the suspension is dissolved to produce a colorless solution. The red colored stage represents the conversion of the starch molecule into a simpler form, *erythrodextrin,* which is still very complex in structure. This then passes into a colorless form, *achrodextrin,* and then into a soluble disaccharide. In the body, this reduction of starch (our principal source of carbohydrate) requires two different enzymes. More will be said about this later.

As we have just said, starch will turn blue in the presence of iodine; glycogen, on the other hand, gives a mahogany-red color with iodine. These are very simple tests, but the identification of the disaccharides is much harder and requires either special apparatus or more complex chemical procedures. Accurate tests for some disaccharides involve their hydrolysis into their individual hexoses and examination of these in a *polariscope.* This instrument makes use of the fact that the hexoses will rotate a beam of polarized light either to the right or to the left. Since this rotating power is known for each hexose, it is possible to identify the disaccharide from which each has been derived. The monosaccharides themselves can only be distinguished from each other by the use of the polariscope, although they all give certain definite reactions with some copper-containing solutions. The two most commonly used tests for the presence of a monosaccharide employ Fehling's solution and Benedict's solution. Both are prepared by mixing a solution of copper sulfate (blue in color) with a strong alkali such as sodium hydroxide. When either is heated with a solution containing a hexose, its blue color changes to yellow, red, or reddish brown depending on the amount of sugar present. These tests are of great importance from a medical standpoint. Glucose, for example, will appear in the urine of a person suffering from untreated diabetes, and its detection there will aid in the diagnosis.

The carbohydrates play several important roles in the body chemistry. In the first place, their oxidation quickly releases a large amount of energy, so it is no wonder that these materials are stored in nearly all of the tissues of the body except the brain. The liver and muscles are the principal storehouses, although the kidneys also contain a sizable quantity. In a well-nourished individual, glycogen may account for approximately 20 percent of the dry weight of the liver. Carbohydrate (as glucose) is carried by the blood to the liver. Here glucose is converted into glycogen. In times of need, the glycogen is reconverted into glucose and enters the blood stream to be carried to those regions where it will be used.

Carbohydrates may also be converted into fats and stored in that form. The farmer who fattens pigs by feeding them corn, which contains a high percentage of starch, is simply applying this fact in a practical manner. Likewise, people who overindulge in carbohydrates may become overweight by storage of fats.

One of the most important functions of carbohydrates is their action as protein savers. Proteins, as we will see later, are essential to the processes of growth and repair of tissue. If the diet consists mainly of proteins the cells must oxidize them to provide energy. If carbohydrates are available, however, these will be oxidized rather than the proteins. Proteins produce as much energy per unit of weight as carbohydrates, but use of proteins for energy rather than repair and growth is obviously uneconomical.

Lipids

The term lipids includes fats, which are solid at ordinary room temperature (20°C.), and oils, which are liquid at that temperature. The lipids also include waxes and other fatlike substances present in many foods. Some lipids contain nitrogen and thus supplement the proteins in their ability to form protoplasm. Because lipids differ chemically among themselves, they are grouped in three subdivisions: the *fats and oils,* the *phospholipids,* and the *steroids.*

The molecule of a fat or oil, like the carbohydrate molecule, contains only the elements carbon, hydrogen, and oxygen, but the two molecules are very different in structure. A typical fat is beef fat, stearin ($C_{57}H_{110}O_6$). This molecular formula does not show the definite relation of the elements to each other. A molecule of a fat or oil is formed by the reaction of one molecule of *glycerol* ($C_3H_5(OH)_3$) with three molecules of a *fatty acid.* See Fig. 13-2. In the case of a stearin, the fatty acid is

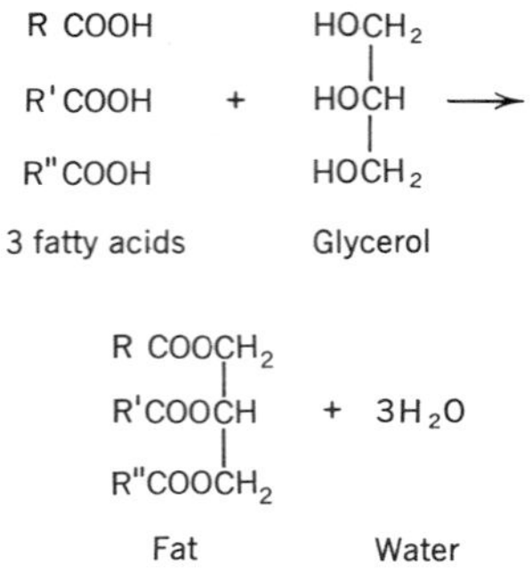

$$
\begin{array}{l}
\text{R COOH} \qquad \text{HOCH}_2 \\
\qquad\qquad\qquad\quad |\\
\text{R' COOH} \quad + \quad \text{HOCH} \longrightarrow \\
\qquad\qquad\qquad\quad |\\
\text{R''COOH} \qquad \text{HOCH}_2 \\
\text{3 fatty acids} \qquad \text{Glycerol}
\end{array}
$$

$$
\begin{array}{l}
\text{R COOCH}_2 \\
\qquad\qquad\quad |\\
\text{R'COOCH} \quad + \quad 3H_2O \\
\qquad\qquad\quad |\\
\text{R''COOCH}_2 \\
\text{Fat} \qquad\qquad \text{Water}
\end{array}
$$

Fig. 13-2 The formation of a fat from glycerol and three fatty acids (all different or all the same). COOH is the acid radical and R,R', and R'' represent the hydrocarbon groups of the fatty acids.

stearic acid. When a fat is digested, its molecules are broken down into glycerol and fatty acids. Some common fatty acids are in the table (page 191).

Many of the fats in our diet contain more than one kind of fatty acid. Since these acids all have distinctive flavors, the mixture of acids gives the food its particular flavor or odor. Butter fat, for example, has the following approximate composition:

Butyric acid	3.0%
Caproic acid	1.4%
Caprylic acid	1.8%
Capric acid	1.8%
Lauric acid	6.9%
Myristic acid	22.6%
Palmitic acid	22.6%
Stearic acid	11.4%
Oleic acid	27.4%

Fats and oils can be subdivided into two groups depending on the amount of hydrogen that is present in the fatty acid molecule. If you examine the formulae of the fatty acids given above,

you will notice that all except lauric and oleic acids have twice as many hydrogen atoms as carbon atoms. Such acids are called *saturated* fatty acids, while those with a smaller number of hydrogen atoms are known as *unsaturated.* It is possible to convert an unsaturated acid into a saturated one by chemical reaction with hydrogen. This is done on a commercial scale in the manufacture of certain shortening compounds used in baking and in butter substitutes. The process is called *hydrogenation.*

The *phospholipids* are a group of chemically complex substances that contain nitrogen and phosphorus in addition to the carbon, hydrogen, and oxygen present in the molecules of the true fats. The phospholipids are abundant in brain, heart, kidneys, eggs, soy beans, etc. The best known phospholipid is *lecithin,* a normal constituent of cells. Egg yolk is especially rich in it, and a crude preparation of lecithin can be obtained from that source.

The *steroids* are represented by *cholesterol* $(C_{27}H_{45}OH)$, the best known member of the group. It is normally present in the brain and spinal cord, and plays an important role in the formation of bile by the liver. In some individuals the cholesterol in the liver is converted into hard bodies, gallstones. These may cause serious difficulty if they obstruct the flow of bile.

Cholesterol has been given considerable publicity because of its role in one of the most common types of cardiovascular (heart and blood vessels) disease. Deposits of cholesterol may form in the walls of the coronary arteries, which supply the heart muscles. These deposits roughen the walls and encourage blood clot formation

Fatty Acid		Occurrence
Butyric acid	$C_4H_8O_2$	Butter
Caproic acid	$C_6H_{12}O_2$	Butter
Lauric acid	$C_8H_{12}O_2$	Spermaceti, coconut oil
Myristic acid	$C_{12}H_{24}O_2$	Butter, coconut oil
Palmitic acid	$C_{16}H_{32}O_2$	Animal and vegetable oils
Oleic acid	$C_{18}H_{34}O_2$	Animal and vegetable fats
Stearic acid	$C_{18}H_{36}O_2$	Animal and vegetable fats
Cerotic acid	$C_{26}H_{52}O_2$	Beeswax, wool fat (lanolin)

which may result in coronary thrombosis. General weakening of the arteries may also occur due to cholesterol and calcium deposits in their walls.

It has been recognized for a long time that overweight people are more prone to cardiovascular disease than those who do not eat fat-producing foods to excess. It has also been observed that in countries where the population eats a large amount of fat, as we do in the United States, the number of deaths from cardiovascular disease is greater than in countries where the average diet contains little fat. The body's supply of cholesterol is not gained solely from dietary sources, as it can by synthesized by various body tissues. However, there is some evidence that the presence of a large amount of saturated fatty acids in the diet may be responsible for the formation of excess cholesterol deposits in the arteries.

In the process of digestion, the fats are reduced to glycerol and fatty acids. They are then carried to the cells and stored in the form of fat. Most of our digested fat is oxidized to produce energy. Pound for pound, fats produce more than twice as much energy as either carbohydrates or proteins. The body stores fats chiefly in the vacuoles of specialized cells that make up adipose tissue.

A simple test for the presence of fats is to put the food in question on a piece of paper. If fat is present, the food will leave a greasy spot on the paper. You see an example of this test when you carry doughnuts in a paper bag. Chemical tests for specific fats are rather complex. First the fats are dissolved from the food by solvents such as ether, chloroform, acetone, or benzene. This is necessary because fats are not soluble in water. Then chemical tests are used to identify the specific fatty acids present.

Proteins. The proteins, as has been said, are among the most chemically complex of all of the organic compounds. The molecules of some of these substances are the largest known and have an extremely high molecular weight. Hemoglobin from the blood has a molecular weight of 66,700. (The molecular weight of water, by comparison, is 18.) Actin and myosin have respective molecular weights of approximately 80,000 and 850,000. Other proteins have lower or higher molecular weights, depending on the sizes of their molecules.

In the same way that carbohydrates and fats are composed of simpler compounds, so the proteins are made up of smaller units that are relatively simple. These are the *amino acids,* of which some 40 have been discovered. Only 23 of them have been accepted as common "building blocks" of the proteins and only about 20 of these are found widely distributed in living organisms. No one protein has been found thus far that contains all of these, but casein of milk, as shown in the table on page 193, contains 18 of them.

Proteins differ from each other depending on their source, and those found in one tissue of an animal's body differ from those found in another tissue. These differences depend on the number and kind of amino acids present, the ratio by weight and number, of one to another, and their arrangement within the protein molecule both linearly and spatially. The table shows two of these characteristics, the type of amino acid and the ratio of each. The arrangement of each can only be shown by a structural diagram of the particular protein.

The amino acids are not acids in the generally accepted sense of the term; they are more properly called amphoteric or *amphiprotic* compounds because they have properties which may be either acidic or basic. Thus under one set of conditions they will give a slightly acid reaction, while under another they are slightly alkaline. The reason for this is that each amino acid contains both amino (NH_2) and carboxyl (COOH) radicals. The amino radical has a basic reaction while the

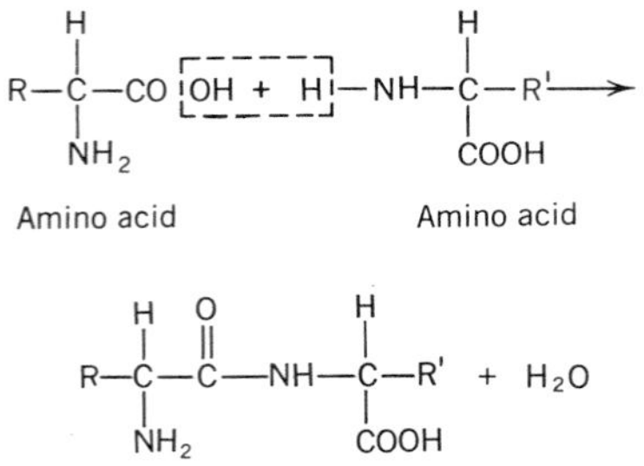

Fig. 13-3 The formation of a dipeptid from two amino acids. R and R' have a different composition for each amino acid.

carboxyl radical is acidic. In a protein molecule the amino acids are joined together through bonds between the carboxyl group of one amino acid and the amino group of another by the removal of a molecule of water, just as fats and carbohydrates are built up from their subunits with the elimination of water. See Fig. 13-3.

It will be noted from the form of the table that amino acids may be divided into two groups: those that are essential and those that are not essential. This simply means that ten of the amino acids listed here should be present in the diet because the body either cannot synthesize them or does so too slowly. Arginine for example can be synthesized by the body, but the rate is too slow to be of value. Histidine may be formed by the action of certain intestinal bacteria. The nonessential amino acids are those which the body can synthesize out of other materials that are present in our food. They are just as important as the so-called essential ones for proper body growth and maintenance, but they do not have to be in the diet. For example, a rat kept on a diet in which gelatin is the principal source of protein will lack tryptophan (see table). The rat will be able to maintain its protoplasm but it will not grow.

The principal use of proteins is to repair and replace protoplasm and to form new living material. Nitrogen is the essential element in the protein molecule. Without this element no protoplasm could be produced. It is interesting to note that although nitrogen makes up about 80 percent of the atmosphere we are unable to use it in this form. Nor can our bodies use this nitrogen in simple compounds such as ammonia and nitrates. We must depend on chemically complex combina-

tions of nitrogen in our food for this vitally important element.

A second use of proteins is to supply energy. There is some doubt as to whether they are oxidized directly or first converted into sugars and fats. If they are oxidized directly, their production of energy is equivalent to an equal amount of carbohydrate. The fuel value of each of the classes of organic nutrients, in terms of the heat they are able to produce when oxidized in the body, is as follows:

1 gram of carbohydrate will yield	4.1 Calories
1 gram of fat will yield	9.3 Calories
1 gram of protein will yield	4.1 Calories

AMINO ACID CONTENT OF SOME PROTEINS

Amino Acids	Approximate Percentages in Dry Protein				
Essential	Gelatin	Casein (milk protein)	Egg white	Pepsin	Insulin
Arginine	8.0	4.3	6.0	1.0	3.4
Histidine	0.8	2.9	2.9	0.8	5.3
Isoleucine	1.3	6.4	7.0	10.8	2.8
Leucine	2.9	7.9	9.9	10.4	13.2
Lysine	4.1	8.9	6.5	1.5	2.4
Methionine	0.9	2.5	5.3	1.7	—
Phenylalanine	2.6	4.6	7.2	6.4	8.3
Threonine	2.2	4.9	4.0	9.6	2.1
Tryptophan	—	1.6	1.2	2.4	—
Valine	2.3	6.3	8.8	7.1	7.8
Nonessential					
Alanine	10.0	3.8	7.6	—	4.5
Aspartic acid	7.5	8.4	9.3	16.0	6.8
Cystine	0.1	0.4	2.8	2.1	12.5
Glutamic acid	10.8	22.5	16.5	11.9	18.6
Glycine	26.0	2.3	3.6	6.4	4.3
Proline	16.5	7.5	3.8	5.0	2.5
Serine	3.4	6.3	8.2	12.2	5.2
Tyrosine	0.4	8.1	4.1	8.5	13.4

The production of heat in the body is measured in units called Calories (capital *C*) which are 1000 times greater than the calorie (lower-case *c*) used in the physical sciences. The Calorie is defined as *the amount of heat required to raise the temperature of 1 kilogram of water from 15° to 16°C*. Since foods must be oxidized to produce energy, any excess food not actually used for energy is largely converted into fats for storage.

Although excess carbohydrates and fats may be stored in the body, there is no provision for protein storage. Therefore, proteins that are not immediately utilized in the formation of new protoplasm are built into the protoplasm of existing cells or are transformed into sugars and fats.

The identification of amino acids is frequently made by a rather interesting process of elimination. There are certain bacteria and fungi whose growth requires one or more definite amino acids. It is possible, therefore, to prepare synthetic food materials containing the protein to be tested. If the bacteria or fungi fail to grow, we know which amino acids are lacking in the protein.

The presence of a protein in any material can be determined by the *xanthoproteic test*. This consists of heating a food in a solution containing nitric acid. If protein is present, a yellow color develops which changes to orange on the addition of an alkali such as sodium hydroxide.

The Diet

The term *ingestion* is applied to the process of taking solid or liquid foods into the body through the mouth.

Green plants manufacture their own nutrients, but animals, as a group, must take in complex foods and then break them down into the simpler products that their bodies can use. This second process is *digestion*. The simplified products pass into the circulating blood and are carried through the body to those areas where they are needed for energy or growth, or are deposited in storage depots. In the first part of this chapter we discussed the role played by each of the nutrients. Let us now consider how the intake of these nutrients is balanced.

A person's food requirements vary with his age and size, the amount of work he does, and the climate in which he is living. All of these factors affect the normal, healthy individual. To meet any pronounced change in one or more of them, he has to alter his diet.

For an average person twelve years of age and over, a basic intake of 2400 Calories per day is considered adequate. This should be supplemented by extra quantities of food to meet the special requirements of various types of work. The suggested additions are:

Type of Work	Addition per Hour of Work
Light	50 Calories
Moderate	50–100 Calories
Hard	100–200 Calories
Very hard	200 Calories or more

These values are based on the needs of a person of average size: for men, a weight of 68 kilograms and a height of 1.70 meters; for women, a weight of 60 kilograms and a height of 1.65

meters. As we will show in a later chapter, most of the body's heat is lost through the skin. If the surface area of the skin is known, this factor can be considered in determining the energy expenditure of the body. The average man has a total body surface area of about 1.8 square meters, while the average woman has a body surface area of about 1.6 square meters. In general, persons with smaller surface areas require fewer Calories, while those with greater surface areas need more Calories.

The diet of a healthy person should include certain general types of foods. Each of these furnishes one or more important food elements. A rough grouping of foodstuffs may be made as follows:

MEAT, FISH, and CHEESE supply proteins. Fats are also present in varying amounts. Organs such as liver, kidneys, and sweetbreads, as well as shellfish, are rich in minerals and vitamins.

MILK is an excellent source of proteins, carbohydrates, fats, vitamins, calcium, and phosphorus.

EGGS contain proteins, fats, vitamins, and minerals.

VEGETABLES and FRUITS are satisfactory sources of vitamins, minerals, and roughage. Root vegetables and legumes (beans, peas, etc.) contain carbohydrates and proteins. Raw fruits and green vegetables are needed as sources of vitamins and minerals.

FATS and SWEETS are sources of energy-producing materials, fat-soluble vitamins, and various fatty acids.

A balanced diet is one that contains all of the essential food elements in sufficient quantities to supply the needs of the body. As we have said, these needs may vary somewhat, with the result that more of certain nutrients are required at one time and less at another. Protein requirements definitely change during the life of the individual. When he is actively growing, the requirement is comparatively higher than it is later in life, because there is a greater demand for the basic materials for protoplasm formation. For example, a child of 3 to 5 years of age will require daily 3 grams of protein for each kilogram of body weight (one kilogram = 1000 grams = 2.2 pounds). Between the ages of 5 and 15, the basic requirement is 2.5 grams per kilogram; from 15 to 17 it drops to 1.5 grams; and above 21 it remains at 1.0 gram. During pregnancy the protein intake should be increased from the normal 1.0 gram to 2.0 grams per kilogram of body weight. A list of some common foods and their energy values in Calories is included in the table on pages 196–197.

Raw vegetables should be eaten in moderate amounts only. All plant materials contain cellulose, a complex carbohydrate that cannot be digested. This forms the greater part of the roughage which is essential for the proper elimination of intestinal wastes. However, large quantities of cellulose are irritating to the lining of the digestive tract. Cooking of vegetables breaks down the cellulose cell walls of plant tissues, thereby freeing some of the nutrients within the plant cells that otherwise might not be digested.

A food is considered to be acid when on oxidation it produces compounds such as uric and phosphoric acids. When used in this connection, the word does not refer to the sour flavor of fruits like

FOOD VALUE OF SOME OF OUR COMMON FOODS

Food	100-Calorie Portion Measure	Distribution of Calories		
		Protein	Fat	Carbohydrate
Apples	One large	3	5	92
Apple pie	1½ in. at circumference	3	41	56
Apple sauce	⅜ cup	1	3	96
Apple tapioca	¼ cup	1	1	98
Apricots	3 large halves	5		95
Asparagus	20 large stalks	32	8	60
Asparagus soup	½ cup	17	56	27
Bacon	4 or 5 slices (small)	13	87	
Banana	One medium	5	6	89
Beans, baked	⅓ cup	21	18	61
Beans, lima (buttered)	¼ cup	16	36	48
Beef, Hamburg	1 cake 2 in. diam. ¾ in. thick	38	52	10
Beef, pot roast	Slice 4¾ × 3½ × ⅛ in.	62	38	
Beef, sirloin	Slice 1¾ × 1½ × ¾ in.	47	53	
Beets, greens	2¼ cups	17	58	25
Beets, fresh	Four 2 in. diam.	14	2	84
Biscuits	Two small	11	27	62
Blackberries	½ cup	9	16	75
Bread, Boston brown	¾ in. slice 3 in. diam.	10	10	80
Bread, graham, w.wh.	2½ slices 3¾ × 3¼ × ¼ in.	14	6	80
Bread, white	2 slices 3 × 3½ × ½ in.	14	6	80
Butter	2 sq. 1¼ × 1¼ × ¼ in.		100	
Buttermilk	1⅛ cups	33	13	54
Cabbage, shredded	Four cups	20	9	71
Cantaloupe	One 4½ in. diam.	6		94
Carrots, buttered	½ cup	4	67	29
Celery	4 cups of ¼ in. pieces	24	5	71
Cheese, American	1⅛ in. cube	26	74	
Cheese, cottage	5 tbsp.	76	9	15
Cherries, fresh, stoned	1 cup	5	9	86
Chicken, white meat	3 slices 3½ × 2½ × ¼ in.	80	20	
Chocolate, beverage (made with milk)	⅓ cup	13	49	38
Chocolate cake	2½ × 2½ × ⅞ in.	5	41	54
Chocolate nut fudge	1¼ × 1 × ¾ in. piece	4	36	60
Clams	Twelve clams	56	8	36
Cocoa, beverage (made with milk)	½ cup	16	44	40
Codfish, creamed	½ cup	32	46	22
Cod liver oil	1 tablespoon		100	
Cod steak	3¾ × 2½ × 1 in. piece	94	6	
Corn bread	Slice 2 × 2 × 1 in. slice	10	24	66
Corn, fresh	½ cup	12	10	78
Crackers, saltines	Six, 2 in. square	10	26	64
Cream, thick	1⅔ tbsp.	2	95	3
Cream, thin	¼ cup	5	86	9

FOOD VALUE OF SOME OF OUR COMMON FOODS (Continued)

| Food | 100-Calorie Portion Measure | Distribution of Calories | | |
		Protein	Fat	Carbohydrate
Cream pie	2 in. at circumference	10	37	53
Cup cakes	$\frac{1}{2}$ cake 2 in. diam. 2 in. thick	8	32	60
Custard, boiled	$\frac{1}{3}$ cup	13	44	43
Doughnuts	$\frac{1}{2}$	6	45	49
Eggs, in shell	$1\frac{1}{3}$ eggs	36	64	
Grapefruit	$\frac{1}{2}$ large	7	4	89
Ham, boiled	Slice $4\frac{3}{4} \times 4 \times \frac{1}{8}$ in.	29	71	
Honey	1 tbsp.	1		99
Ice cream	$\frac{1}{4}$ cup	4	63	33
Lamb chops	One $2 \times 1\frac{1}{2} \times \frac{3}{4}$ in.	40	60	
Lettuce	2 large heads	25	14	61
Macaroni and cheese	$\frac{1}{2}$ cup	7	39	44
Maple syrup	$1\frac{1}{2}$ tbsp.			100
Milk, whole	$\frac{5}{8}$ cup	19	52	29
Milk, skim	$1\frac{1}{8}$ cups	37	7	56
Muffins, bran	$1\frac{3}{4}$ muffins $2\frac{3}{4}$ in. diam.	20	16	64
Oats, rolled—cooked	$\frac{1}{2}$ to $\frac{3}{4}$ cup	17	16	67
Onions	3 to 4 medium	13	6	81
Orange juice	1 cup			100
Oranges, whole	One large	7	2	91
Oyster stew	$\frac{1}{2}$ cup	16	63	21
Parsnips	One 7 in. diam., 2 in.	10	8	83
Peaches, fresh	Three medium	6	3	91
Peanut butter	1 tbsp.	19	69	12
Peanuts, shelled	20 to 24 nuts	19	63	18
Pears, canned	3 halves, 3 tbsp. juice	1	4	94
Peas, green	$\frac{3}{4}$ cup	28	4	68
Plain cookies	Two $2\frac{1}{4}$ in. diam.	6	33	61
Pork chops	$\frac{1}{2}$ chop	32	68	
Potatoes, white	One medium	11	1	88
Prunes, stewed	Two, 2 tbsp. juice	2		98
Raisins	$\frac{1}{4}$ cup	3	9	88
Rice pudding	$\frac{1}{2}$ cup	18	32	50
Salmon, canned	$\frac{1}{2}$ cup	45	55	
Sauerkraut	$2\frac{1}{2}$ cups	25	17	58
Spinach, cooked	$2\frac{1}{2}$ cups	12	8	80
Sponge cake	Piece $1\frac{1}{2} \times 1\frac{1}{2} \times 2$ in.	11	19	70
Squash	1 cup	12	10	78
Squash pie	2 in. at circumference	10	25	65
Sugar, granulated	2 tbsp.			100
Tapioca cream	$\frac{2}{5}$ cup	12	28	60
Tomatoes, whole	2 to 3 medium	16	16	68
Veal cutlets, breaded	$\frac{2}{5}$ serving	30	52	18
Walnuts, English	8 to 16 nuts	11	82	7
Watermelon	Slice $\frac{3}{4}$ in. $\times$ 6 in. diam.	5	6	89
Yeast, compressed	Six cakes	32		68

lemons. Most meats are acid-producing foods, while many fruits are alkaline regardless of their flavor. The reason for this is that oxidation of fruits and vegetables frequently produces salts of magnesium, sodium, and potassium. These eventually combine with other chemicals to produce basic (alkaline) compounds. The flavor of a food does not indicate whether it will have an acid or alkaline reaction in the body.

Physiologically, a distinction can be drawn between appetite and hunger. The former means the craving for a certain food or drink, which can be satisfied quite easily. Appetite is usually based on some previous pleasurable experience. Hunger, on the other hand, is a complex condition that is accompanied by vigorous contractions of the empty stomach, a feeling of emptiness in the abdomen, and a generalized discomfort that cannot be localized in any particular area. It is unwise to confuse these two terms, especially when considering the diet of young children. If appetite is mistaken for hunger, a child may be able to convince his elders that a diet unbalanced in favor of some well-liked item is more satisfactory than one that contains all of the necessary food factors. Hunger is a constant reminder of the primary needs of the body for energy and specific nutrients.

VOCABULARY REVIEW

Match the phrase in the left column with the correct word in the right column. *Do not write in this book.*

1 A type of chemical reaction involving water.
2 Final product of the digestion of starch.
3 Contains the NH_2 group.
4 A measure of heat.
5 The process resulting in the break-down of food.
6 Cholesterol belongs to this group of organic substances.

a amino acid
b Calorie
c carbohydrate
d digestion
e hexose
f hydrolysis
g ingestion
h nutrient
i lipids

TEST YOUR KNOWLEDGE

Group A

Select the correct statement. *Do not write in this book.*

1 The term organic nutrients includes (a) water (b) mineral salts (c) carbohydrates (d) carbonates.
2 Carbohydrates do *not* contain the element (a) oxygen (b) hydrogen (c) nitrogen (d) carbon.

3 One of the monosaccharides is (a) sucrose (b) glucose (c) maltose (d) lactose.

4 Carbohydrates contain (a) equal amounts of hydrogen and oxygen (b) more hydrogen than oxygen (c) less hydrogen than oxygen (d) no definite proportion of hydrogen and oxygen.

5 A carbohydrate *not* obtained from plants is (a) glucose (b) starch (c) glycogen (d) cellulose.

6 Amino acids are derived from (a) starches (b) fats (c) oils (d) proteins.

7 One product of the oxidation of glucose in the body is (a) water (b) salt (c) carbon monoxide (d) glycerol.

8 One chemical test for a monosaccharide is (a) copper nitrate (b) sodium hydroxide (c) Fehling's solution (d) the xanthoproteic test.

9 The organic nutrient with the highest calorific value is (a) sugar (b) starch (c) fat (d) protein.

10 Of the following, the *best* source of protein is (a) fish (b) beans (c) apples (d) butter.

Group B

1 What is the importance of the chemical process of hydrolysis in nutrition?

2 Using as an example a recent meal you have eaten, briefly discuss the nutritional value of each of the foods that composed it.

3 What is the difference between the molecular formula of a compound and its structural formula?

4 Why do some people have to take into consideration the number of Calories that are in their food?

5 What are the differences and similarities between inorganic and organic materials?

Chapter 14 Enzymes and Vitamins

ENZYMES

The growth and activity of every cell in the body is dependent on the presence of a group of chemical compounds, the *enzymes*. Although it is most convenient to discuss them in relation to digestion, the role of enzymes in living processes is not limited to digestion, as was seen in Chapter 3. Their chemical activities are so varied and complex that enzymes accomplish with ease processes that a chemist can reproduce only with difficulty in the laboratory. From a chemical standpoint, life consists of a series of chemical reactions which are either hastened or retarded by the activities of the enzymes.

An enzyme is a complex protein compound capable of a chemical reaction without itself being changed in composition. The chemical name for this is an *organic catalyst*. Many commercial processes depend on the use of inorganic or organic catalysts to break large molecules into simpler ones or combine elements and simpler compounds to form large molecules. Some of these catalysts are relatively simple compounds while others may be highly complex, but they all differ from enzymes in that they are not produced by living cells. A few industries do make use of enzymes, but in each case it is necessary to grow the appropriate organisms to produce them. Thus the brewing industry makes use of the enzymes of yeast. Cheese manufacturers use enzymes of molds and bacteria, and farmers use bacterial enzymes in the preparation of silage.

The outstanding characteristic of an enzyme is its ability to accelerate chemical change without itself entering into the reaction. The late Sir William Bayliss once likened the action of a catalyst to that of a drop of oil beneath a metal weight on an inclined glass plate. The oil does not start the weight moving down the plate, nor does it combine chemically with either the metal or the glass. It merely hastens the sliding process.

The action of catalysts can be illustrated by the following simple experiment with cane sugar. A molecule of cane sugar is formed by the combination of two molecules of simple sugars with the loss of a water molecule. This action is carried out in the plant under the influence of enzymes. Now, if we make a solution of cane sugar in water and allow this to stand for a long period of time, water will be taken up by the molecules of the cane sugar and they will break down into the same simple sugars from which the cane sugar was originally formed. The presence of these simple sugars can be detected by Fehling's Test. This chemical change is a fairly slow one involving a period of weeks, but we can obtain the same results in a few minutes by the use of a catalyst. A few drops of dilute hydrochloric acid are added to a solution of cane sugar in a test tube. The solution is then boiled and allowed to cool. When tested with Fehling's solution, it shows the presence of simple sugars. Here the hydrochloric acid is the catalyst and can be recovered unchanged at the end of the experiment. In this case it is necessary to use a strong acid and heat to bring about the change. If, however, we use the enzyme *sucrase,* that is secreted by the walls of the small intestine, we find that the change occurs rapidly without application of heat.

All enzymes that have been isolated and chemically identified thus far are proteins. In many instances these cannot function by themselves but require the presence of some other substance. Some of these accessory materials are relatively simple, like calcium or magnesium ions, while others are highly complex. A good example of the need for these auxiliary materials is found in the action of *rennin,* an enzyme secreted by the stomach of a calf. This enzyme digests the protein of milk (casein), but only in the presence of calcium ions. Since milk normally contains calcium, this process occurs regularly in the calf's stomach. If, however,

the calcium is removed from the milk, no digestion by rennin is possible.

Another important characteristic of enzymes is their ability to carry on their activities outside the cells that produce them. Even though they are manufactured by living cells, their structure is such that they do not require the actual presence of protoplasm to make them effective. One of the simplest demonstrations of this characteristic is the effect of saliva on starch. Saliva contains an enzyme which decomposes starch into double sugars. If a starch solution is mixed with some saliva in a test tube, and then placed in an incubator at body temperature, the solution will slowly change from an opaque condition to a clear one as the transformation of starch to sugar takes place. The enzyme in the saliva is an amylase called *pytalin.* Its ability to convert starch to a simpler form in a test tube indicates that the presence of living protoplasm is not required to make an enzyme active. The specialized cells in the walls of other digestive organs also form enzymes and liberate these into the alimentary canal where their action takes place.

Enzyme Action

The activity of an enzyme depends on the structure of its surface. Since they are highly complex proteins, their surfaces bear an intricate design determined by the spatial as well as linear arrangement of their amino acids. Each enzyme evidently has a highly individualistic pattern of surface structure. It is as though one enzyme had a pattern of curved pits, another of angular pits or projections, and a third of both curves and angles. Each particular pattern

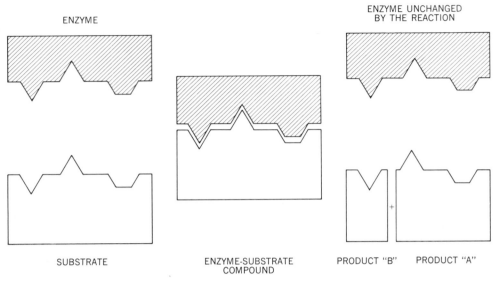

Fig. 14-1   Model for enzyme action. When the enzyme-substrate compound is formed, the substrate breaks into simpler compounds, but the enzyme remains unchanged.

would have its opposite configuration on the surface of some other type of molecule that the enzyme could affect— the *substrate*. Experimental evidence suggests that when these two materials, the enzyme and the substrate, come in contact with each other, they form briefly an enzyme-substrate complex. In one case, this complex was found to exist for only 1/85,000 of a second. The combination of enzyme and substrate lowers the activation energy necessary for a given reaction and the substrate breaks into simpler compounds or two molecules of substrate are joined to form one. See Fig. 14-1.

Some of the most important enzymatic reactions in cells cannot occur unless an additional substance is present to aid the principal enzyme (the *apoenzyne*) in splitting or linking substrates. This auxiliary material is called a *coenzyme*. The reactivities of these coenzymes are especially noticeable in processes concerned with the release of energy within cells, although they function in other reactions as well. One of the most important and widely spread of these substances is known as *coenzyme A (CoA)*. The molecule of this material contains a vitamin, pantothenic acid, which will be discussed later. CoA enters into several reactions within cells which result in the formation of high-energy bonds similar to the ones concerned with the release of energy during the contraction of a muscle fiber. Another substance which acts as a coenzyme is vitamin B_1 (thiamine). The role of vitamins in the chemical reactions occuring within cells is frequently that of coenzymes. The primary effects of vitamin deficiency are therefore at a molecular level although they may subsequently become evident as visible tissue changes.

Enzymes are easily affected by various conditions in their environment. Changes in temperature, acidity, and the accumulation of the products of their activities; all these things have a tendency to influence the rate at which they work.

Effects of temperature. Temperature plays an important part in the rate at which enzyme reactions occur. The rate of most chemical reactions increases from two to three times with each rise of 10°C (18°F) degrees. Enzyme reactions obey this general rule. At 0°C (32°F), their action generally ceases, although a few plant enzymes function below this temperature. The rate of activity increases as the temperature rises until heat destroys the enzyme. The destruction of most animal enzymes occurs in the temperature range between 40° and 50°C (approximately 104°–122°F). In all animals and plants there is a rather definite temperature at which an enzyme functions most efficiently. This is called the *optimum temperature* for the enzyme. In man, this temperature is approximately 37.5°C (98.6°F), as determined by a thermometer placed under the tongue.

Certain light-producing organisms demonstrate the effect of temperature on enzyme action. Whether the light is produced by fireflies, bacteria, or the fungus growing on a decaying stump, the process depends on the presence of an enzyme, *luciferase.* A protein, *luciferin,* also must be present. This protein is oxidized in the presence of luciferase to produce the light. As Fig. 14-2 shows, when the temperature of a

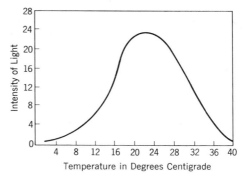

Fig. 14-2 The effect of temperature on an enzymatic process in light-producing organisms. Note that low temperatures slow down the process, but that light intensity increases with increase in temperature only until an optimum temperature of about 23.5°C is reached. Temperatures higher than this retard the reaction. The enzyme is destroyed at about 40°C.

mixture of luciferin and luciferase is low, the amount of light produced is is low. As the temperature is raised, the intensity of the light increases and then decreases until it is totally extinguished. At this point the enzyme responsible for the production of light is destroyed, although the organism may still be alive.

Effect of pH. The action of many enzymes depends on the pH of the solution in which they react. A good example of the effect of hydrogen ion concentration is found in the action of ptyalin on starch. In the mouth, ptyalin acts in a solution that has a pH of approximately 6.7 (very close to the neutral point). When the food is swallowed and enters the stomach, it encounters the gastric juice, which has a pH varying between 1 and 3. This greatly increased acidity tends to stop the action of ptyalin and arrest further digestion of starch. In the same manner, the gastric enzyme, pepsin, will act in a highly acid medium, but its activity is stopped when the food enters the small intestine, where the pH varies from about 5 to approximately 8.

Accumulation of end-products. The materials produced by the activities of enzymes are called their end-products. It is difficult to demonstrate how the accumulation of end-products affects human activities. However, laboratory experiments with various enzymes have indicated that the rate at which the enzyme activities proceed may be affected.

One of the most carefully studied enzymes is the *zymase* of yeast. This is responsible for the splitting of starch into sugars. Other enzymes then convert sugar into alcohol and carbon dioxide. If the end-product (alcohol) is not removed, its gradual accumulation will slow down the action of the enzymes and eventually bring the process to an end. The most active enzymes of brewer's yeast can therefore produce only 13 to 15 percent alcohol from a grain or fruit substrate before their own products stop their activities.

VITAMINS

Some of the most important and impressive advances in the field of physiology have been made in the study of vitamins. These materials are essential food substances although they contribute no material for production of energy or building of protoplasm. They are required in only minute amounts, but without them the body cannot use some of the foods it takes in or carry on numerous vital functions that are necessary for the maintenance of normal conditions.

The body's need for certain food factors has been recognized for many years, but it is only within this century that we have learned the relation of some of these to specific body conditions. In the sixteenth century, the British sea rover, Richard Hawkins, observed that lemons and oranges prevented the appearance of *scurvy* among his sailors. About two hundred years later a British physician, James Lind, recommended that the sailors of the Royal Navy be given regular rations of lime juice to stop the ravages of scurvy. This course was adopted and from it arose the nickname of Limey or Lime-juicer for the British sailor.

Beriberi is a vitamin deficiency disease that plagued the people of the Far East. In 1882 Admiral Takaki of the Japanese Navy improved the diet of his men by including a greater amount of meat, barley, and fruit than had been used previously. He held the mistaken idea that beriberi was caused by a diet low in proteins. Five years later, a Dutch physician, Eijkman, discovered that a disease similar to beriberi could be induced in birds fed on polished rice, a staple in the diet of Oriental peoples. He later cured the birds of the disease by feeding them the husks of the rice that had been removed in the polishing process. He formulated the hypothesis that the lack of some factor present in the husks but lacking in the polished grain might be the cause of beriberi. Eijkman's hypothesis aroused the interest of physiologists who were studying the human diet and led to intensive research in the field of food deficiency diseases. In 1911, Dr. Funk developed a general theory to account for these diseases. He used the term *vitamine* as a name for the factors lacking in deficiency diseases, because his work indicated that they all contained the amino (NH_2) group which is characteristic of all amines. It was assumed that this chemical group was characteristic of these vital food factors and essential to their proper functioning. Later work has shown that this is not always true, so the spelling has been changed to *vitamin*.

In the early days of research on vitamins, each of the newly identified factors was designated by a letter of the alphabet. This was not only a convenient method of naming them, but it was also noncommittal about their chemical structure. As the methods of chemistry improved, new information on vitamins was obtained, and the use of letters was found to be inadequate in many cases. Some of the older names have been kept because of their long-established usage, but many of the more recently discovered compounds have been given chemical names.

All of the known vitamins fall into one of two categories: those that are soluble in *oil* and those that are soluble in *water*. The oil-soluble vitamins are known by letters, A, D, E. etc.; while the water-soluble ones are named. An exception to this general rule is water-soluble ascorbic acid, which is still commonly called vitamin C. This difference in the solubility of the vitamins is of considerable dietary importance because it indicates how vitamins will behave when foods are cooked. The loss of oil-soluble vitamins in cooking water, generally speaking, will be less than may be expected of a water-soluble vitamin under the same conditions.

Oil-soluble Vitamins

Vitamin A ($C_{20}H_{29}OH$). Vitamin A is a product of animal metabolism and is not found in plants. It is a slightly yellowish substance that is stable in air at ordinary temperatures but is rapidly destroyed in the presence of oxygen at high temperatures. This explains why foods cooked in open pans frequently lack vitamin A while those that are prepared in pressure cookers retain it.

One source of vitamin A in human nutrition is the ingestion of animal products. The human body is also able to convert a plant product, *carotene,* into the vitamin. Three different chemical forms of carotene are found in plants; one of these can be converted into two molecules of vitamin A, while the others produce only one molecule each. Carotenes are all yellowish materials that give the typical color to carrots and other plant foods. Another yellow pigment *cryptoxanthine,* is found in corn and can also serve as a source of vitamin A. If an excess of carotene is eaten, the skin may take on a yellow discoloration and form excess freckles.

The recommended daily intake of vitamin A is 5000 International Units per day, which is actually an extremely small quantity. For instance, one ounce of pure crystalline vitamin equals 30 million units. Overdosing with vitamin A can have serious consequences, such as fragility of the bones, drying and peeling of the skin, nausea, headaches, and loss of serum proteins. Recovery from these effects is usually slow.

An individual who has a normal, adequate diet can store enough vitamin A in his liver to meet his needs for several weeks or months if his usual supply is cut off. Retention of the vitamin, however, is apparently dependent on the presence of a second vitamin, E. Animals which have been deprived of vitamin E are unable to store vitamin A properly because it is changed by oxidation.

Vitamin A is essential to the health and well-being of epithelial tissues and makes them less susceptible to diseases. Since the lining of the digestive and respiratory tracts (as well as many glands) is made up of epithelium, the normal function of many organs is upset if vitamin A is lacking. The tissues become dry and the cells slough off. Glands cease to produce their normal secretions. As a result of this effect on epithelium, the secretion of tears may be lessened and the outer layers of the eye become dry and opaque. This condition, known as *xerophthalmia* (dryeye), is a cause of blindness. Vitamin A plays another important role in vision. As pointed out in Chapter 11, vitamin A is essential for the proper formation of rhodopsin in the rods of the retina of the eye. If the vitamin is partially lacking, recovery from glare is slow and the ability to see in a dim light is impaired.

The proper formation of tooth enamel depends on the action of vitamin A.

Fig. 14-3   Two rats from the same litter, both 11 weeks old. One (top) was fed no vitamin A and is suffering from xerophthalmia and rough fur. It weighs only 56 grams. The other (bottom) was fed sufficient vitamin A and weighs 123 grams. (U.S. Department of Agriculture)

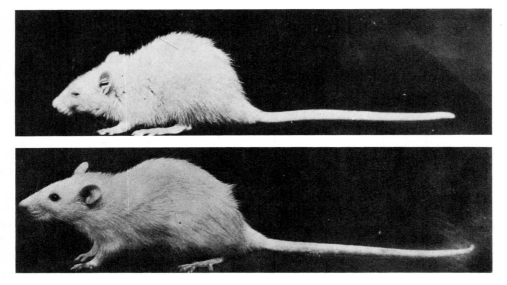

The cells which form the enamel are a highly specialized type of epithelium. If there is a deficiency of vitamin A in a child's diet, or in the diet of the mother before the birth of the child, the minute hexagonal pillars of enamel are not properly formed. Pits may result in the surface of the enamel which will lead to early tooth decay.

Sources: The common foods which are rich in vitamin A contain a large amount of carotene. These are all yellow vegetables and the outer leaves of cabbage and lettuce. Many leafy vegetables are also rich in carotene, although the yellow color may be hidden by the green chlorophyll. Spinach, kale, collard, and mustard greens all contain carotene. Animal foods, such as whole milk, butter, eggs, liver, kidney, and some fish, contain the vitamin itself.

Vitamin D ($C_{28}H_{44}O$) or ($C_{27}H_{44}O$). There are two principal chemical forms of this vitamin, both of equal value to humans. They are white, odorless crystals that are stable to heat, oxidation, alkalies, and acids.

Vitamin D promotes the proper growth and development of bones and teeth and other structures when calcium is present. It promotes the absorption of calcium and phosphorus from food, as well as improving the reabsorption of phosphorus in the kidneys. If this vitamin is lacking, rickets may develop even with a diet that contains adequate amounts of necessary minerals.

Rickets is characterized by failure of bony materials to form normally at the epiphyseal cartilages. An excess of improperly ossified cartilages accumulates, which results in bowed legs, enlargement of the joints, abnormally

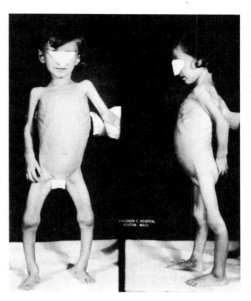

Fig. 14-4 An advanced case of rickets. At the left, note the enlargement of the joints, the shape of the femur (sickle thigh), and the bowing of the lower leg bones. Some enlargement of the ribs can be seen in this view, but this feature is more evident in the photograph at the right. Advanced cases of this type are becoming rare because of the general improvements in the diets of children. (F. R. Harding, Children's Hospital, Boston, Mass.)

flattened bones, irregularly formed ribs, and skull deformities (Fig. 14-4). Teeth are also affected and develop with thin, poorly calcified enamel. Fortunately, advanced cases of rickets are becoming rare in this country due to advances in the knowledge of proper diets.

Human skin normally contains a sterol compound known as *cholesterol.* There is now definite evidence that cholesterol can be changed by a series of complex chemical reactions into another compound, 7-dehydrocholesterol.

This new compound is then converted into vitamin D_3 ($C_{27}H_{44}O$), by exposure to certain wave lengths of ultraviolet radiations. Not all such radiations are capable of producing this vitamin, as indicated in Fig. 14-5, some merely cause sunburning. Another substance, *ergosterol,* which was first found in the fungus from which ergot is obtained, can also be converted into a vitamin D by ultraviolet irradiation. This is vitamin D_2 ($C_{28}H_{44}O$), which is produced in commercial quantities from yeast.

Rickets is primarily a disease of the temperate regions of the earth. In these areas there is much cloudiness, fog, and dust in the air. Also, the wearing of heavy clothes prevents the formation of the vitamin in the skin by shielding it from the ultraviolet radiations in sunlight. Window glass also filters out those ultraviolet radiations which are most effective in producing the vitamins. Many of these conditions do not exist in tropical regions, with the result that rickets seldom occurs in these areas. It should be pointed out, however, that continued overexposure to sunlight can have a very serious effect on the skin and, in some individuals, may cause skin cancers. Overdosing with vitamin D will result in nausea, loss of appetite, headaches, and a lack of the retention of calcium and phosphorus.

The recommended daily intake of vitamin D is between 300 and 400 International Units for infants and adolescents. Pregnant women should receive more because of the demands made on their calcium and phosphorus supplies by the developing child.

Sources: Very few of the common foods contain significant amounts of vitamin D. Dairy products and eggs vary in their amounts depending on the exposure of the cows and hens to sunlight. It is possible, however, to increase the vitamin D content of these foods by feeding the vitamin to the animals. Milk may also have its vitamin D content increased by exposure to ultraviolet radiations from carbon or mercury-vapor arc lamps. Since vitamin D, like A, is stored in the liver, oils pressed from the livers of cod, halibut, and sharks is a commercially important source. Saltwater fish, such as herring, mackerel, salmon, and sar-

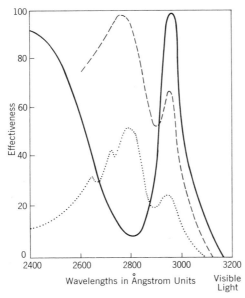

Fig. 14-5   The effects of various wave lengths of solar radiations outside the visible spectrum. The solid line represents wave lengths most able to cause sunburn; the dashed line, those having the greatest power to prevent rickets; and the dotted line, the wave lengths at which ergosterol is formed in the greatest quantities.

dines contain relatively large amounts of vitamin D.

Vitamin E. One group of fat-soluble vitamins has been designated by the letter E. Chemically, these belong to the group of organic compounds known as *tocopherols,* a name derived from the Greek: *tocos* (childbirth) + *phero* (to carry) because they are necessary for normal reproduction. There are four different tocopherols which have approximately the same effects, but one, alpha tocopherol ($C_{29}H_{50}O_2$), is the most potent.

Very little is known about the relation of this vitamin to human health. It is known that the tocopherols prevent fats from oxidizing and becoming rancid. This antioxidation effect makes vitamin E essential for the proper use and storage of vitamin A. Many claims have been made for the supposed effects of vitamin E on different abnormal conditions in humans, but they have not been backed-up by later research.

Sources: The tocopherols are widely distributed in common foods, and this may account for the fact that so little is known about their deficiency. Green leafy vegetables, whole cereals, and egg yolk are excellent sources.

Vitamin K. There are two naturally occurring forms of this vitamin: K_1 ($C_{31}H_{46}O_2$) and K_2 ($C_{41}H_{56}O_2$). The former is present in food while the latter is normally formed in the intestinal tract by bacteria. Both can be absorbed properly only if bile is present in normal quantities.

In addition to these natural forms, there are several synthetic water soluble forms of vitamin K which are more effective in their action than the normally occurring materials. The most efficient one is vitamin K_3, or *menadione* ($C_{11}H_8O_2$). This is the standard against which all other vitamin K products are checked. Menadione, when injected, is more effective than the oil-soluble forms in natural foods because the latter must be absorbed through the walls of the intestine, and lose some of their efficiency in this process.

In cases of jaundice, bile does not reach the small intestine and normal absorption of vitamin K is inhibited. Until recently, surgeons hesitated to operate on people suffering from jaundice because their blood did not clot rapidly enough. This failure of the blood to react normally is due to a decrease in the amount of prothrombin, a blood protein which requires vitamin K for its formation. An inherited condition, known as hemophilia, in which the blood also fails to clot properly is caused by the lack of another blood protein, thromboplastin. Vitamin K has no effect on the production of this substance.

Newborn infants occasionally die from uncontrollable hemorrhages. Such babies do not receive enough vitamin K during prenatal development because their mothers' diet is inadequate. Injections of one of the water-soluble forms of vitamin K protect the baby from excessive bleeding during the first few days of life. He soon acquires the necessary intestinal bacteria to form the vitamin for himself.

Sources: Vitamin K is stable in the presence of oxygen and heat. It is widely distributed in plant foods, especially the green parts. Cabbage, cauliflower, spinach, and soy beans are especially rich in it, with tomatoes, the cereal grains, peas, and mushrooms being less well provided. Pork liver contains a considerable quantity, although there is no indication that the vitamin is stored in any quantity in animal liver.

The Water-soluble Vitamins

The vitamin B complex. In the early days of vitamin research, a substance was discovered which prevented beriberi in humans. Later work in the field has shown that what was originally thought to be a single substance is actually a group of compounds to which the name vitamin B complex has been given. Each of the substances composing this group has its own specific action in preventing some deficiency disease. Not all of these materials can be related directly to specific human conditions, but studies of the diet of animals whose food requirements are similar to man's indicate that they play some role in human nutrition. It is known that some of them can be synthesized by bacteria in the intestine and then absorbed. This would greatly lessen the appearance of deficiency symptoms due to their absence from the diet.

Thiamine hydrochloride (vitamin B₁)($C_{12}H_{17}ClN_4OS \cdot HCl$). This vitamin is also known as the *antineuritic vitamin* because its inclusion in the diet will help prevent the appearance of certain types of nervous disorders. If birds are kept on a thiamine-deficient diet, partial paralysis of the muscles may result. This condition is known as polyneuritis. In man, there is retardation of growth, loss of appetite, degeneration of the nervous system, muscular difficulties, enlargement of the heart, and eventual death.

A person's ability to absorb and use this vitamin properly is important. Increased intake of thiamine may be required to offset the effects of diarrhea. Thiamine is often used in cases of severe gastro-intestinal disorders or extensive surgery of the digestive tract. Occasionally beriberi has appeared in people who suffer from tuberculosis, dysentery, or cancer of the liver. Chronic alcoholism may lead to beriberi because the diet of a heavy drinker is often lacking in thiamine.

The staple diet associated with human beriberi is one based on so-called polished grain. In the process of polishing grain, the outer husk is removed and with it goes the thiamine. Thus the ordinary commercial rice is white because the husk has been removed in the process of refining it. The same process is used in making white flour. If polished grains constitute the principal article of the diet, a deficiency in thiamine may result. This deficiency is particularly notable in the Far East, where polished rice is eaten in large quantities. This diet tends to exclude other foods that might replace the thiamine.

Thiamine forms a phosphate salt, a coenzyme known as cocarboxylase, within the cells. This coenzyme is essential for complete oxidation of carbohydrates to carbon dioxide and water, the normal end products. In its absence various intermediate substances accumulate and upset the metabolism of the cells. One of these is pyruvic acid. Since the brain derives its energy almost exclusively from the oxidation of carbohydrates, any disturbance in the oxidation of these compounds makes itself evident in abnormal function of the nerve cells of the brain. Fortunately, symptoms of an upset metabolism become evident early and the administration of thiamine prevents actual damage to the cells.

Very little thiamine is stored in the body. Some small amounts are normally present in the liver, kidneys, and muscles, but these are quickly used up. Bacteria in the intestine may be able to synthesize thiamine, but not in sufficient quantities to satisfy the body needs.

Sources: Thiamine is found in many foods. Its richest sources are: the husks of grains, wheat germ, milk, potatoes, cabbage, meat (especially pork), eggs, and brewer's yeast.

Vitamin B_1 is easily destroyed by heat. In dry heat there may be a loss of from 10 to 25 percent, whereas cooking in water may dissolve or destroy as much as 50 percent. To meet the body's daily needs, one should drink three glasses of milk per day or have one serving of meat or one egg. Since thiamine can be added to flour, six slices of vitamin-enriched white bread may supply the needed quota.

Riboflavin (vitamin B_2 or G) $(C_{17}H_{20}N_4O_6)$. This is a yellow compound that is stable in air and dry heat but easily destroyed in light.

Riboflavin aids in the release of energy from cells, and also regulates some of the vital activities. It is able to form members of a group of substances known as flavo-protein enzymes. These belong to a large group of enzymes, the oxidases, which are the active agents in releasing energy within the cells. Riboflavin may also play a role in

Fig. 14-6   Top: a rat, 28 weeks old, fed on a diet free of riboflavin. It weighs 63 grams. Bottom: the same rat six weeks later following a diet rich in riboflavin. It weighs 169 grams. (U.S. Department of Agriculture)

vision. It is present in the pigment of the retina of the eye, where it aids in light adaptation.

A deficiency of this vitamin results in retarded growth in some experimental animals, such as the rats in Fig. 14-6. In man, a lack of the vitamin causes sores in the region of the lips and corners of the mouth, a scaly skin, increased sensitivity of the eyes to bright light, and the growth of fine blood vessels over the surface of the eyes. The last condition may eventually lead to blindness.

Sources: Riboflavin is quite widely distributed in foods. It is present in milk, eggs, wheat germ, green leafy vegetables, peas, lima beans, and many kinds of meat. The richest sources are pork or beef liver, and brewer's yeast.

Niacin ($C_6H_5O_2N$). Other names for this vitamin are *nicotinic acid* and the *pellagra preventative vitamin,* in reference to the disease pellagra.

The name *pellagra* comes from words meaning dry skin, because this condition is one of the most evident signs of niacin deficiency (Fig. 14-7). The chief features of pellagra are red, dry skin which is irritated by sunlight and heat, and a sore mouth. The nervous system may be affected, resulting in walking difficulties, jerky movements, and finally paralysis.

This disease, like beriberi, is very frequently associated with the overbalance of one staple grain in the diet. In this case it is corn, which is notoriously poor in niacin but is the principal grain in the diet in certain parts of this country. Its low cost, in those regions where corn, cotton, and tobacco are the main crops, accounts for the appearance of the disease among low-

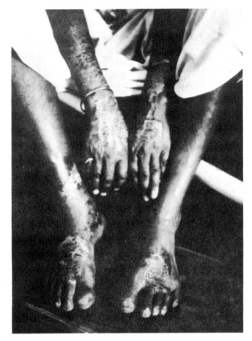

Fig. 14-7 The skin rash of pellagra. (C. G. King, Columbia University)

income farmers in southern parts of the United States.

Sources. Most meats are well supplied with niacin. Liver, whole wheat, peanuts, and brewer's yeast are especially rich in it. Fresh peas and cabbage are good sources, and many green leafy vegetables contain an adequate amount to prevent the appearance of pellagra.

Pyridoxine (vitamine B_6) ($C_8H_{12}O_3NCl$). In common with some of the other vitamins, pyridoxine plays a role in cellular metabolism. It was noted in a study of rats that had been kept on a pyridoxine deficient diet that there was a pronounced failure of some of the sulfur containing amino acids

THE HUMAN BODY
IN ANATOMICAL TRANSPARENCIES

By GLADYS McHUGH, Medical Artist
Associated with the University of Chicago Clinics

The "Trans-Vision" process presents the human body in a unique manner in which you can perform a "dissection" and proceed through the depths of its structures by turning transparent pages. You can see organs overlying other organs in a three-dimensional effect. As you turn the pages, a layer of anatomy is removed and a deeper layer comes into full view. The right pages give you a front view of the structures. To see the same structures from the back side, you turn the page. Thus, you can see the relation of organs to the body as a whole and to each other. You can single out an individual part for more detailed observation.

The pages preceding and following the anatomical transparencies give you a description of each view—how it was made and what it shows. The numbers you find on many of the structures refer to an identification key on the back of Plate 6. This will identify any structure you wish quickly and easily. Numbers have been omitted where they would detract from structural detail. Many such structures are referred to in the description of the various plates.

The structures shown are detailed and accurate. This presentation of the human body will serve as an adequate basis for anatomical study in any degree of thoroughness and complexity you may desire.

Key identifying numbered structures on back of Plate 6.

1

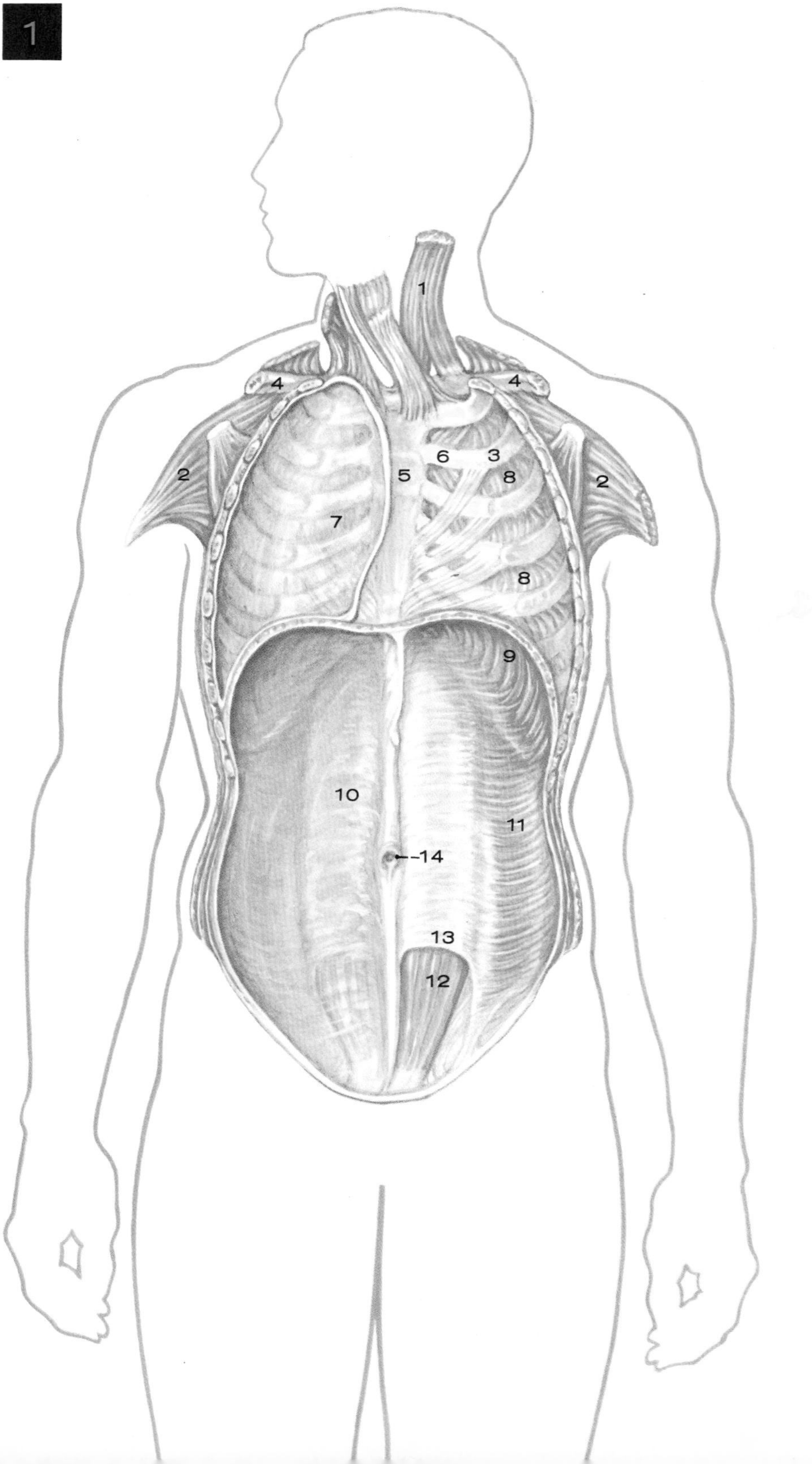

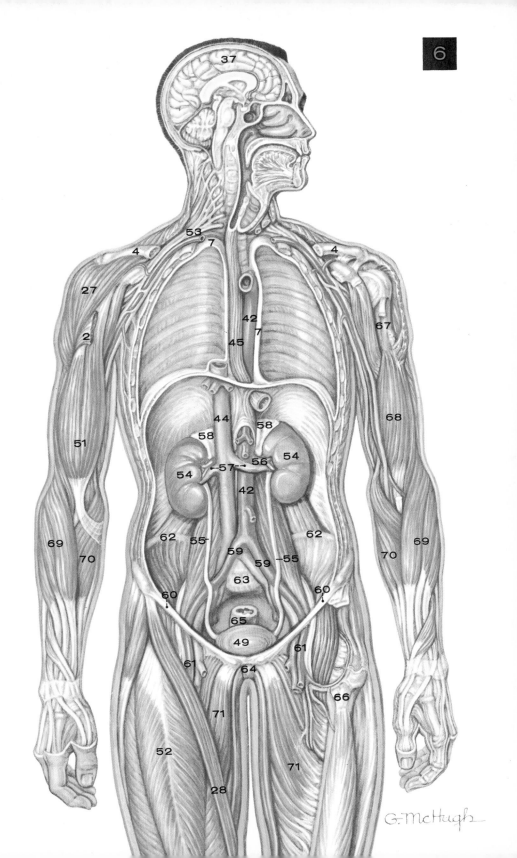

Key to the Structures of the Human Body

1. Sterno-mastoid muscle, used in turning the head
2. Pectoral muscle, used in moving the arm across the chest
3. Rib
4. Clavicle
5. Sternum
6. Rib cartilage
7. Pleural membrane
8. Intercostal muscles, used in breathing
9. Diaphragm
10. Peritoneum
11. Transverse abdominal muscle, supporting the abdominal wall
12. Rectus abdominal muscle, used in flexing the trunk
13. Rectus sheath
14. Umbilicus
15. Temporal muscle, used in chewing
16. Masseter muscle, used in chewing
17. Parotid salivary gland
18. Thyroid cartilage
19. Thyroid gland
20. Lung
21. Pericardium
22. Liver
23. Gall bladder
24. Stomach
25. Colon
26. Small intestine
27. Deltoid muscle, used in raising the arm
28. Sartorius muscle, used in crossing the leg
29. Cranial cavity
30. Spinal canal
31. Nasal cavity
32. Tongue
33. Pharynx
34. Larynx
35. Superior vena cava
36. Root of the mesentery
37. Cerebrum
38. Cerebellum
39. Nasal septum
40. Trachea
41. Heart
42. Aorta
43. Pulmonary artery
44. Inferior vena cava
45. Esophagus
46. Duodenum
47. Spleen
48. Pancreas
49. Urinary bladder
50. Cerebral septum
51. Biceps muscle, used in flexing the arm
52. Flexor muscles of front of thigh
53. Brachial plexus
54. Kidney
55. Ureter
56. Renal artery
57. Renal vein
58. Adrenal gland
59. Iliac artery
60. Inguinal ligament
61. Femoral artery
62. Crest of hip bones
63. Lumbo-sacral joint
64. Pubis
65. Rectum
66. Femur
67. Humerus
68. Brachial muscle, which works with biceps as a flexor of the arm
69. Extensor muscles of hand
70. Flexor muscles of hand
71. Adductor muscles which bring legs together

such as tryptophan to function properly. If this condition were permitted to continue, the apoenzyme that controlled the reaction was destroyed and no recovery was possible following administration of the vitamin to the rats. In human beings, especially infants who are fed on a highly artificial formula, the most apparent symptoms of a lack of pyridoxine are cerebral seizures resembling epilepsy. These may be the result of the lack of the vitamin in the brain tissues and the reduction of the chemical reactions that are normally dependent on a compound of pyridoxine, pyridoxal phosphate.

During experiments with normal, healthy human volunteers who were kept on a diet deficient in vitamin B_6, pronounced changes in the electroencephalogram (EEG) patterns of the volunteers were noted. The onset of the abnormal condition was hastened by also reducing the protein intake of the subjects. These experiments helped to support the idea that there is some direct connection between the presence of the vitamin and the proper use of amino acids in tissues, especially those of the central nervous system.

There has been sufficient data gathered in Thailand and Burma to link the formation of "stones" in the kidney with lack of vitamin B_6. The exact relationship of pyridoxine to this condition is not currently known.

The amount of dietary pyridoxine required by an individual is unknown; some pyridoxine is supplied to the body by the activity of the intestinal bacteria. The National Research Council has recommended 1–2 milligrams per day. Quite recent work apparently indicates that this is not entirely suffi-cient, especially for the group of older people whose bodily chemistry has changed over the years. It has been suggested in this connection that an adequate supply could be assured the general population by adding pyridoxine to bread thereby "enriching" it with another important food material, the significance of which is just becoming apparent.

Sources: This vitamin is abundant in common foods such as milk, fresh vegetables, whole wheat, and many meats. It is not easily destroyed by dry heat, but readily dissolves when food is cooked in water.

Biotin (vitamin H) $(C_{10}H_{16}O_3N_2S)$. The exact relation of this vitamin to human nutrition is not well understood. Recently however it has been shown to play an important role in fatty acid synthesis. If rats are kept on a biotin-free diet, they develop severe itching of the skin. Then the hair around their eyes falls out (spectacle eye), and eventually baldness results. In man, skin changes, loss of weight, and muscular pain have been ascribed to a lack of biotin, but there is little conclusive evidence that this is the case.

A biotin-free diet contains raw egg white as the principal source of protein. Raw egg white contains a substance, *avidin,* which combines with biotin and makes it unusable by the body. Cooking evidently changes the composition of egg white so that it no longer will combine with biotin, and therefore the vitamin is available to regulate bodily processes.

Although biotin may not function like some of the other vitamins in regulating a definite process, it plays an important role in the body's chemistry.

It appears to be an essential part of a coenzyme which is required to form citrullin, a substance with is necessary for the production of urea by the liver.

Sources: Biotin is present in greatest quantities in liver, yeast, and poultry. It is found in lesser amounts in egg yolk, tomatoes, and carrots. Biotin can be synthesized by some bacteria in the large intestine.

Pantothenic acid ($C_9H_{17}O_5N$). Like biotin, pantothenic acid has an important role in cellular metabolism. It is a constituent of the compound called coenzyme A which is essential in the formation and breakdown of fatty acids and the liberation of energy from fats and carbohydrates. Also, it is required for the formation of hemoglobin and various sterols, such as cholesterol and some of the reproduction hormones. No deficiency has been discovered in man that can be related to a lack of pantothenic acid from the diet due, perhaps, to its wide distribution in foods.

Sources: Three rich sources of pantothenic acid are brewer's yeast, liver, and kidney. Other good sources are eggs, whole milk, sweet potatoes, tomatoes, and roasted peanuts.

Folic acid ($C_{19}H_{19}N_7O_6$). This vitamin is involved in many bio-synthetic reactions and appears to be closely associated with proper blood formation. At one time it was believed to be a specific preventative of pernicious anemia, but investigation shows that folic acid alone is not able to repair the damage this disease does to the nervous system. The chemical structure of folic acid indicates that it is closely related to vitamin B_{12}.

Sources: Liver, kidney, mushrooms, and yeast are especially rich in folic acid. The word *folic* is derived from the Latin *folium,* a leaf, because green leaves and grass were first recognized as an excellent source of this vitamin.

Vitamin B₁₂ (cyanocobalamin) ($C_{63}H_{90}N_{14}O_{13}PCo$). In its pure form, this vitamin is a red crystalline compound that is not affected by heat in a neutral solution, but is rapidly destroyed by heating in acids or alkalis. It contains the element cobalt (Co), an infrequent constituent of protoplasm.

The discovery of the action of vitamin B_{12} required careful detective work. It was known that folic acid helped in the treatment of pernicious anemia, but did not actually cure the disease. Further investigation showed that an extract of fresh liver could be used to treat pernicious anemia. Also, it was found that victims of this disease improved after treatment with an extract of muscle tissue plus normal gastric juice. However, gastric juice from a person ill with pernicious anemia was not effective in this treatment. These facts indicated that two necessary factors must be present. Careful chemical analysis of muscle and liver tissue showed that both tissues contained vitamin B_{12}, the so-called *extrinsic factor.* Normal gastric juice contains the *intrinsic factor,* whose function is to assure the proper absorption of the vitamin. Vitamin B_{12} is normally formed in sufficient quantities by bacteria in the large intestine. As coenzyme B_{12} it takes part in synthetic reactions in conjunction with folic acid.

Sources: Good sources of the vitamin are liver and kidney; moderate amounts can be obtained from milk, lean meats, and fish. It is in very low concentrations in plant products.

Fig. 14-8 Two New Hampshire cockerels, $3\frac{1}{2}$ weeks old. The one at the left was fed a diet deficient in vitamin B_{12} and weighs 157 grams. The one at the right was fed sufficient vitamin B_{12} and weighs 280 grams. (U.S. Department of Agriculture)

Vitamin C ($C_6H_8O_6$). Vitamin C is now commonly known by either the term *antiscorbutic vitamin* or *ascorbic acid*. The former name is used because this vitamin acts as a specific agent in the prevention and cure of scurvy (scorbutus). For many years this disease was prevalent among sailors and explorers who were deprived of fresh fruit and vegetables for long periods. Scurvy is still common in regions where the diet is limited by custom or occupation, as in isolated fishing villages. There the main article of winter diet is salt or dried fish, because there is no adequate supply of fresh foods. In such places scurvy is known as "spring sickness" because the symptoms appear in late winter or early spring before home-grown vegetables are available. Strangely enough, farmers may show some signs of scurvy. When vegetables are prepared for home canning by boiling them in open pots, vitamin C is destroyed. If such vegetables are used to the exclusion of fresh ones, the symptoms of scurvy appear. Foods that have been canned commercially by pressure cooking keep much of their vitamin C content.

The absence of vitamin C from the diet will result in fragility of the capillaries and subsequent loss of blood. This condition is due to a weakening of the cement material that holds the endothelial cells together. Tooth structure may also suffer from improper formation and maintenance of the dentine. Also the cementum that holds the tooth in its bony sockets weakens and the tooth becomes loose. These effects are accompanied by loss in weight and constant and annoying pains in the joints. Extreme deficiency of vitamin C in the diet may sometimes result in death.

Although the antiscorbutic action of vitamin C is its most widely known effect, it has other important properties. The amount of ascorbic acid required by pregnant and lactating women is considerably higher than at other times. This increased requirement is due to the demands made on the mother by the unborn child for enough vitamin C to build skeletal structures and maintain a normal condition of the blood vessels. Also, an increase of vitamin C in the diet appears to be necessary following severe injury or a surgical operation. It has been found in these cases that the amount of ascorbic acid in the serum of the blood drops, and it becomes concentrated in the scar tissue that forms at the site of the wound. An inadequate amount of vitamin C slows the healing process. This relationship is due to the action of the vitamin in speeding up the formation of collagen fibers. These elongated fibers of protein form the basis of connective tissue, and its formation will be slowed or stopped in the absence of vitamin C. A process that is accelerated by the presence of an adequate amount of ascorbic acid is the absorption of iron through the intestinal walls. It has been found that certain types of nutritional anemias can be improved by the addition of extra vitamin C to the diet.

This vitamin evidently plays a number of roles in cellular metabolism that are not clearly understood at present. Some of these appear to be associated with the utilization of the amino acid tyrosine, but the exact method of this action has not been determined. Unlike some animals (mice, rats, chickens, and pigeons), man has no method of synthesizing vitamin C; he is dependent on his diet for his supply. Nor does he have the capacity to store sufficient quantities to allow him to remain in a healthy condition during a long period when he might be deprived of it.

Sources: Vitamin C is found in large quantities in fresh tomatoes, turnips, green leafy vegetables, and the majority of fruits. One of the best sources is the juice of oranges, lemons, and grapefruit. Fresh meats contain vitamin C in moderate amounts.

VOCABULARY REVIEW

Match the phrase in the left column with the correct word in the right column. *Do not write in this book.*

1 Any substance that affects the rate of a chemical reaction.
2 Bacterial growth is retarded by refrigeration to this temperature.
3 If an acid is formed during a chemical reaction there is an increase in this.
4 The material affected by an enzyme.
5 A general name for an organic catalyst.
6 The digestion of starch in the mouth is due to the action of this enzyme.

a catalyst
b end product
c enzyme
d hydrogen ion concentration
e maximum temperature
f minimum temperature
g optimum temperature
h ptyalin
i substrate
j zymase

TEST YOUR KNOWLEDGE

Group A

Select the correct statement. *Do not write in this book.*

1 An enzyme is a complex (a) protein compound (b) carbohydrate (c) fatty acid (d) inorganic compound.
2 An enzyme may be called a(n) (a) oxidizing agent (b) reducing agent (c) catalyst (d) disaccaride.
3 An enzyme that digests milk is (a) ptyalin (b) gastric juice (c) rennin (d) sucrase.
4 The optimum temperature for most human enzymes is about (a) 0°C (b) 50°C (c) 3.5°C (d) 98.6°C.
5 Saliva contains an enzyme that hydrolyses (a) sugars (b) starches (c) proteins (d) fats.
6 Of the following, the one that is *not* a member of the vitamin B complex is (a) thiamine (b) riboflavin (c) folic acid (d) ascorbic acid.
7 One oil-soluble vitamin is (a) vitamin A (b) vitamin B (c) vitamin C (d) vitamin G.
8 The vitamin associated with the clotting of blood is (a) vitamin A (b) vitamin C (c) vitamin K (d) vitamin B_1.
9 The vitamin associated with rickets is (a) vitamin A (b) vitamin D (c) ascorbic acid (d) thiamine.
10 The best source of vitamin C is (a) meat (b) potatoes (c) oranges (d) butter.

Group B

1 (a) What is an enzyme? (b) What factors in its environment affect the activities of enzymes?
2 In what ways are enzymes alike, and in what ways are they different?
3 (a) How may vitamins be classified? (b) Give two examples of vitamins from each of the groups you have mentioned. (c) What characteristic, or characteristics, place each of these in the group you mentioned in (a)?

Chapter 15 **The Mouth**

Structures of the Mouth

The mouth (Fig. 15-1) is the first part of the digestive system with which food comes in contact. It is bounded above by the *hard* and *soft palates,* on the sides by the *cheeks,* and in front by the *lips*. The *tongue* serves as part of the floor of the cavity. The lining of the mouth is a highly modified type of stratified squamous epithelium. It is

Fig. 15-1 The mouth cavity showing the opening into the pharynx.

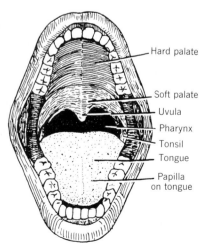

Hard palate

Soft palate

Uvula

Pharynx

Tonsil

Tongue

Papilla on tongue

similar in structure to the outer skin of the body but lacks the keratin that makes the skin a strongly protective organ. This lining contains nerve endings that respond to pain, pressure, heat, and cold just as the skin does. The mouth, therefore, has a degree of sensitivity that is lacking elsewhere in the digestive system.

The mouth opens into the *pharynx* (Fig. 15-2), a region shared by both the digestive and respiratory systems. The pharynx can be divided into three general parts: the *nasopharynx,* above the opening of the mouth and communicating with the cavity of the nose; the *oral* pharynx, behind the mouth; and the *laryngeal* pharynx, leading downward to the opening of the windpipe. The opening of the nasopharynx is protected by a small flap of tissue, the *uvula,* which hangs from the soft palate into the opening at the back of the mouth (Fig. 15-1). When a person swallows food, this flap is drawn upward and effectively closes the entrance to the nasal region. The *tonsils* (Fig. 15-1), masses of lymph tissue, lie on either side of the opening, and occasionally other groups of similar cells, the *adenoids,* grow in the nasopharynx. If very large, these may cause difficulty in nasal breathing. The tonsils have pitted surfaces and may easily become infected by the growth of bacteria in the depressions.

The tongue. Not only does the tongue play a primary role in speech, but it also aids in swallowing and contains nerve endings that give us the sense of taste. It is a highly muscular organ; the muscle bundles are arranged in a very complex manner so that they lie in several different planes. The result is that the tongue can be moved in many different directions.

The epithelial surface of the tongue contains nerve endings for receiving sensations of pressure, heat, and cold, and for the detection of those chemical substances that give flavor to food. If you examine your tongue, you will see it is covered by small projections, the *papillae,* which give parts of it a velvet-like appearance. Some of the papillae are very small and inconspicuous, but toward the back of the tongue they become large and are raised above the surface. The sense organs of taste, or *taste buds,* are located in the small depressions in the epithelium of the tongue, and are more heavily concentrated in the areas surrounding the papillae. A substance must be in solution before its flavor can be recognized. The solution passes through the opening of the taste bud and stimulates the taste cells, giving rise to impulses

Fig. 15-2   *A longitudinal section through the head showing the mouth cavity and related structures. (Model designed by Dr. Justus F. Mueller. Photograph, Ward's Natural Science Establishment)*

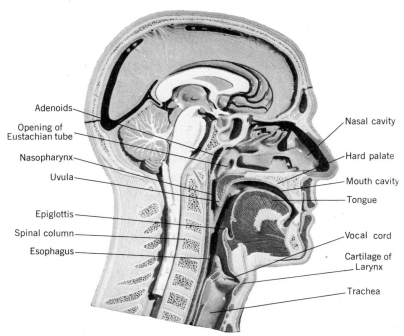

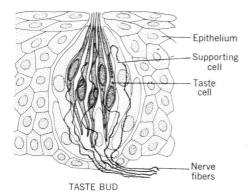

Epithelium

Supporting cell

Taste cell

Nerve fibers

TASTE BUD

Fig. 15-3   A taste bud with its nerve connections. There are approximately 9000 taste buds. (From Marsland, Principles of Modern Biology, *Holt, 1957)*

which are carried to the brain through nerve fibers at the base of the taste bud. See Fig. 15-3.

Not all surfaces of the tongue are sensitive to all flavors to the same degree. Taste buds are sensitive to four basic flavors: sweet, sour, bitter, and salt. Taste buds especially sensitive to sweet flavors are located near the tip of the tongue; those for sour flavors lie along the sides. The receptors for bitter flavors are on the surface of the tongue toward the back. Salty flavors are detected by taste buds at the front and sides of the tongue. If these various groups of taste buds are stimulated individually by a very weak electric current, the stimulus will produce in each group the same sensations as a chemical substance of the flavor for which the taste buds are specialized.

Many complex flavors are the result of smelling foods rather than tasting them. Odors are carried to the olfactory nerve endings located in the upper part

of the nasal cavity, and we are apt to confuse the odor of the substance with its flavor simply because the material is present in the mouth at the time. You can seldom get the real flavor of a food when you have a heavy head cold. At that time the increased mucus secretions cover the olfactory nerve endings and you cannot detect the odors of the food.

The teeth. Dividing the food into smaller bits by chewing is called *mastication*. This primary function of the teeth is an important one because it increases the surface area of the morsel of food. As a result, the digestive juices can function more efficiently and rapidly than they could if the food were swallowed without being chewed. In man, as in all mammals, there are two sets of teeth. The first set begins to appear within a few months after birth and is later replaced by the set that is designed to remain throughout life. These first teeth are called the *deciduous teeth* (Latin, *decidere,* to fall down) and they are replaced by the *permanent teeth.* Since the deciduous teeth appear early in life, their formation begins some months before birth. The same is true of some of the permanent set; the buds from which they develop are formed in the tissues of the jaws before birth. The approximate ages at which the teeth appear are shown in Fig. 15-4.

When a tooth makes its appearance through the gum it is said to *erupt.* In the case of the deciduous teeth, this process may cause the young child some pain. It is also usually accompanied by an increase in the amount of saliva formed because of the irritation to the gums and the resulting stimulation of the nerve endings. The eruption of the permanent set, with the possible exception of the last molars, is usually not accompanied by similar discomfort because there is not the same amount of irritation present. These second teeth do not simply push the earlier teeth out of their sockets, as many people believe. Instead, as they grow within the jaw, a group of cells, the *odontoclasts,* form in front of the tip of the tooth and dissolve the base of the first tooth. In this respect, the odontoclasts behave much as the osteoclasts do (Chapter 5). Finally, the first tooth is held in place by only the tissues of the gum and final separation from these may occur spontaneously, or the tooth may need a little coaxing.

There are four different types of teeth in the adult set, but only three in

Fig. 15-4 The types of teeth, their arrangement, and the ages at which they appear.

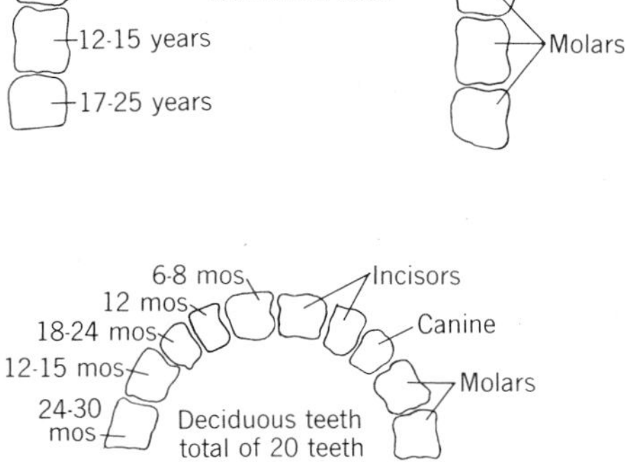

the deciduous set (Fig. 15-4). In front, in each jaw, there are four *incisor teeth* with edges adapted for biting. On each side of these is a *canine tooth* with a slightly pointed tip to aid in tearing food, and in back of these are four *premolar teeth*—two on each side. (These are sometimes spoken of as the *bicuspid* teeth because of the presence of two points, or *cusps.)* Since the cusps of the upper and lower teeth mesh when biting, their action is one of cutting or shearing food. There are three *molar teeth* on each side of the jaw behind the premolars. These are characterized by relatively flat surfaces that permit the food to be ground between them. The last of the molars to appear are the *wisdom teeth.* Normally, the total number of permanent teeth is 32, as compared with the 20 in the deciduous set. Occasionally these numbers vary because of the failure of some of the teeth to erupt.

During the formation of a tooth (Fig. 15-5) two separate groups of tissues add to its structure. The *enamel* is formed from *ameloblasts,* specialized types of epithelial cells. At the same time, the *dentine* is formed by *odontoblasts* which are derived from mesenchyme. Both of these regions are formed in a manner that is similar to the formation of bone (Chapter 5). The enamel is laid down in the form of microscopic hexagonal pillars that are held together by a cementing substance. This binding material is attacked first by mouth acids in the initial stage of tooth decay. Acids dissolve the cement and loosen the pillars, making a pathway for the invasion of the softer dentine by bacteria. The enamel covers only the crown of the tooth and

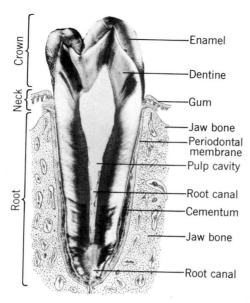

Crown

Neck

Root

Enamel

Dentine

Gum

Jaw bone

Periodontal membrane

Pulp cavity

Root canal

Cementum

Jaw bone

Root canal

Fig.15-5 Photomicrograph of a section of a human tooth under polarized light. The surrounding tissues have been sketched in.

extends a short distance below the gum line.

The dentine is the somewhat softer part of the tooth that lies below the enamel. It contains a large number of very fine canals, the *dentinal tubules,* which are similar to the small bony tubes in which the osteocytes lie (Fig. 15-5). Within each of these tubules is a cellular process from an odontoblast, the body of which lies against the dentin wall of the *pulp cavity.*

Below the surface of the gum, the tooth is surrounded by the *periodontal membrane,* which supports it in a sling-like manner. The fibers of this membrane are firmly attached by the *cementum,* at one end to the tooth socket in

the jaw bone, and at the other end to the root. The slinglike arrangement prevents the tooth from being pushed inward by the considerable force exerted on the tooth in chewing. However, in those cases where the teeth do not oppose each other at the correct angle (malocclusion), there is a lever action of one tooth against another which pushes the teeth out of their normal position. The abnormal pressure may eventually result in disease of the periodontal membrane. After a tooth has been removed from the jaw, bone fills in the cavity that is created.

The pulp cavity contains the blood vessels and nerves essential for the nourishment of the teeth. If bacteria invade these soft tissues, infection can spread rapidly downward through the *root canal* to the bone. An abscess then forms at the base of the root.

Digestion

Mastication. Four pairs of muscles are involved in the process of chewing. Because of the manner and place of their origins and insertions, they give to the lower jaw a slightly rotary motion as well as the more obvious up-and-down movement. The lips, cheeks, and tongue also have their roles in the process of mastication. When the teeth come together, the food is squeezed out from between them into the mouth cavity. The lips and cheeks are then drawn inward as the jaw is lowered for the next biting motion and the tongue spreads outward toward the sides. These motions tend to push the food back between the teeth again so that it can be rechewed. The potential force of the muscles responsible for the motion of the jaws is much greater than that actually required to chew the food. By the use of an electronic device, it has been found recently that the incisors can exert a force of between 11 and 25 newtons while the molars can develop forces of between 33 and 90 newtons, depending on the individual (1 newton = 1 kilogram × 1 meter per second per second). It is not the force of the teeth on the food that is important, but rather the grinding motion mentioned above that serves to break the food down.

An important result of proper mastication of food is the breakdown of the cellulose fibers of plant materials. This allows the digestive juices to act on parts of the food that would otherwise be inaccessible because of their enclosure by cellulose walls. Persons who have lost their teeth or are equipped with poorly fitting dentures sometimes suffer from nutritional disorders because they cannot chew their food sufficiently. On the other hand, excessive chewing of food, as some food faddists recommend, has no special merit.

Salivary glands. The process of mastication is helped by the saliva, which is formed in and secreted by the salivary glands. This fluid plays several roles. It softens and lubricates the food mass, which can then be more easily chewed and swallowed. It also dissolves some of the food so that it can be tasted. Saliva partly digests the starch in foods. In addition, saliva washes the teeth, helps neutralize mouth acids, and keeps the inside of the mouth flexible and moist. This last function of the saliva is especially important for speech.

Saliva is a mixture of different chemical compounds, the composition chang-

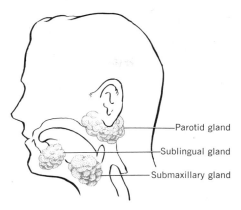

Fig. 15-6   A diagram of the location of the three pairs of salivary glands.

ing as the activity of the gland increases or decreases. It has an average pH value of 6.7, which means that it is slightly acid in its reaction. Normally, the saliva contains about 99 percent water. In this are dissolved minute traces of several different salts of sodium, potassium, and calcium, as well as organic materials like the enzyme *ptyalin* (salivary amylase). Mucin also forms a part of the saliva.

The three pairs of salivary glands (located on either side of the mouth) empty into the mouth by separate ducts. In Fig. 15-6, the *parotid gland* may be seen lying just in front of and slightly below the level of the opening of the ear. These are the largest of the glands and the ones that usually become enlarged during an attack of mumps. Below it and near the angle of the lower jaw is the *submaxillary* (submandibular) *gland,* and under the side of the tongue is the *sublingual gland.*

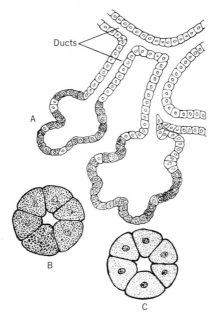

Ducts

A

B

C

Fig. 15-7 At A is a diagram of a compound gland. The darker cells are those that secrete. At B is a group of cells following a period of inactivity. At C is shown a group of cells after a period of active secretion. Note the difference in the number of granules in B as compared with C.

The saliva-forming cells are at the ends of small branches leading off from the main duct. They are a compound type of gland composed of serous and mucous cells (Fig. 15-7A). When the gland has been inactive for a time, the cells become filled with granules (Fig. 15-7B) which disappear after a period of activity (Fig. 15-7C). This indicates that the materials in saliva, other than water, are formed as solid substances within the cells and then converted into a fluid that can pass through the cell membrane, and thence by ducts into the mouth.

All of us are aware of the fact that if we become excessively thirsty, our mouths become dry. This is the result of water being withdrawn from the glands in order to replenish some of the fluid that has been lost from the blood by sweating, or as a result of strenuous physical exertion.

Thirst is a physiological signal to replenish the body's water supply and the intensity of the drive is directly related to water deficit.

The sensation of thirst is a quite complex reaction. It may be a response to some stress affecting the body, such as an infection, prolonged heat or cold, or nervous tension. Under these conditions, the pituitary gland (Chap. 30) secretes a hormone which helps the body withstand stress. This hormone evidently stimulates a center in the hypothalamus of the brain that has been called the drinking center. When this area is stimulated in an experimental animal, the animal shows all of the symptoms of thirst. Also, animals which have been placed under stress during the course of an experiment become thirsty much more rapidly than they do when they are in their home cages and relieved of the perplexing situation.

Quite recently a mechanism has been found which evidently accounts for thirst in human beings. It appears that there are highly sensitive nerve cell endings in the walls of the carotid artery, the artery that carries blood to the brain. These nerve endings respond very quickly to changes in the osmotic pressure of the blood. If it becomes too great, impulses are sent to the hypothalamus, and indirectly to the pituitary gland. The pituitary then secretes a hormone which reduces the amount of urine formed by the kidneys. This has a tendency to lower water loss. At the same time the thirst center is stimulated and the person drinks water which will be absorbed by the blood and lower its osmotic pressure.

The secretion of saliva is entirely controlled by the nervous system. As we have pointed out in an earlier chapter, this secretion may be the result of a conditioned reflex brought about by the sight, smell, or even the thought of food. There is also a simple reflex secretion of saliva when food is placed in the mouth, especially if the food is in small particles, as is the case when it is thoroughly chewed. These small particles stimulate nerve endings in the mouth, and in solution also stimulate the taste buds. Through fibers of the autonomic nervous system, there is a dilation of the blood vessels passing to the glands, and saliva is produced. The daily flow of saliva in a normal individual amounts to between 950 and 1420 milliliters. Much of this saliva is produced during periods when there is no food in the mouth.

Digestive action of ptyalin. Food remains in the mouth for such a relatively short time that little digestion can occur there. Nevertheless, the action of the enzyme ptyalin should not be underrated. Ptyalin can decompose starch into simpler products, the dextrins, and break off some of the disaccharide maltose. Although little of this activity occurs in the mouth, digestion of food by ptyalin may continue in the stomach. The ball of food (bolus) does not break up immediately upon reaching the stomach, so digestion by the action of ptyalin may continue for as long as half an hour. Eventually, the acid in the gastric juice will stop this digestion, but by then, as much as 75 percent of the starch of potatoes and bread may have been broken down. The digestive process in the small intestine completes the breakdown of carbohydrates by the action of numerous enzymes. This will be discussed in detail in Chapter 17.

Deglutition

The act of swallowing, called *deglutition,* is a complex one involving many of the muscles that form the walls of the mouth and pharynx. It is usually started as a voluntary process, but shortly becomes involuntary as the food comes under control of the smooth muscles of the esophagus. When we swallow, the tip of the tongue arches slightly and starts to push the food toward the back of the mouth. Then a wave of muscular contraction passes over the tongue which forces the food against the hard palate. At the same time, the uvula and soft palate close the opening to the nasopharynx, preventing food from going into that region. Once the bolus reaches the pharynx, involuntary control takes over the process. As the food moves backward and downward, the soft palate is drawn slightly down, just as the *larynx* (voice box) is drawn upward toward the *epiglottis* (the covering of the opening of the windpipe). As added insurance against the entry of food into the respiratory system, the nervous impulses responsible for breathing stop so that it is almost impossible to breathe and swallow at the same time. Action of the involuntary muscles sends the bolus down the *esophagus* to the stomach.

The series of involuntary muscular contractions that move food along the digestive tract is known as *peristalsis* (Fig. 15-8). These contractions appear

Fig. 15-8   These two diagrams show the passage of a bolus of food through the esophagus. A series of nerve impulses (black dotted lines) causes the muscles (white diagonal lines) to contract in succession so as to force the food toward the stomach. (Encyclopaedia Britannica Films)

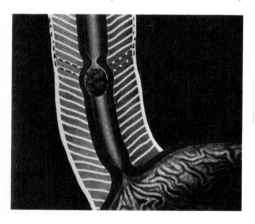

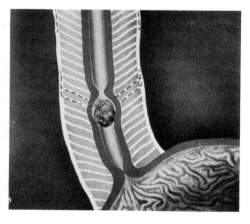

behind the bolus and push it forward. At the same time, the muscles of the walls of the esophagus in the region of the food and preceding it relax, thereby relieving pressure on the bolus so that it can go forward. Solid food of an average consistency requires between five and six seconds to pass along the ten inches of the esophagus. When the peristaltic wave reaches the stomach, it causes the muscle fibers that guard the cardiac opening of the stomach to relax. This opening allows for the passage of food from the esophagus to the stomach.

When fluids are drunk, they pass quickly through the slightly dilated esophagus and outrun the peristaltic wave. They then collect at the cardiac sphincter and are able to enter only after the wave of muscular contraction has reached that point. If, however, there is a rapid succession of swallowing movements, as in drinking a glass of water, the opening may remain relaxed. The fluid then passes directly from the esophagus into the stomach.

It is possible to hear food enter the stomach. If a stethoscope is placed just under the lower margin of the left ribs and a piece of cracker swallowed with the help of a small amount of water, the entry of the food can be heard as a distinct sound. This occurs about five or six seconds after the food has been swallowed indicating that the descent has been quite gradual.

The esophagus is covered by a tough protective tissue. Beneath this are two layers of muscle fibers, the outer of which runs longitudinally (lengthwise) and the other in a circular direction. It is through the alternating contraction and relaxation of these two layers that the peristaltic wave is set up. Internally, the esophagus is lined by a mucous membrane that is held loosely to the circular muscles by connective tissue. The mucous secretion of the lining serves as a lubricant.

VOCABULARY REVIEW

Match the phrase in the left column with the correct word in the right column. *Do not write in this book.*

1 The hardest substance in the body.
2 Contains the nerves and blood vessels of the tooth.
3 The part of the tooth containing the processes of the odontoblasts.
4 The material which attaches the periodontal membrane to the tooth.
5 A region common to both the digestive and respiratory systems.
6 A structure in the neck that moves upward when you swallow.
7 A process resulting from the contraction of smooth muscle fibers.

a cementum
b peristalsis
c dentine
d enamel
e incisor
f larynx
g mastication
h pharynx
i ptyalin
j pulp cavity

TEST YOUR KNOWLEDGE

Group A

Select the correct statement. *Do not write in this book.*

1 The parotid glands produce an enzyme that digests (a) protein (b) sugar (c) starch (d) fats.
2 Incisor teeth are best fitted for (a) tearing (b) biting (c) grinding (d) chewing.
3 The mouth opens into the (a) uvula (b) soft palate (c) pharynx (d) larynx.
4 A basic flavor to which the taste buds are *not* sensitive is (a) sweet (b) sour (c) salt (d) pepper.
5 A full set of permanent teeth consists of (a) 20 teeth (b) 24 teeth (c) 28 teeth (d) 32 teeth.
6 Tooth enamel is formed from (a) odontoclasts (b) odontoblasts (c) ameloblasts (d) orthodontists.

7 Of the following, the one that is *not* a salivary gland is (a) parotid (b) sub-lingual (c) submaxillary (d) subcutaneous.

8 The covering for the opening of the windpipe is called the (a) glottis (b) epi-glottis (c) trachea (d) larynx.

9 The series of muscular contractions that moves food along the digestive tract is called (a) perimysium (b) pericardium (c) periosteum (d) peristalsis.

10 In the structure of a tooth, the enamel covers (a) the whole tooth (b) only the crown (c) only the root (d) only the dentin.

Group B

1 Discuss the statement, "Digestion is both a chemical and mechanical process," from the standpoint of the activities occurring in the mouth.

2 How is the secretion of saliva controlled?

3 Briefly describe the process of deglutition.

4 Make a carefully drawn and fully labeled sketch of a human tooth.

5 Why will proper care of the teeth of a young child help to insure better teeth in the adult? What factors are involved in this care?

Chapter 16 **The Stomach**

Structure of the Stomach

The human stomach lies to the left of and just below the diaphragm, as shown in the Trans-Vision insert which follows page 312. Actually the shape of the stomach is highly variable, depending on several factors. As Fig. 16-1 shows, the stomach is J-shaped in the living individual, with the upper portion considerably greater in diameter than the lower end. If a thin person is standing, the stomach becomes quite elongated, but if he lies down, the organ becomes shorter although it retains its general J-shape. The amount of food present in the stomach will alter its shape, as will the effect of the pressure of the intestines that lie below it. It is therefore impossible to say that the stomach has a definite shape in the same way that other organs of the digestive system retain their form.

The esophagus opens into the stomache at the *cardiac orifice.* The outlet of the stomach is at the *pyloric orifice,* or *pylorus,* (Fig. 16-1) which connects with the *duodenum* of the small intestine. Sphincter muscles (muscles which encircle tubes and openings and control their opening and closing) control the size of both orifices. The stomach is divided into three general parts: the *fundus,* which is the upper portion; the *pyloric portion* at the lower curve of the J; and the *body,* which lies between these two regions.

Fig. 16-1 X-ray photograph of a normal stomach. The fundus is at the top right, below it is the body, and to the lower left of this is the pyloric portion. The pyloric orifice is marked by the white arrow. (Jerry Harrity)

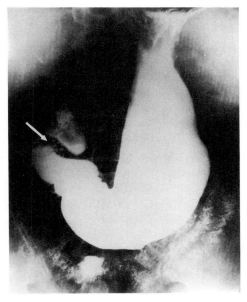

The outside layer of the stomach is a thick serous coat which is continuous with the *peritoneum,* the membrane that lines the entire abdominal cavity. At various points, the peritoneum grows forward and surrounds the organs of the abdomen, holding them, as it were, in a sling. From the larger, or left, curve of the stomach, this membrane grows downward to form the *greater omentum* which hangs like an apron over the front of the intestines. It contains large deposits of fat. Three layers of smooth muscle form the walls of the stomach (Fig. 16-2). As in other organs of the digestive tract, there are layers of both circular and longitudinal muscles, but in addition, the fundus and part of the body contain an inner layer of oblique muscle tissue. The inner lining of the stomach is a mucous membrane, the *gastric mucosa,* which lies in folds called the *rugae* (see Fig. 16-3). When the stomach is distended with food, these folds stretch so that the interior of the stomach is quite smooth. The capacity of the stomach is 0.95 to 1.4 liters.

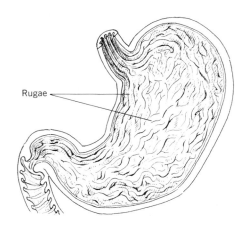

Fig. 16-3   A section of the stomach to show the mucous lining and the rugae.

The Gastric Glands

The gastric mucosa contains the long tubular *gastric glands* that secrete the *gastric juices* used in digestion. It has been estimated that some 35,-000,000 of these glands line the normal stomach. As Fig. 16-4 shows, they contain three principal types of cells: *parietal cells* which secrete hydrochloric acid, *chief cells of the body of*

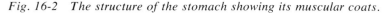

Fig. 16-2   The structure of the stomach showing its muscular coats.

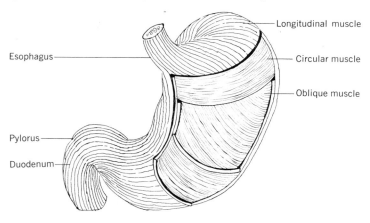

the gland which secrete pepsinogen, and *chief cells of the neck* which secrete mucin.

Hydrochloric acid (HCl). Gastric juice is the most acid fluid of the body. This characteristic is due to the presence of hydrochloric acid, which in a concentrated form will destroy tissue. In the concentration in which it is present in the gastric juice (about 0.6 percent) it does not affect the normal, living stomach in this way. Following death, the mucous lining of the stomach is rapidly destroyed, so there is evidently some protective agent present in the living organ that prevents destruction of the tissue. It has been suggested that protection is afforded by the salts of sodium and potassium, especially when they are combined with chlorides and bicarbonates. The range of acidity of gastric juice is from a pH of 1.7 to 0.3.

The HCl functions in several different ways during gastric digestion. In the first place, its presence is necessary for the formation of pepsin from the proenzyme, pepsinogen. Secondly, pepsin is unable to act in the digestion of proteins unless the gastric contents remain acid. A third activity of the HCl is to destroy bacteria that might have entered with the food. This action is so efficient that the material that leaves the stomach is virtually sterile under normal conditions. A fourth property of hydrochloric acid is that it dissolves some of the salts in the food we eat.

Pepsin. The active enzyme of the stomach, pepsin, digests proteins only. Purified crystals of pepsin are shown in Fig. 16-5. This enzyme is able to break down the large molecules of proteins into simpler molecules which, however, are still too large to enter the blood stream. These intermediate products, the *proteoses* and *peptones,* are groups of amino acids that have been separated from the original protein molecule. This preliminary action of pepsin converts the insoluble forms of protein into soluble compounds which pass into the small intestine for final digestive processes.

For many years, a second protein-splitting enzyme was thought to be present in the human gastric juice. This enzyme was called *rennin* and was considered to act only on the protein of milk, *casein.* Now it is known that this enzyme does not exist in adult humans, although it may be present in the secretions of infants and other young mammals. The action of rennin is truly amazing. One part of the enzyme is

Fig. 16-4 A gastric gland showing the three principal types of cells.

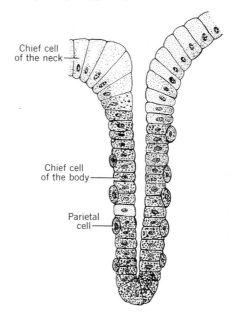

Chief cell of the neck

Chief cell of the body

Parietal cell

Fig. 16-5 Crystals of pepsin, the active protein-digesting enzyme of the gastric juice. (John H. Nothrop)

able to convert 4,550,000 parts of milk into a curd in approximately 10 minutes at a temperature of 40°C and a pH value of 6.2. Its action hastens the digestion of milk because it changes a liquid form of the protein into a solid form and thereby permits the digestive enzymes to function more effectively on it. Pepsin has a similar action on the milk we drink. It is able to convert 800,000 times its own weight of milk protein into a curd under the same temperature and time conditions. The fact that the pH value of the adult stomach is considerably below that necessary for the proper function of rennin is a strong argument against the presence of this enzyme in our digestive system. Commercially, rennin is used extensively in the manufacture of cheese and milk desserts.

Gastric lipase. Another enzyme in the stomach about which there has been much debate among physiologists is *gastric lipase.* This has the ability to break fats down into fatty acids and glycerol. There appears to be a small amount of this enzyme formed in the stomach, but its action is very weak because it is destroyed when the acidity of the stomach becomes higher than 0.2 percent. The extent to which it affects fats is unknown, but any action it might have would occur early in the digestive process and in connection with the breakdown of fats such as those found in milk and ice cream.

Phases of Gastric Secretion

The quantity of gastric juice produced during an average meal is about 0.5–0.7 liters. The secretion of gastric juice is partly under nervous control and partly under the control of a chemical regulator belonging to the group of secretions known as *hormones.* When solid food enters the stomach, cells in the pyloric region are stimulated to produce a hormone, *gastrin.* This is absorbed by the blood and carried to the fundus of the stomach where it stimulates the glands to produce gastric juice.

The sight, odor, taste, or even the thought of food will start the activity of the gastric glands. This is the *psychological phase* of gastric secretion. In this respect the glands are showing the same type of conditioned reflex that the salivary glands exhibit under similar stimuli. The gastric juice produced in this phase is rich in pepsin, which has been called "appetite juice." When food is introduced experimentally directly into the stomach, without the influence of sensory stimuli, secretion is only three-fourths normal. The question then arises as to what type of mechanism continues the secretion of gastric juice in sufficient quantities to complete the digestion of the meal. Nervous stimulation caused by the presence of food in the stomach is not the answer, because when all gastric nerves are cut, secretion is hardly affected.

Experimental animals were used to work out the problem of how digestion continues after the psychic phase has stopped. The cardiac end of the stomach was tied off while these animals were under anesthesia, and salt solution was introduced into the cavity of the stomach through a tube placed in the pyloric portion. After an hour the solution was removed and tested for the presence of hydrochloric acid and pepsin. None was found. However, if small bits of the lining of the pyloric portion were ground up and an extract of this injected into the stomach's blood supply, pepsin and hydrochloric acid both appeared in the salt solution. This indicated that some chemical substance present in the extract of the pyloric lining was responsible for the stimulation of the gastric glands and the formation of more gastric juice. Later work has confirmed this idea. It is now believed that the presence of food in the stomach stimulates the formation of the hormone of *gastrin,* which is discharged into the blood stream. Carried by the blood, gastrin arouses the gland cells to further activity. This phase of secretion is called the *gastric phase.* Gastrin belongs to that group of hormones known as *secretagogues* because of their ability to stimulate secre-

tion in a gland. We will have occasion to refer to some more of these in the next chapter.

Stomach Movements

From X-ray studies of the stomach during digestion it is seen that the fundus plays the role of a reservoir for the reception of food from the esophagus. Fig. 16-6 shows a series of waves passing over the pyloric end of the stomach; note how the body and the fundus show little activity. A succession of slower waves gradually passes the food from the fundus and body of the stomach toward the pylorus, where the food is churned into a thick liquid state called *chyme.* As the stomach empties at the pyloric orifice, it again resumes its J-shape.

Fig. 16-6 X-ray photograph of the stomach showing a wave of muscular contraction passing over it. (Jerry Harrity)

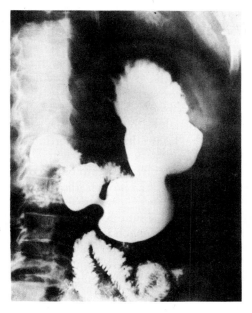

If a small rubber balloon is swallowed and then inflated in the stomach, a record can be made of the intensity and frequency of the movements of the stomach under various conditions. Such records are made with the aid of a kymograph similar to that used to record a muscle twitch. An apparatus of this type shows that when the stomach is empty, hunger contractions occur at irregular intervals. These contractions may be of sufficient intensity to arouse hunger pangs. It was thought at one time that when a person was suffering from these hunger pangs, the sight, odor, or thought of food would increase the contractions of the stomach. However, recent work on gastric motility has shown that the opposite is true. The stomach slows down in anticipation of the food it is to receive.

The passage of food from the stomach to the small intestine is controlled by the peristaltic rhythms of the stomach muscles and by the consistency of the stomach contents. The presence of food in the first part of the small intestine (duodenum) causes a slowing down of the gastric peristaltic movements, and the pyloric valve remains closed. Experiments have demonstrated that the pyloric sphincter closes when solid particles touch it. These are moved back into the pyloric portion for further liquefaction. Only when the food mass is converted into chyme is it allowed to pass through the pyloric sphincter.

The length of time food normally remains in the stomach is determined by the nature of the material. Foods that are rich in proteins remain longest in the stomach and produce the high acidity necessary for the action of pepsin. Thus, meats may stay from three to four hours. Foods with high carbohydrate content, such as many fruits and vegetables, leave the stomach within 1.5 and 2 hours. Cereal foods have an intermediate position. The difference in emptying time for these various foods is probably related to the length of time required to liquefy them. Carbohydrates and fats are more easily reduced to a semifluid state than are proteins; thus they can be released into the duodenum sooner.

Very little material is absorbed into the blood stream through the walls of the stomach because most foods have not been decomposed into sufficiently simple materials to permit this. Some water, glucose, a few salts, and alcohol are the principal materials passing through the stomach membranes into the blood stream.

Vomiting

Under certain conditions, the normal path of the peristaltic waves of the stomach is reversed and vomiting occurs. This process is controlled by a nerve center lying in the medulla oblongata of the brain. The stimuli capable of affecting this response may arise from several different areas, the back of the mouth and the pharynx being especially sensitive to stimuli that give rise to vomiting. This explains why the process can be started by tickling the back of the throat. The lining of the stomach and especially the duodenum are so highly sensitive that vomiting may be initiated by emetics such as strong salt solutions and mustard in warm water. Rhythmic stimulation of the semicircular canals of the inner ear as the result of the motion of a ship, an airplane, or a swing, may cause a person to vomit.

The movements involved in vomiting are primarily those of the abdominal muscles. The feeling of nausea that precedes an attack is accompanied by an increase in the amount of saliva. This is swallowed and, with air that is also gulped down, causes a distension of the lower end of the esophagus. The stomach is pressed against the diaphragm and the force required to eject the material from the stomach is then supplied by the contraction of the abdominal muscles.

Psychosomatic Effects

Since the movements of the stomach are closely controlled by the nervous system, it is not surprising that a person's emotional outlook may be reflected in the way his stomach behaves. If a person is of a calm disposition and does not permit small annoyances to bother him unduly, his stomach will probably respond in a normal manner. On the other hand, if he tends to be emotionally upset, his stomach will reflect his attitude.

A very graphic demonstration of this control of the nervous system over the activities of the stomach was made by two New York doctors, Wolf and Wolff. A patient had swallowed some boiling hot chowder when he was nine years old. The scar tissue that formed over the burned area closed off the esophagus, and it was necessary to make an opening through the abdominal wall directly into the stomach so that he could feed himself. Such a permanent opening is called a *fistula*. Through this opening the doctors were able to observe the action of the stomach on various types of food as well as its behavior in states of emotional stress. When the patient reacted to a situation with fear, the gastric mucosa became pale and the normal movements of the stomach were reduced or stopped entirely. If, however, he became angry and showed resentment toward a situation, the mucosa responded by becoming filled with blood and turning a bright red color. The motions of the stomach were also greatly increased in rate and force. There was an increase in acid production. These findings bore out observations made in 1833 by an Army surgeon, Dr. William Beaumont, on Alexis St. Martin, a man who had been wounded in the abdomen and had developed a fistula.

Continued attacks of indigestion may reflect both emotional and physical problems. Some serious diseases produce symptoms which may be confused with those arising from simple indigestion. Stomach or duodenal ulcers, cancer of the stomach, and some types of liver diseases may be the underlying causes for these symptoms.

Some individuals who are under emotional strain of one type or another may react by developing ulcers either in the lining of the stomach itself or in the lining of the duodenum, the first part of the small intestine. The former are gastric ulcers, and the latter are duodenal ulcers (see Fig. 16-7). In both cases they are the result of the over-secretion of gastric juice. Duodenal ulcers are usually the result of the nervous overstimulation of the gastric glands. Since the empty duodenum has little protection against the activity of pure gastric juice, chronic progressive ulcers may develop. Gastric ulcers, on

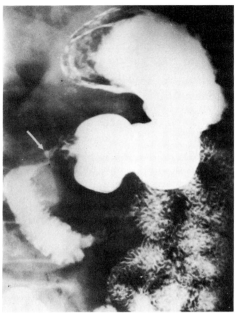

the other hand, appear to be the result of the oversecretion of gastrin resulting from the continued stimulation of the glands that produce it. A cause for this may be the retention of food in the pyloric part of the stomach for an abnormally long time due to faulty emptying. The presence of duodenal ulcers which may affect the pyloric sphincter will also cause the food to remain too long in this region.

Unfortunately, many deaths each year are caused by stomach cancer. Many of these are undoubtedly unnecessary. Often people do not heed the vague preliminary symptoms. The cause of stomach cancer is unknown, but if it is detected early enough, the chances for complete recovery are increasing each year.

Fig. 16-7   The arrow in this X-ray shows the location of a duodenal ulcer. Note that the fundus of the stomach is almost empty. Some of the barium compound which is fed to a person before X-rays are taken to make the organs opaque, has left the stomach and is seen in the intestine beyond. (Dr. John A. Campbell)

VOCABULARY REVIEW

Match the phrase in the left column with the correct word in the right column. *Do not write in this book.*

1 Is at the lower end of the esophagus.	**a** cardiac orifice
2 Its opening is controlled by the consistency of chyme.	**b** duodenum
3 The lining of the abdominal cavity.	**c** gastrin
4 A hormone produced in the pyloric region of the stomach.	**e** hydrochloric acid
5 Flexible folds that line the stomach.	**f** parietal cell
6 Secretion of the parietal cells.	**g** pepsin
7 The rounded upper part of the stomach.	**h** peritoneum
8 An enzyme that has been crystallized.	**i** pyloric sphincter
	j pylorus
	k rugae

TEST YOUR KNOWLEDGE

Group A

Select the correct statement. *Do not write in this book.*

1 The esophagus opens into the stomach at the (a) cardiac orifice (b) pyloric orifice (c) duodenum (d) pharynx.
2 The enzyme pepsin digests (a) starch (b) sugar (c) protein (d) fat.
3 Adult human gastric juice contains (a) ptyalin (b) rennin (c) hydrochloric acid (d) casein.
4 Cells in the pyloric region of the stomach secrete the hormone (a) pepsin (b) amylase (c) gastrin (d) lipase.
5 Of the following, the one that is *not* a general region of the stomach is the (a) fundus (b) body (c) pyloric (d) gastric.
6 The inner lining of the stomach is called (a) gastric mucosa (b) rugae (c) greater omentum (d) parietal cells.
7 Of the following, the one that is *not* a muscle coat of the stomach is (a) longitudinal muscle (b) circular muscle (c) oblique muscle (d) transverse muscle.
8 Gastric lipase breaks down large molecules of fats into (a) glucose (b) peptones (c) glycerol (d) proteoses.
9 The pH of gastric juice is about (a) 4 (b) 1 (c) 7 (d) 10.
10 The lining of the stomach contains (a) parotid glands (b) parietal glands (c) peptic glands (d) gastric glands.

Group B

1 (a) Briefly describe the structure of the stomach, taking into consideration the nature of its walls and the functions of its various regions. (b) How does the stomach move, and what are the results of this activity?
2 Briefly describe the structures responsible for the production of gastric juice and how its secretion is controlled.
3 Outline what happens to a piece of beef (protein and fat), a slice of bread (carbohydrates, proteins, fat, and mineral salts), and a glass of milk during gastric digestion.
4 Briefly describe the entry of food into the stomach and the factors that control its passage into the small intestine.

Chapter 17 **The Intestines**

Once the food material (chyme) is released from the stomach, it is subjected to a variety of changes during its passage through the small intestine. The relation of the various organs responsible for these changes can be more clearly understood with the help of the Trans-Vision following page 312, and the longitudinal section as shown in Fig. 17-1. Reference should be frequently made to these drawings to clarify the concepts discussed.

The great length of the small intestine, some 7 meters, is evidence of its importance in the digestive process. From the small intestine digested material is absorbed into the blood stream for distribution through the body.

Food material passes from the stomach through the pylorus into the *duodenum,* which is the first part of the small intestine. The name of this region is derived from the Latin word, *duodeni,* meaning twelve each, because it was originally identified as being as long as the breadth of twelve fingers (about 250 millimeters). The duodenum curves upward, forming a horseshoe shape around the head of the pancreas. Its diameter, slightly greater than that of the remainder of the small intestine, is about one and three-fourths inches. Structurally, it is quite inflexible compared to the rest of the small intestine and lacks their supporting membranes.

The second section of the small intestine is the *jejunum,* which is approximately 2.7 meters long. The third region, the *ileum,* is about 4 meters long. Total length of the small intestine may vary from the average of 7 meters to between 4.7 and 9.5 meters. The difference has no apparent relation to the age, height, or weight of the individual.

These parts are supported by thin membranes which are folds of the *peritoneum,* the lining of the abdomen. These *mesenteries* not only support the folds of the intestine, but they also hold the numerous blood vessels and nerves that pass to and from it. Unlike the duodenum, the jejunum and ileum both have considerable freedom of movement and shift their positions as a result of changes in posture and of the peristaltic activity of the intestine.

For a full understanding of the digestive process that occurs in the small intestine, it is necessary to consider two important glands that pour digestive juices into the duodenum. These are the liver and the pancreas.

THE LIVER

The liver is the largest gland in the body and weighs 1.1–1.6 kilograms. It lies in the upper right quarter of the abdominal cavity and occupies a major part of that region, with the left lobe projecting downward over part of the stomach. There are four lobes in the liver; two of these can be seen from the front and the other two, much smaller lobes, are visible on the back surface. Under the large right lobe is the pear-shaped *gall bladder,* the lower margin

Fig. 17-1   A longitudinal section through the abdominal cavity showing the relative positions of the mesenteries and the peritoneum. Some of the organs, such as the pancreas, are at first covered by the peritoneum, but this later disappears.

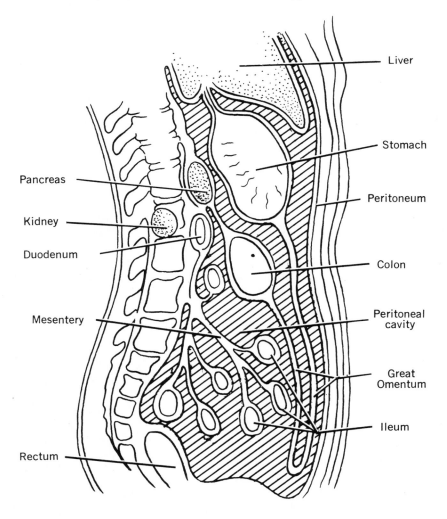

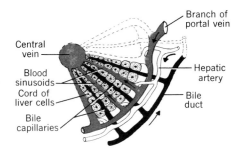

Fig. 17-2   A diagram of a section of a liver lobule. The central vein carries blood away from the blood sinusoids.

of which projects slightly below the edge of the lobe. In consistency, the liver is a soft solid of reddish-brown color that is easily broken and cut. The entire liver is surrounded by a serous coat and a thin fibrous coat, but the peritoneum covers only part of it.

Structure of the Liver Lobule

Internally, the liver is divided into *lobules,* each one to two millimeters in diameter (1 millimeter = 0.039 inch). These, in turn, are made up of *cords of liver cells* that radiate out from a *central vein* in an irregular manner, the columns frequently joining each other. Each cord consists of two adjacent rows of cells (Fig. 17-2). Within these cells the many chemical activities that make the liver an intricate laboratory concerned with digestion and metabolism take place.

Our chief interest here is the function of the liver in secreting bile. The other important liver functions are discussed later in this book.

Portal circulations. The blood supply of the liver is extensive. The *portal vein* (Fig. 17-3) brings blood to the liver from the gastrointestinal tract, the pancreas, and the spleen. A relatively small *hepatic artery* also carries blood rich in oxygen to it. The portal vein, once it enters the liver, breaks down into small branches which form a network around the lobules. These branches, joined by capillaries of the hepatic artery, give rise to the *sinusoids* (Fig. 17-2) which pass between the cords to enter the central veins. The central veins of several lobules join and eventually form the *hepatic veins,* which conduct the blood from the liver on its way to the heart.

Fig. 17-3   Diagram of the general distribution of the portal vein. It arises as capillaries in the digestive organs and flows to the liver, where it again breaks down into capillaries. Generally, veins arise from capillaries and flow to the heart without forming another set of capillaries.

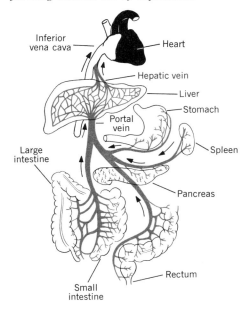

The sinusoids are unusual parts of the circulatory system in that their capillary-like walls are so extremely thin and irregular that the blood is often in direct contact with the cells. This permits an easy exchange of material between the cells and the blood.

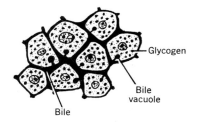

Bile Formation

The liver cells obtain from the blood the materials needed to form *bile,* which is collected in minute *bile vacuoles* within the cells (Fig. 17-4). These empty their secretions into the *bile capillaries* (Fig. 17-2), which pass between the two layers of cells that make up the cords. The bile capillaries run together to form increasingly large ducts and these eventually leave the liver as the *hepatic ducts.* The bile passes from the hepatic ducts through the *cystic duct* into the gall bladder where it is stored until needed for the digestive process. When food reaches the small intestine, the gall bladder contracts, expelling bile which flows through the cystic duct into the *common bile duct.* As Fig. 17-5 shows, this duct is joined by the *pancreatic duct* at a common opening in the duodenum about three inches beyond the pylorus.

When food is not present in the small intestine, the bile is kept from flowing into the duodenum by a sphincter valve at the opening of the duct into the intestine. Closure of this valve and the resulting pressure changes cause the bile in the ducts to back up into the gall bladder to await the digestive requirements of the duodenum.

There is a curious anatomical feature found in the cystic duct of the mammal group to which we belong, the Primates, and to no other. The mucous mem-

Fig. 17-4   Diagram of liver cells showing bile formation. Glycogen is stored as granules in the cells.

brane lining this tube is folded in a series of spiral ridges that make the duct quite rigid and prevent its collapse or distension as the pressure in the gall bladder changes when a person stands or lies down. The ridges, known as the *spiral valves of Heister* in honor of the man who discovered them, are probably a response to the upright method of locomotion that characterizes the Primates.

Composition of bile. Bile, though it contains no enzymes, plays a highly

Fig. 17-5   Diagram showing the relation of the liver and the pancreas to the small intestine.

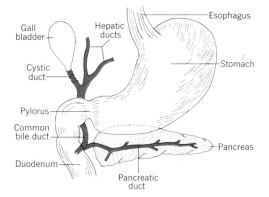

important role in the digestion of fats. The most important constituents are three substances, of which the first two are responsible for the characteristic color and action of the bile. These are *bile pigments, bile salts,* and *cholesterol.*

Human bile is a yellowish-orange fluid that owes its color to two different bile pigments: *bilirubin,* a red compound, and *biliverdin,* which is green. In man, bilirubin is the chief pigment of bile; biliverdin is only the product of oxidation of bilirubin. These pigments are formed as a result of the breakdown of the blood's red pigment, hemoglobin. This worn-out hemoglobin is broken down into bilirubin in various organs and tissues, such as the spleen, liver, connective tissue, and bone marrow. The iron content of hemoglobin is extracted for storage in the body, and the bilirubin is carried into the portal circulation and thence to the liver cells. In the liver, the cells that effect the conversion of hemoglobin are the specialized *Kupffer cells* which lie in the walls of the sinusoids. The bilirubin passes from these cells directly into the nearby liver cells which form the bile.

Within the small intestine, the bilirubin undergoes further chemical changes and is converted into *urobilinogen,* a yellowish substance that accounts for the yellow color of the feces (intestinal wastes). Some of the urobilinogen is absorbed through the walls of the large intestine and is returned to the liver for reuse.

In the formation of bile the liver has both a secretory and excretory activity. A *secretion* may be defined as a product of cells that is used by the body in its performance of a definite function. As we will see shortly, the bile aids in the digestion of fats, and it can thus be considered a secretion. An *excretion,* on the other hand, is a substance that the body eliminates as a waste material. In this capacity the liver plays an excretory role because it gets rid of worn-out hemoglobin and other products, all of which could cause damage to the body if they accumulated.

Digestive action of bile. The bile salts are *sodium glycocholate* and *sodium taurocholate.* These are the constituents of bile concerned with the digestive process. Just as we use soap and water to wash grease off our hands, the digestive system uses these salts as a detergent. Droplets of fat are surrounded with a layer of "soap" molecules so that they do not fuse together again. By separating fats and oils into very small droplets of this type, an emulsion is formed. The breaking of the large droplets of oil into tiny droplets greatly increases the surface area of the oil so that it can be subjected more completely to the action of the digestive enzymes. This action occurs in the small intestine as will be discussed later. The bile salts also form water soluble complexes with fatty acids and cholesterol since many do not readily dissolve in the intestinal fluids. A third function of these salts is to activate the enzyme *lipase* produced by the pancreas (and, to some extent, the lipase produced by the stomach). They also help neutralize the acidic gastric chyme. After their use in the digestive process, about 90 percent of the bile salts are reabsorbed by the portal circulation and returned to the liver for reuse.

Jaundice. We have not discussed cholesterol in connection with the digestive action of bile because its purpose is not known. Possibly it aids the bile salts in the emulsification process. However, most of our knowledge of biliary cholesterol is concerned with its role in gall bladder disease. In the bile ducts and gall bladder this material may collect in the form of solid masses, gall stones, which may block the ducts and seriously interfere with the normal secretion of bile. This condition results in jaundice. The whites of the eyes and the skin become slightly yellow due to the accumulation of bilirubin in the blood. The feces are pasty-colored as a result of the absence of urobilinogen. Jaundice due to gall stones is the most common type and is known as obstructive jaundice. Another type results from a viral (virus) infection of the liver cells which reduces the liver's capacity to secrete bile. The viral type, infectious hepatitis, is highly contagious and can be transmitted by contaminated food or water. Serum hepatitis, another viral type, may be transmitted by blood transfusions in which the organism is present in the transfused blood.

Other Liver Functions

Glycogen. The formation of glycogen from glucose is an important function of the liver. A sample of blood taken from the portal vein following a meal rich in carbohydrates is found to contain a great deal of sugar. On the other hand, a sample of blood taken at the same time from some other part of the circulatory system shows a much smaller rise in sugar content. Since the blood from the digestive system flows through the liver, it is evident that the liver can regulate the amount of sugar in the general circulation. If this condition did not exist, great waste of this energy-producing food would result. The kidneys would then remove the excess and give rise to a condition of *glycosuria,* or sugar in the urine. After a period of fasting, blood from the portal vein contains very little sugar, but the sugar in the general circulation remains at about its normal level. This is because the liver has converted some of its stored glycogen into glucose needed to keep the normal general level.

There is one normal condition with which the liver cannot cope. Following a meal in which excessive amounts of carbohydrates have been eaten, the blood and urine of a normal individual may show an increase in glucose. This simply means that such large quantities of sugars or starches have been ingested that the liver cannot take care of all of it.

As we shall see in a later chapter, the glands that produce hormones also play a role in maintaining the sugar content of the blood. *Adrenaline,* a hormone secreted by the adrenal glands, will increase the conversion of glycogen into glucose to meet the demands of an emergency. *Insulin,* on the other hand, prevents too rapid hydrolysis of glycogen and also promotes the storage of glycogen in the liver and its utilization in the muscles.

Deamination. This is a highly important function of the liver. Following a meal rich in proteins, a very high concentration of amino acids is found in the blood of the portal vein. In the liver, these acids are removed and the carbon, hydrogen, and oxygen are split off from them to form other compounds, especially glycogen. The amino groups (NH_2) are then split off, a process called deamination, and converted into *urea* which is the principal nitrogen-containing waste of the body. The urea enters the circulatory system and is carried to the kidneys, from which it is excreted in the urine.

Another liver function takes place in the Kupffer cells of the sinusoids. These cells take in the droplets of fat that appear in the blood in a form too large to permit their passage through the cell membranes. The Kupffer cells chemically change the fat into a form that the liver cells can use, but just how is unknown.

THE PANCREAS

The pancreas is an elongated, somewhat triangular gland, the head of which lies in a loop of the duodenum to the right of the pylorus (Fig. 17-5). The remainder of the gland extends to the left for about 14–15 centimeters and tapers gradually to a narrow tail region. The microscopic structure of the compound gland shows that it is similar to the salivary glands; it is a racemose type of gland, the subdivisions resembling a bunch of grapes.

The pancreas contains two different types of glandular tissue (Fig. 17-6). One type is the more common and is the source of the pancreatic juice which contains digestive enzymes. The cells of this tissue liberate their secretions into ducts which all join up to form the pancreatic duct. The other glandular tissue is composed of isolated groups of cells that lie scattered among the enzyme-secreting cells and liberate their secretions into the blood stream.

These form the *islets of Langerhans* and produce *insulin,* a hormone which plays an important role in carbohydrate metabolism. Within the islets are two different types of cells. The majority of cells are small and store insulin, or its forerunner, in the form of granules. These are called the *beta cells*. The other cells are larger and secrete *glucagon* which like insulin affects carbohydrate metabolism. These are called the *alpha cells.*

The secretion of pancreatic juice and bile is not a nerve reflex activity, but is under the control of secretagogues similar to that controlling the secretion of gastric juice. If all the efferent nerves to the gall bladder and the pancreas are cut, there is little reduction in the amount of digestive juices entering the intestine. As soon as the chyme passes from the stomach into the duodenum, the cells of the latter produce two hormones: *cholecystokinin* and *secretin.* Both secretagogues enter the blood stream, but their actions are highly specific. Cholecystokinin stimulates only the gall bladder, causing the muscles of its walls to contract. This expels bile into the ducts and thence into the duodenum. Secretin stimulates only the pancreas in its production of juices.

The pancreatic juice flows through the *pancreatic duct* to join the bile in the common bile duct as it enters the duodenum. The juice is alkaline, with an average pH value of 7.5 which may increase to 8 as the rate of secretion increases. Because it is so alkaline, the pancreatic juice stops activity of the acidic gastric juice, and digestion of the chyme in the intestine proceeds under entirely different chemical conditions than those found in the stomach.

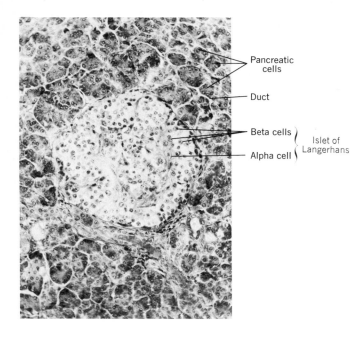

Pancreatic cells

Duct

Beta cells ⎫
Alpha cell ⎭ Islet of Langerhans

Fig. 17-6 Micrograph of section through the pancreas. Note in this photograph the islet of Langerhans with its two types of cells, lying among the enzyme-secreting cells. (General Biological Supply House)

The Pancreatic Enzymes

The main digestive enzymes of the pancreatic juice are *trypsin, steapsin,* and *amylopsin.* There are two other enzymes, *chymotrypsin* and *carboxypeptidase,* that play minor roles in digestion. Trypsin (pancreatic *protease)* is an enzyme that digests proteins, steapsin (pancreatic *lipase)* acts on lipids and amylopsin (pancreatic *amylase)* affects carbohydrates. The italicized names in parentheses are general terms designating the function of each enzyme. A protease is any enzyme that will digest a protein; a lipase will digest fats and oils, and so forth. You will note that many of the names of enzymes end in "-ase." This suffix was adopted by chemists to designate an enzyme, just as "-ose" designates a carbohydrate. Some of the digestive

enzymes that were identified before this modern method of naming enzymes was adopted retain the old "-in" suffix. Thus we have ptyalin, pepsin, and amylopsin.

In the intestine, the proenzyme, *trypsinogen,* which is formed in the pancreatic cells, is changed into trypsin by the action of an intestinal enzyme, *enterokinase.* Trypsin can act on raw proteins, just as pepsin does in the stomach, but if the protein has been cooked, or has already been acted on by pepsin, trypsin is more effective. As stated in Chapter 16, in the stomach the proteins are broken down into proteoses and peptones by the action of pepsin. Trypsin reduces these molecules to simpler chains of amino acids, the *polypeptids.* Chymotrypsin continues the breakdown of the polypeptids into still simpler chains. Carboxypeptidase acts on these smaller polypeptids to reduce them to dipeptids (two amino acids). The final breakdown into single amino acids takes place by the action of enzymes in the intestinal juice.

Chymotrypsin also has a forerunner, *chymotrypsinogen,* which is converted into chymotrypsin by the action of trypsin. Chymotrypsin curdles milk with about the same effectiveness as pepsin, so that any milk that has escaped action in the stomach will be digested in the small intestine.

The ability of steapsin (the pancreatic lipase) to digest fats is remarkable; an extremely small amount of the enzyme is needed to bring about normal digestion of fats in the intestine. If the pancreas is removed from an experimental animal so that steapsin is lacking, the amount of fat excreted in the feces is greatly increased. However, if only a small piece of the pancreas is allowed to remain, little fat is lost. Steapsin acts on the fats that have been emulsified by the bile salts and changes

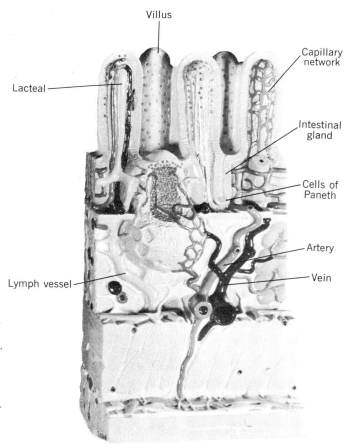

Fig. 17-7 A section through the intestinal mucosa. Note the structure of the villi and of the intestinal glands. (Model designed by Dr. Justus F. Mueller. Photograph of Ward's Natural Science Establishment Inc.)

them into fatty acids and glycerol, the forms in which fats can be absorbed.

The amylopsin of the pancreas is sometimes referred to as the pancreatic ptyalin because its action so nearly parallels that of the saliva. Amylopsin is able to break down the starch molecule into dextrins and split off the maltose much more rapidly than ptyalin, but it is unable to carry the reduction farther. These complex carbohydrates cannot be absorbed directly by the blood but must await reduction into hexoses by the intestinal juices.

THE INTESTINES

The cells of the intestinal walls produce a digestive fluid that is quite distinct from either bile or pancreatic juice. It is formed by special cells lying at the bottom of the simple tubular glands that dip below the surface of the intestinal lining. These glands, the *crypts of Lieberkuhn,* are scattered quite generally over the inner surface of the intestine between the villi (Fig. 17-7). At the bottom of each gland are cells that contain large granules in their cytoplasm. These are the *cells of Paneth* which produce the enzymes of the intestinal juice. Along the sides of the glands are cells that show mitotic figures indicating rapid division. This condition has given rise to the belief that these dividing cells are able to move toward the surface and replace those cells of the lining of the intestine that are worn away by the normal activity of the organ.

The Intestinal Enzymes

Digestion of the chyme is completed by the intestinal enzymes. Thus far, the digestive enzymes—with the exception of steapsin—have only reduced the foods to intermediate products which cannot be absorbed because of the size of their molecules. Therefore, to render the foods suitable to pass into the blood stream, these more complex substances must be further reduced. The *intestinal juice* contains six enzymes that bring about this final reduction: *erepsin* (currently known to be more than one enzyme, i.e. aminopeptidase and dipeptidase), *maltase, sucrase, lactase,* and *lipase.* The juice also contains enterokinase which, as we have seen, activates the pancreatic trypsinogen into the active enzyme trypsin.

Erepsin completes the digestion of proteins by splitting the remaining polypeptids and dipeptids into their component amino acids. In this form they are finally absorbed into the blood stream.

Since the action of ptyalin and amylopsin on carbohydrates is relatively slight, the major responsibility for carbohydrate digestion falls to the intestinal enzymes. Maltase converts maltose into glucose which can be absorbed. One of the principal carbohydrates in our diet, sucrose, remains unchanged until it encounters the intestinal enzyme, sucrase, which breaks it down into glucose and fructose. Milk has a special carbohydrate, lactose, which can only be digested by lactase in the small intestine, where it is converted into glucose and galactose. In the meantime the dextrins have been further reduced to maltose by the continued action of amylopsin, although little is known about the steps in this process.

The small amounts of the intestinal lipase are important in digestion of fats and oils. Lipase functions in addition to the steapsin of the pancreas.

The action of the various digestive juices are shown in the table.

Intestinal Movements

The structure of the walls of the small intestine is much the same as that of the other digestive organs. On the outside is a serous coat which is surrounded by the mesenteries. Inside this is a layer of longitudinal muscle fibers, and then a layer of circular muscles. These muscle layers, which are responsible for the movements of the intestines, are attached to the mucosa by a submucosa of connective tissue. The mucosa contains the intestinal glands and the food-absorbing mechanisms.

In the small intestine the partly digested food is subjected to three differ-

Material(s) Digested	Place of Digestion	Digestive Juice	Enzyme	Material(s) Formed	Material(s) Absorbed
Starch	Mouth and stomach	Saliva	Ptyalin	Dextrins Maltose	None
Starch	Small intestine	Pancreatic juice	Amylopsin	Dextrins Maltose	None
Dextrins Maltose	Small intestine	Intestinal juice	Maltase	Glucose	Glucose
Sucrose	Small intestine	Intestinal juice	Sucrase	Glucose Fructose	Glucose Fructose
Lactose	Small intestine	Intestinal juice	Lactase	Glucose Galactose	Glucose Galactose
Fats and oils	Stomach	Gastric juice	Lipase	Fatty acids Glycerin	Fatty acids Glycerin
Fats and oils	Small intestine	Bile	None	Emulsions of fat	None
Emulsified fats and oils	Small intestine	Pancreatic juice	Steapsin	Fatty acids Glycerin	Fatty acids Glycerin
Emulsified fats and oils	Small intestine	Intestinal juice	Lipase	Fatty acids Glycerin	Fatty acids Glycerin
Proteins	Stomach	Gastric juice	Pepsin	Proteoses Peptones	None
Proteoses Peptones	Small intestine	Pancreatic juice	Trypsin Chymo-trypsin	Polypeptids	None
Polypeptids	Small intestine	Intestinal juice	Carboxy-peptidase Erepsin	Amino acids	Amino acids

ent types of intestinal movement. First, by *peristaltic contraction* food is moved along the tube at a fairly constant rate of about 7.6 centimeters in eight or nine minutes (Fig. 17-8). This motion, however, follows a spiral course and makes a complete revolution in about every 25 or 28 centimeters of its forward progress because of the slightly spiral arrangement of the longitudinal muscles. A second type of motion, *segmental action* (Fig. 17-8), appears as the food flows along but does not help it in its forward progress. This action constricts the tube every few seconds, thus churning its contents so that a new surface is constantly being presented to the digestive juices. A third type of movement is the *pendular movement,* in which separate sections of the intestine are involved. The muscular coats contract in such a way that the contents appear to be flung along the loop. Actually, the food does not progress but is mixed thoroughly with the secretions. As we shall see shortly, there is another type of motion in the intestine: that of the swaying villi. These tiny fingers stir the fluids and the chyme into a more thorough mixture.

In Chapter 4, we discussed some of the properties of smooth muscle tissue. Not only does smooth muscle move more slowly than striated muscle fibers, but it is capable of remaining contracted for longer periods of time. It can also contract to a greater degree without suffering permanent injury. These characteristics make the smooth muscle tissue admirably suited to its function in the intestines. As a moving force in the digestive system, a wave of contraction passes over these muscles. The violence of the contractions is

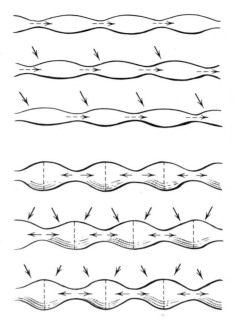

Fig. 17-8 A region of the small intestine during successive periods of peristalsis (above) and segmentation (below).

evidenced by the fact that the walls turn pale in the area of the contraction as the blood is squeezed out of the tissues momentarily. Another characteristic of the muscle is its ability to contract rhythmically as a result of the steady flow of nervous impulses to it. Cardiac muscle can also contract in a rhythmic manner, but it does this without nervous stimulation; if skeletal muscles contract in this way, they do it for only a short time before becoming fatigued. Smooth muscle cells are also more sensitive to mechanical and chemical stimuli than either skeletal or cardiac muscle cells. Thus the mere presence of food in the digestive system starts their contraction and maintains it. These intestinal movements mix the chyme with the digestive juices and also aid in absorption of the end-products of digestion.

As indicated in Chapter 9, the autonomic nervous system controls the movement of the smooth muscle of the digestive tract.

Absorption of Nutrients

Both the mucosa and submucosa of the small intestine are thrown into a series of *circular folds* that greatly increase the surface area of the intestine. They form permanent and rather hard ridges that do not disappear when the intestine becomes distended with food.

Over the surface of the circular folds are small fingerlike projections which extend into the opening of the intestine. These are the *villi* (Fig. 17-9). They are so small that they can barely be distinguished by the naked eye, but they give the lining a velvety appearance. Smooth muscle tissue in their walls enables the villi to move with a swaying motion or to stretch or shorten slightly. This movement continually exposes new surfaces over which the digested materials can be absorbed.

The structure of the villi is shown in Fig. 17-7. The outer surface of a villus is composed of a single layer of epithelial cells, while inside there is a network of capillaries and a branch of a lymph vessel, the *lacteal*. Through the outer layer of cells pass the digested proteins, carbohydrates, and fats.

The lacteals contain an emulsion of fats. How fats get into the lacteals in this form has been the subject of much research. We know that the emulsified fats are broken down in the intestine into fatty acids and glycerol, which enter the villi. Apparently these substances recombine to form fats after they pass through the cell membranes of the villi. The emulsion gives a milky appearance to the fluid in the lacteals. (The word "lacteal" means milky.) This theory also accounts for the *chylomicrons,* extremely minute droplets of fat, that may be found in the blood stream following a meal that contains a normal amount of fat.

Water, glucose, and amino acids are absorbed by the blood capillaries of the villi and carried to the liver and thence into the general circulation. Part of the

Fig. 17-9 A photomicrograph of the lining of the small intestine, showing the villi. (Encyclopaedia Britannica Films)

fat is carried to the liver through the portal circulation and part enters a lymph vessel, the *thoracic duct,* from which it finally enters the blood stream. The thoracic duct empties its contents into the left subclavian vein in the region of the shoulder after being joined by other lymph vessels. The return of some fat particles to the blood stream in this manner accounts for the presence of the chylomicrons in the blood.

THE COLON

Structure

The large intestine or colon, a tube approximately 1.5 meters long, forms about one fifth of the total length of the intestinal canal. At the point where the small intestine enters it, the large intestine is approximately 76 millimeters in diameter, but from this point on, it tapers finally to little more than 25 millimeters. The large intestine forms three sides of a square beginning at the lower right-hand quarter of the abdomen. From the *caecum,* it passes upward along the right side of the abdominal cavity to form the *ascending colon.* The *transverse colon* crosses toward the left side of the body at about the level of the third lumbar vertebra, rising slightly before turning downward. The part that passes downward along the left side is the *descending colon* (see Trans-Vision). This third part makes a sharp S-shaped curve, known as the *sigmoid colon.* The rectum is continuous with the sigmoid colon; its adjoining *anal canal* eventually opens to the exterior at the *anus.*

The differences between the small and large intestine are more than a mat-

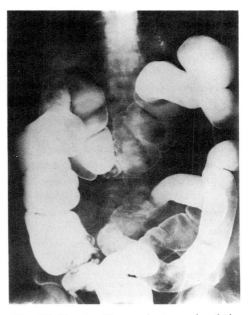

Fig. 17-10 An X-ray photograph of the large intestine. The haustrae are clearly shown. The ascending portion of the colon is on the left of the photograph and the descending and sigmoid portions on the right. (Jerry Harrity)

ter of size. Externally, the small intestine has a relatively smooth surface. This is not true of the large intestine. While it contains the usual four layers of tissue, the layer of longitudinal muscle tissue does not completely surround the intestine, but is confined to three narrow bands placed almost equidistant from each other. These long strips of muscle pucker the wall of the intestine into pouches, called the *haustrae* (sing., haustra). Unlike the circular folds of the small intestine, the haustrae involve all the layers of the colon. These segments can be seen in Fig. 17-10. The cells that line the colon do not secrete digestive enzymes. Intestinal glands are present, but they secrete mucus. There are no villi. The opening between the large and small intestines is guarded by the *colic valve*. This sphincter controls the passage of the intestinal chyme. Below the colic valve is a blind pouch, the caecum. In a few individuals the caecum may be conical or pyramidal in shape, but the average appearance is that shown in Fig. 17-11. Just beneath this colic valve is the *vermiform* (worm-like) *appendix* which projects into the abdominal cavity. This may vary in length from 19 millimeters to almost 200 millimeters, but its average length is slightly over 76 millimeters, and its diameter approximately that of the little finger.

As far as has been ascertained, the appendix plays no role in the process of digestion. Being a blind sac, it fills easily but empties sluggishly so that materials may remain in it for unusu-ally long periods. Hard or rough sub-stances may irritate the inner walls of the appendix, making it a favorable place for the growth of bacteria. An in-flammatory infection of the walls (ap-pendicitis) results.

Water Absorption

The material that enters the caecum is, at least theoretically, lacking in us-able nutrients. These have all been digested and their products absorbed during their passage through the small intestine. There are, however, still some substances that the body can use and that the large intestine will re-trieve before the waste materials are eliminated. As the contents of the small intestine enter the caecum, they are in a very liquid state. Because water ac-counts for a large percentage of the to-tal fluids of the body, its loss would represent a serious depletion of fluid reserves. One of the principal functions of the large intestine is to recover this water so that the body's reserves are not lowered.

The large intestine's capacity to ab-sorb water is very great. Some people who suffer from constipation have the idea that their condition can be helped by drinking large quantities of water to make the evacuation of the bowel's content easier. Even though a person drinks as much as three liters of water per day above his usual intake, there is no indication that the nature of the feces is changed or that constipation is relieved. This additional water is absorbed through the walls of the large intestine, and the amount of urine formed and excreted is proportionally increased.

Fig. 17-11 Junction of the small and large intestine. Note the saclike nature of the caecum.

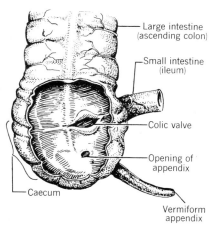

Large intestine (ascending colon)

Small intestine (ileum)

Colic valve

Opening of appendix

Caecum

Vermiform appendix

Bacterial Action

Throughout the entire life of the individual from a few hours after birth until death, the large intestine contains bacteria. These enter with ingested food, and although many are destroyed by the action of the digestive juices, a sufficient number escape and find their way to the large intestine where they multiply rapidly and in huge numbers. Without these bacteria, life would be supposedly impossible.

Bacteria in the large intestine are largely harmless to man. They act on the undigestible materials and break some of them down into gases, acids, amines, and other waste products. Some of these products of decomposition are excreted through the intestinal tract. Potentially toxic materials are absorbed and are carried by the portal vein to the liver for *detoxification.* Special enzymes produced by the liver cells convert these toxic substances into nontoxic materials which are excreted by the kidneys. This process of detoxification is one of the chief functions of the liver. The poisons of the body are thus kept from reaching a dangerous level in the circulation.

Formation of Feces

After the absorption of water from the contents of the large intestine, bacterial action changes the consistency of the intestinal contents from a liquid to a semisolid state, the *feces.* Some of the contents of the feces are bacterial; the other substances are waste products brought by the blood, the products of bacterial action in the intestine, many inorganic salts, mucus, and the undigestible components of food, such as cellulose. Cellulose, as we have seen, is the fibrous part of plant food that man is unable to digest. It forms a small percentage of the feces and contributes to its bulk. This bulk stimulates the lining of the intestines and induces peristalsis. Fruits and vegetables provide this type of roughage.

Movement

There is considerable question as to just how the colon moves its contents along its tube. No frequent, rhythmic peristalsis or other movement has been observed except in the transverse colon, where there are slow, weak accordion-like movements of the haustrae rather than actual peristalsis. At certain intervals two or three times in 24 hours, mass peristalsis occurs in the large intestine. This massive movement often sweeps through the entire length of the colon, pushing the feces before it. These waves are sometimes preceded by peristalsis in the small intestine and by the taking of food into the stomach. The sigmoid colon, which serves as a storehouse for the feces, becomes filled so that during a wave of mass peristalsis the feces may enter the rectum. The lower end of the rectum is equipped with stiff longitudinal folds of tissue that lie just beneath the lining mucosa. These delay the further progress of the contents into the anal canal. Nervous stimuli are set up which result in *defecation,* or the act of excreting intestinal wastes through the anus. This process is voluntarily controlled by the action of two sets of sphincter muscles which surround the anus. In an infant, the ability to control the act of defecation is lacking, but with patient training it can be established.

Constipation is a condition in which normal, regular defecation is lacking. The feces are retained within the bowel for a longer than normal period of time. This permits continued bacterial activity with production of large amounts of gas. During longer retention in the intestine, the feces lose more water, forming a very solid mass difficult to evacuate. Ignoring the desire to defecate at the habitual time is often the beginning of constipation. Otherwise, if a normal person follows a well-balanced diet and takes a moderate amount of exercise, there are few physiological reasons for constipation.

Sometimes a person becomes "bowel conscious" and worries about his inability to have bowel movements at regular intervals. The individual may seek aid from laxatives which eventually reduce the muscle tone of the bowel so that he becomes more and more dependent on them. Thus habitual use of laxatives may replace the normal functions of the large intestine. For people whose activities and diet are limited by conditions such as long periods of illness, the use of laxatives is required, but healthy individuals seldom have a need for them. The regularity of bowel movements varies from two or three a day to one every other day.

VOCABULARY REVIEW

Match the phrase in the left column with the correct word in the right column. *Do not write in this book.*

1 A sac-like appendage of the colon.
2 A storage form of carbohydrates.
3 A function of the liver connected with protein conversion.
4 Fats are reformed in its walls from fatty acids and glycerol.
5 May collect deposits of cholesterol.
6 Carries bile to the gall bladder.
7 Important in the control of sugar metabolism.
8 The principal absorptive structure in the small intestine.
9 The longest part of the small intestine.

a caecum
b common bile duct
c cystic duct
d deamination
e gall bladder
f glycogen
g ileum
h islets of Langerhans
i jejunum
j lacteal
k villus(i)

TEST YOUR KNOWLEDGE

Group A

Select the correct statement. *Do not write in this book.*

1 The liver receives blood through the (a) portal veins (b) portal arteries (c) hepatic veins (d) renal arteries.
2 The stomach connects directly with the (a) jejunum (b) ileum (c) duodenum (d) peritoneum.
3 The largest gland in the human body is the (a) pancreas (b) liver (c) thyroid (d) gastric.
4 Bile passes into the gall bladder through the (a) hepatic duct (b) cystic duct (c) common bile duct (d) pancreatic duct.
5 Gall stones are mainly composed of (a) bilirubin (b) biliverdin (c) cholesterol (d) bile salts.
6 The intestinal enzyme that digests proteins is (a) maltase (b) pepsin (c) rennin (d) erepsin.
7 The digested nutrient that commonly enters the blood through the thoracic duct is (a) starch (b) sugar (c) protein (d) fat.
8 Peristalsis does *not* occur in the (a) ileum (b) duodenum (c) caecum (d) liver.
9 The term that is *not* closely related to the others in the following group is (a) villus (b) vermiform appendix (c) lacteal (d) lymph vessel.
10 Narrow muscle bands in the small intestine produce pouches called (a) rugae (b) feces (c) haustrae (d) lobules.

Group B

1 (a) What are the functions of the liver? (b) Briefly describe how each of these is performed.
2 Describe how each of the products formed by the digestion of each of the classes of organic nutrients enters the circulatory system.
3 (a) Describe the structure of the small intestine, including its relation to the mesenteries. (b) What is the function of each of the parts mentioned?
4 What is the relation of the portal circulation to the digestive system?

Chapter 18 **Respiratory Structures**

Respiration is a combination of two distinct processes, one mechanical, the other chemical. *External respiration* is a mechanical process controlled by the pressure of the atmosphere and the action of certain body muscles. Its function is to bring air into the lungs where oxygen is absorbed by the blood and to remove gaseous wastes brought from the body by the blood. In one form or another, all animals and plants carry on external respiration, exchanging their waste products for the usable materials present in the environment. While the methods may vary widely, the fundamental principles are the same. *Internal respiration* is the exchange of oxygen and carbon dioxide between the body's cells and the fluids which circulate around them. The cells obtain energy for their chemical activities by oxidizing nutrients. The waste products of this oxidation, carbon dioxide and water, are picked up by the blood and returned to the lungs for removal. Chemically, these are highly complex reactions in which numerous enzymes present in the cells act on various substrates.

In a way, external respiration may be compared to the action of a bellows. As the bellows is expanded, the cavity within it becomes enlarged so that the internal air pressure is less than that of the surrounding air. The outside air therefore rushes in to equalize the air pressure, thereby inflating the bellows. If the handles of the bellows are then pressed together, the inside volume is reduced and the pressure within becomes greater than that of the outside air, so that the air within is forced out. The difference in air pressure between the outside and the inside of the bellows is the essential principle of the breathing movements.

THE RESPIRATORY SYSTEM
The Nasal Cavity

The first part of the human respiratory system with which air normally comes in contact is the nasal cavity, whose inner walls are folded into three ridges on each side—the *nasal conchae* or *turbinate bones*. Like the rest of the lining of the nose, the conchae are covered by a layer of mucous membrane that is richly supplied with blood. As Fig. 18-1 shows, the air passes over the series of ridges, the arrangement and relationship of which are evident in the diagram of a vertical section through this region (Fig. 18-2). This folding of the inner surface of the nasal cavity is of great importance, because here the air is warmed by the blood, moistened by the mucus, and cleansed of small particles of dirt which are drawn in with the air. This last function is carried on by the sticky mucus which traps the particles. Small hairs are located in the front of the nostrils, and they aid in preventing larger bits of dirt from entering. By the time the air reaches the lungs it is saturated with water vapor, cleaned, and warmed to body temperature.

The nose cavity is divided internally into two smaller cavities by a wall,

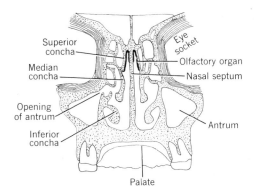

Fig. 18-2   Vertical frontal section through the nasal cavities to show the arrangement of the nasal conchae. Note that the antra are paranasal sinuses which have no respiratory function.

Fig. 18-1   Sagital section of the head showing the relation of the nasal and other air passages. (Model designed by Dr. Justus F. Mueller. Photograph, Ward's Natural Science Establishment Inc.)

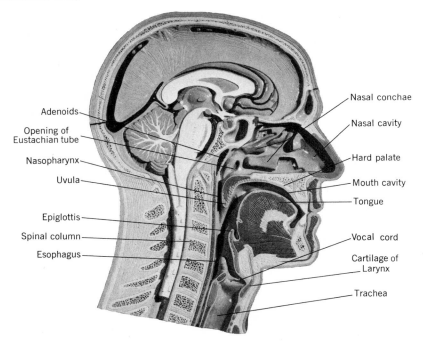

called the *septum*. The external openings of these two cavities are called the *nostrils*. The cartilage of the outer nose is firmly attached to the *nasal bones* that make up the bridge of the nose. The conchae are located on the walls of the nasal bones. In the covering of of the uppermost of these three ridges, the superior concha, we find the nerve endings for the olfactory sense, smell. The cells which receive olfactory stimuli are elongated and supported by larger cells. Their free ends have hairlike projections against which the air brushes as it flows along the nasal passage (Fig. 18-3).

Although our olfactory sense is poorly developed as compared with that of some animals, we still have a remarkable ability to distinguish odors

Fig. 18-3 Olfactory cells and supporting structures. The nerve fiber from each olfactory cell joins with others to form the olfactory nerve, which carries impulses to the olfactory center in the brain.

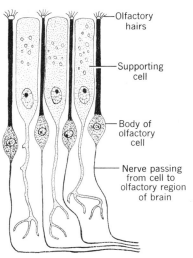

Olfactory hairs

Supporting cell

Body of olfactory cell

Nerve passing from cell to olfactory region of brain

even when they are present in surprisingly small concentrations. Next to sight, smelling is our most acute sense, and it is possible that proper training would greatly improve it. We can detect the odor of ethyl alcohol in a concentration of approximately 0.0000003 percent by weight, but it must be present in about a 14 percent solution before we taste it, and between 25 and 50 percent before it produces a burning sensation on the lining of the mouth.

The Pharynx

From the nasal cavity, the air enters the *pharynx*. This is the region in the back of the mouth which serves as a passageway for both food and air. The pharynx is the site of extensive growth in the early embryo. From its walls develop a fingerlike projection that later divides into two parts and gives rise to the lungs and the passages leading to them. Also, from this area develop masses of lymphatic tissue which become the *tonsils* and *adenoids*. The lower end of the pharynx ends at the *glottis*. This is the opening into the *larynx*. The glottis is a narrow slit, with its long axis lying in a front-to-back direction. In men, this opening is about 23 millimeters long; in women, somewhat smaller, being 17 to 18 millimeters long. Around the rim of the glottis are the vocal cords, whose vibrations produce the sounds we make. The *epiglottis* acts as a trap door to close the glottis during swallowing.

The Larynx

The glottis opens into a roughly triangular chamber, the *larynx,* or voice box. The apex of this triangle points

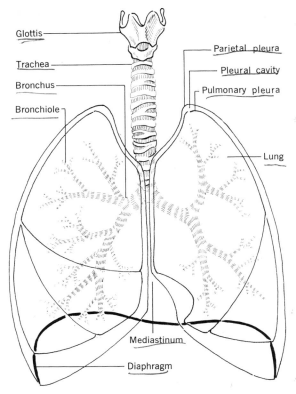

Fig. 18-4 The lungs and associated parts. The size of the pleural cavity is exaggerated; the size between the parietal and pulmonary pleurae is very slight. The only place where there is an appreciable separation of the two layers is below the lungs.

forward and may be quite conspicuous in some men. It is commonly called the *Adam's apple*. The walls of the larynx are composed of plates of cartilage derived from the embryonic structures called the *pharyngeal arches*.

The Trachea

Below the larynx is the trachea. This tube is about 109 millimeters long and 18–25 millimeters wide. It is larger in males than in females. The trachea is not a perfectly round tube, being slightly flattened along the rear surface. Its walls are composed of alternate bands of membrane and cartilage, the former supporting and holding the cartilage in place (Fig. 18-4). The bands of cartilage may be either horseshoe shaped, almost completely encircling the trachea, or short, passing only part of the way around the tube. In either case, their free ends are held together by a tough membrane that contains scattered bands of smooth muscle tissue. The esophagus lies just behind the trachea, in line with the openings in the cartilage bands. It is possible, then, for the esophagus to swell as a bolus of food passes along it, so that part of the swollen region projects into the tracheal cavity. We have all had the unpleasant experience of swallowing a mass of food that was too large to be easily accommodated by the esophagus. When this occurs, there is a momentary feeling of suffocation due to the pressure of the enlarged esophagus on the trachea. This cartilage is quite firm but also somewhat elastic. Thus the partial rings of the trachea maintain an open passage for the air at all times. The lining of the trachea is composed of a ciliated epithelium in which are goblet cells that secrete mucus. The function of the cilia is to carry small particles of

Fig. 18-5 Air sacs. At the left capillaries are shown covering the alveoli. In the center is an air sac with protruding alveoli. At the right the sac is shown in cross section. (From Whaley, Breland, Heimsch, Phelps, and Rabideau, Principles of Biology, *Harper & Bros., 1954)*

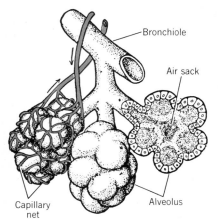

debris from the lungs upward to the mouth or nasal cavity so that they can be eliminated.

The Lungs

The lower end of the trachea divides into two parts, the *bronchi* (singular, bronchus). They are similar in structure to the trachea in that their walls are thickened by cartilage rings, and they are lined by ciliated epithelium. Each bronchus then divides many times, each division having a Y-shaped form. These small branches, the *bronchioles,* eventually form a very large number of minute terminal branches which end in the *air sacs.* According to some recent work on the architecture of the lung, there are about 14 million of these. The total effect is that in appearance the bronchial branches resemble the branching of a tree. This huge number of smallest air passages affords an adequate pathway for the passage of air into the lung and the exit of gases from it.

As shown in Fig. 18-5, the walls of the air sacs are not smooth, but have protruding pouches. These are the *alveoli* through the walls of which the actual exchange of gases takes place. There are approximately 300 million of these alveoli regardless of the size of the lung.

Through the thin walls of the alveoli and capillaries the blood is able to obtain oxygen from the air, and give up waste gases to the space. The blood supply to these vessels is so extensive that they contain approximately one-fourth of the total blood supply at any one moment. If all of the internal surface of the lungs were spread out to form a single sheet of tissue, it would cover an area of between 40 and 80 square meters, the lower figure referring to a child and the higher to an adult.

The texture of the lungs is similar to that of a sponge with very fine openings. In the walls of the air sacs are blood vessels and strands of smooth muscle fibers. The sac is lined by a layer of epithelium so extremely thin that it was not known to exist until 1953, when photographs taken with an electron microscope showed its presence. The finest branches of the bronchioles, less than 1 millimeter in diameter, lack the rings of cartilage that are characteristic of the larger tubes. These fine vessels may therefore occasionally constrict and shut off the passage of air to some small areas of the lungs.

The subdivisions of the lungs, the *lobes,* are shown in the Trans-Vision following page 212. Each lung is covered by a membrane, the *pleura,* made up of tough endothelial cells. The pleura is composed of two layers. The *visceral pleura* completely covers the lungs, dipping between the lobes. The outer layer, the *parietal pleura,* forms the lining of the chest cavity and continues over the surface of the diaphragm. Each lung, therefore, lies in a two-walled sac. Since the lungs almost completely fill the thoracic cavity, the space between the two layers of the pleura is negligible. This potential space, called the *pleural cavity,* is moistened by a thin film of serous fluid which prevents the pleurae from rubbing together when the chest moves with each breath. If this *pleural fluid* increases as the result of a local infection, the distressing symptoms of pleurisy may appear. Although small,

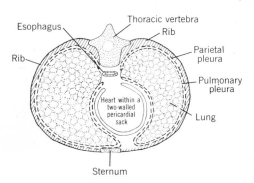

Fig. 18-6   Cross section through the chest cavity to show the relation of the various structures. The space between the pleurae is exaggerated.

the pleural cavity is of great importance in breathing.

The interpleural space, containing the heart, large blood vessels, nerves and other structures of the thoracic cavity, is called the *mediastinum* (Fig. 18-4). The relationship of the main structure is shown in Fig. 18-6.

MECHANICS OF BREATHING

The process of breathing is quite distinct from the chemical reactions of internal respiration. Breathing is a purely mechanical act that can be divided into two stages: the intake of air into the lungs, called *inspiration,* and the passage of air out of the lungs, called *expiration.* Inspiration is the more active of the two processes because it requires the contraction of more muscle bundles than expiration. The latter is largely the result of relaxation of muscles and the return of the viscera to their original position. With each breath of air that we inhale, we expend considerable muscular energy. Exhalation, being more passive, requires less energy.

When we inhale, the major role is played by the *intercostal muscles,* attached to the ribs, and the *diaphragm.* Other groups of muscles also help in the process, but their actions are not so pronounced. The intercostals are divided into two groups: the *external intercostals* and the *internal intercostals.* The fibers of these muscles pass at an angle of almost 90° to each other. The external layer lifts the ribs upward and thereby increases the volume of the thoracic cavity during inhalation (Fig. 18-7). The inner layer draws the adjacent ribs toward each other and reduces the size of the thoracic cavity during exhalation. In the process of inhalation the sternum rises as the ribs rise. At the same time the dome-shaped diaphragm

drops to a flatter condition as its muscles contract. This motion of the diaphragm results in pressure on the viscera of the abdomen, which is transmitted to the front muscular walls so that they protrude slightly. The result is a marked increase in volume of the thoracic cavity, as is evident by comparing the X-ray pictures in Fig. 18-7.

The classic method of demonstrating the mechanics of breathing is by means of an apparatus shown in Fig. 18-8. Here we have a glass bell jar in the top of which is a cork with a glass tube passing through it. The tube acts as the trachea, and divides into two branches, representing the bronchi. At the end of these branches, thin-walled toy balloons are fastened to represent lungs. A rubber sheet, fitted with a device to pull it downward, is stretched over the

Fig. 18-7   X-ray photographs of the lungs during inspiration (left) and expiration (right). Note the differences in position of the diaphragm and the clavicles. (The Children's Hospital Medical Center, Boston)

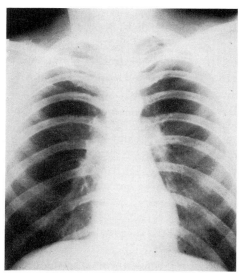
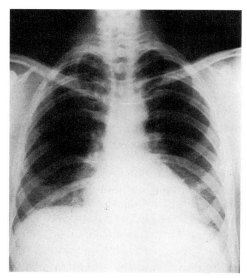

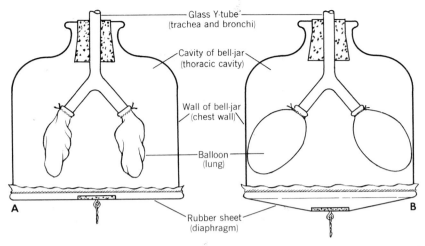

Glass Y-tube
(trachea and bronchi)

Cavity of bell-jar
(thoracic cavity)

Wall of bell-jar
(chest wall)

Balloon
(lung)

A

B

Rubber sheet
(diaphragm)

Fig. 18-8 A mechanical model to demonstrate the process of breathing. In A the balloons (lungs) are collapsed because of the relaxation of the rubber diaphragm. In B the balloons are inflated because the volume of the cavity in the bell jar (chest cavity) has been increased.

bottom opening of the bell jar to represent the diaphragm.

This apparatus has several limitations as a model. First, the walls of the bell jar are rigid, whereas the walls of the chest are moved by the intercostal muscles. Second, the stretched rubber diaphragm is flat, while the normal human diaphragm is dome-shaped in its relaxed position. However, this model is very useful in explaining the mechanics of breathing. The balloons alternately inflate and deflate as the rubber diaphragm is pulled down and released; just as our lungs inflate and deflate following the movements of the muscular diaphragm. Inflation results from a decrease in air pressure within the cavity of the bell jar (or the chest) as the diaphragm is pulled down. Deflation occurs when the diaphragm is re-

leased or relaxed, increasing the air pressure within the cavity. If the diaphragm is pushed up into the cavity of the jar, the balloons will deflate still more because of the further increase in air pressure within the bell jar. Compare this to our forceful exhalation of air, using the abdominal muscles to press the diaphragm further up into the chest cavity.

Pressure of gases is generally measured by the height of a column of mercury that the gas will support. If we fill a long glass tube, sealed at one end, with mercury and invert it in a dish of mercury, the liquid in the tube drops to a point where its weight is balanced by the pressure the air exerts on the surface of the mercury in the dish. Such an instrument is called a *barometer*. At sea level, the height of the column of mercury, measured from the level of the mercury in the dish, is 760 millimeters, or 30 inches. This is taken as the standard pressure of air at 0°C (32°F). However, at a higher elevation, the height of the mercury column drops because the pressure of the air constantly decreases with elevation. The importance of this effect on breathing is discussed later. But it should be clearly understood now that standard air pressure is the weight of a column of mercury that can be supported by the air pressure at sea level.

In the process of inhalation, pressure within the chest cavity decreases by only about two millimeters of mercury. This slight change of pressure is sufficient to cause air to enter the lungs. Exhalation produces a much greater pressure within the chest cavity. Due to relaxation of the elastic diaphragm, the pressure of the abdominal viscera, and the lowering of the ribs, the pressure within the thorax is increased by about four millimeters of mercury. This forces air out against the normal atmospheric pressure.

VOCABULARY REVIEW

Match the phrase in the left column with the correct word in the right column. *Do not write in this book.*

1 Folds in a bone that increase the surface area of the nasal passages.
2 Passes downward from the pharynx anterior to the esophagus.
3 Includes the exchange of gases between the cells and the blood.
4 Space containing the heart and the large blood vessels.
5 Divides the nasal cavity into two halves.
6 The trachea divides into a left and a right of this.
7 Ends in an air sac.
8 The result of increased intrathoracic pressure.
9 A thin-walled structure through the walls of which the blood absorbs oxygen.
10 Sometimes called "Adam's apple."

a alveolus
b bronchiole
c bronchus
d expiration
e glottis
f inspiration
g larynx
h mediastinum
i nasal conchae
j pharynx
k pleura
l respiration
m septum
n trachea

TEST YOUR KNOWLEDGE

Group A

Select the correct statement. *Do not write in this book.*

1   The sense of smell depends on sense organs lying (a) below the septa (b) in the antra (c) on the tongue (d) above the superior conchae.
2   Tonsils and adenoids are masses of (a) glandular tissue (b) muscle tissue (c) epithelial tissue (d) lymphatic tissue.
3   The region in back of the mouth cavity which is a passageway for both food and air is the (a) pharynx (b) larynx (c) trachea (d) esophagus.
4   Rings of cartilage stiffen the walls of the (a) esophagus (b) nasal cavity (c) trachea (d) diaphragm.
5   Windpipe is a common name for the (a) larynx (b) pharynx (c) trachea (d) bronchus.
6   Of the following, the one that is *not* within the chest cavity is the (a) heart (b) lungs (c) esophagus (d) larynx.
7   The membrane that covers the lungs is called the (a) mesentary (b) pleura (c) mediastinum (d) pericardium.
8   Muscles that play a part in breathing are (a) intravenous (b) interstitial (c) intercostal (d) internal oblique.
9   Oxygen passes into the blood of the capillaries chiefly through the walls of the (a) bronchioles (b) bronchi (c) alveoli (d) pharyngeal arches.
10   Of the following terms, the one that is *least* closely related to the others by structure is the (a) larynx (b) trachea (c) glottis (d) air sac.

Group B

1   Trace the path taken by a molecule of oxygen from the time it enters the respiratory system until it is absorbed by the blood. Tell over what structures it passes, and the effect each has on it.
2   What are the differences between external and internal respiration?
3   Briefly describe the mechanics of breathing.
4   Distinguish between the members of the following pairs of terms (a) air sac—alveolus (b) bronchus—bronchiole (c) glottis—epiglottis (d) pharynx—larynx.

Chapter 19 **Breathing**

The Atmosphere

The atmosphere we live in is a mixture of gases rather than a chemical compound. The gases in the air are not combined with each other, nor is their proportion definite and unchanging. The proportions of the various gases in the air differ from one locality to another; over cities or volcanoes, the amount of some substances may be very high, while over the open sea or grasslands and deserts these same compounds may be in low concentrations or entirely absent. The wind flow in a certain locality may bring in traces of unusual gases. The water content of air is also variable from place to place. If moisture is present in large quantities at any given temperature, it may produce a fog, or, if other substances are also present, a smog. Air composition is therefore usually stated in terms of *dry air* and the measurements are made in a place that is not subject to local pollution. The following table gives the percentage by volume of the main constituents of dry air at sea level when the various factors that may contaminate it have been excluded.

Nitrogen	78.03%
Oxygen	20.99%
Argon	0.94%
Carbon dioxide	0.03%
Neon	0.0012%
Helium	0.0004%

The gases that are important in respiration are *oxygen, carbon dioxide,* and *water vapor.* Although nitrogen constitutes the major part of the air we breathe, it plays no part in respiration, nor do the other gases listed. These elements combine with other substances with great difficulty.

Each of the respiratory gases acts independently in the body according to the laws of chemical and physical behavior. A knowledge of the gas laws which govern the behavior of gases is therefore helpful in the study of respiration. The mathematical formulations of these laws are useful in physiological research.

The Gas Laws

The *Principle of Avogadro* (1776–1856) states that equal volumes of all gases at the same temperature and pressure contain the same number of molecules. It has been found that at standard temperature and pressure (STP), that is, 0°C and 760 millimeters of mercury, the molecular weight in grams of any gas will occupy a volume of 22.4 liters. Avogadro's law is the basis for the volumetric method of determining the composition of gaseous mixtures.

Boyle's Law states that when a gas is subjected to pressure, its volume decreases inversely with the pressure, temperature remaining constant. If the pressure on 100 milliliters of gas is doubled, the volume is reduced by one-half. If the pressure is tripled, the volume is reduced to one-third. This relation of volume to pressure can be stated mathematically as $V/V' = P'/P$ where V is the volume of the gas at STP, and where V' is the volume of the gas at the increased, or decreased, pressure; where P is the pressure at STP and P' the pressure at the changed condition. The law is named for the English physicist, Robert Boyle (1627–1691), who discovered and stated this relationship.

Charles' Law (Jacques Charles, 1746–1823) states that the volume of a gas is directly proportional to its absolute temperature. Absolute zero, the temperature at which all molecular motion stops is −273°C. Therefore, for each rise of one degree Centigrade, a gas will expand 1/273rd of its original volume due to the increase of activity of the molecules composing it. This relation between volume and temperature can be expressed as $V/V' = T/T'$. The capital letter T is used to show that the measurements are based on the Absolute, or Kelvin, scale 0°C = 273°K).

Charles' Law explains why oxygen enters the blood more efficiently after the air containing it has been warmed in the lungs. The higher temperature results in increased molecular activity and increased pressure of the gas since the volume of the lung is limited. This hastens the absorption of gas molecules by the blood.

Charles' Law, Boyle's Law and Avogadro's Law are all combined in the *Ideal Gas Law* which can be expressed as $PV = nRT$ where P, V, and T are respectively the pressure, volume, and temperature (°K) of the gas; where n is the number of gram molecules (a gram molecule is the molecular weight in grams) of the gas and R is a constant. An example of the operation of this law as applied to non-living systems is found in the fact that the air pressure inside an automobile tire rises as the tire becomes hot when driving at high speeds.

Dalton's Law of Partial Pressures states that each gas in a mixture exerts its own pressure in proportion to the volume percentage of its occurence in the total volume of the gas. Thus, at sea level, the volume of oxygen present in the air of average composition is approximately 21 percent. Hence, oxygen will exert only 21 percent of the total pressure of the air. If we rise to an altitude of 19,312 meters (12 miles), the percentage of oxygen is reduced to about 18 percent of the total volume of the air and will therefore exert a pressure that is 2 to 3 percent below that at sea level. This means that the molecules of oxygen are so widely separated that they are unable to exert adequate pressure on the cell membranes and therefore cannot enter the blood in sufficient numbers. The relation can be stated as follows: $PV = V(p_1 + p_2 + p_3 \dots)$ where V stands for volume,

P for total pressure of the mixture and p_1, p_2, p_3 . . . . . for the pressure of individual gases. In other words, the total pressure of the mixture of gases in a given volume equals the sum of all of the partial pressures.

Henry's Law plays an important part in the entrance of gases into the blood. The law states that a gas is dissolved by a liquid in direct proportion to its partial pressure, provided that it does not react chemically with the liquid. Henry's Law is important to the understanding of the physical processes in breathing, since all gases that pass through the cell membranes in the lungs must first be dissolved in the cell fluids and in the thin film of fluid that covers the inner surface of the lungs.

A mathematical statement of Henry's Law can be worded as $P/V = K$. In this equation, P is the pressure of a gas in the lungs, V, its concentration in the plasma, and K, a constant, the *Coefficient of Solubility*. The value of this coefficient depends on the units used for expressing P and V and can be obtained by consulting tables in more advanced texts.

Gases of low molecular weight are absorbed into the blood less easily than those of higher molecular weight. For example, a molecule of hydrogen with a molecular weight of 2 is absorbed less easily than a molecule of oxygen with a weight of 32. Carbon dioxide, with a molecular weight of 44, enters the fluid of the blood, the plasma, more readily than oxygen. On the other hand, nitrogen, with a molecular weight of 28, is present in the plasma of the blood in much larger quantities than its weight might indicate because it is four times more common in air than oxygen. The total amount of oxygen absorbed by the whole blood is much greater than the amount dissolved in the plasma. Oxygen combines easily with hemoglobin, and the red blood cells are constantly taking up oxygen from the plasma. Since nitrogen does not combine easily with any of the components of the blood, it is all carried as a dissolved gas in the plasma. The relative solubility in the plasma of these three important gases is shown in the table. The values are milliliters of gas dissolved in 100 milliliters of plasma.

Oxygen	0.25
Carbon dioxide	2.69
Nitrogen	1.04

It is important biologically that the solubility of these gases at low temperatures is considerably greater than at high temperatures. This explains why minute plants flourish in the colder waters where carbon dioxide is more readily available for the process of photosynthesis. These plants, in turn, attract the great whales that use them as their principal food.

Lung Capacity

The capacity of the lungs increases from birth to adulthood. The table at the right shows human lung capacity at various age levels in terms of liters of air.

In the process of normal breathing, a surprisingly large quantity of air is exchanged between the body and its surroundings. When resting, the average adult male exchanges about 8.4 liters of air per minute, while an adult woman exchanges about 4.5 liters. By comparing the values for adults with those for a newborn child, we see the effect of

Age	Total Lung Capacity in Liters	Range in Liters
6	1.6	1.3–2.1
10	2.5	2.2–3.1
18 males	5.9	4.3–8.0
females	4.1	3.1–5.3
25 males	6.4	4.3–9.0
females	4.2	3.1–5.4
45 males	5.9	4.3–8.0
females	4.3	2.9–5.7
65 males	5.6	4.3–7.0
females	3.9	2.8–5.5

the size of the lungs on the amount of air inhaled and exhaled. In a newborn infant in a resting condition (asleep), there is an average exchange of only 0.72 liter per minute.

The moving body of air at each normal breath is called the *tidal air,* or *tidal volume.* If a person inhales as deeply as possible, he can draw in about six times as much air as he does on a quiet breath. This forced inhalation constitutes the complemental air, or *inspiratory reserve volume.*

A forceful expiration which follows a normal inspiration will expel an additional quantity of air constituting the supplemental air or *expiratory reserve volume.* Even after the supplemental air has been expelled from the lungs, there is always about 1 liter to 1.5 liters of air remaining. This is called the residual air, or *residual volume,* and will remain until such time as the chest cavity is opened so that the pressure within is equal to that of the surrounding atmosphere. Even when this apparently last remnant of air has been removed, there is a very small amount

of gas still clinging to the walls of the bronchioles and alveoli. This is termed the *minimal air,* and is scarcely measurable in volume.

The usual method of measuring the capacity of the lungs is by a simple device called a *spirometer.* If a person breathes into a device of this type, his *vital capacity* can be determined. This represents the sum of his tidal, complemental, and supplemental air, and amounts to about four liters for a male of average size and physical fitness. If we consider this in relation to the body surface area of such a person (see Fig. 28-5 on page 381) it is equal to about 2 liters per square meter of body surface for women and 2.5 liters per square meter for men. A well-trained athlete will average about 2.8 liters per square meter of body surface.

During the time when air is in the lungs, a change occurs in its composition. Some of the oxygen that enters the lungs dissolves in the blood plasma and some combines chemically with the hemoglobin of the blood to form oxyhemoglobin. Likewise, some of the carbon dioxide that has been formed by cellular oxidations leaves the blood and enters the air sacs. Therefore a difference is found in the composition of the inhaled air as compared with that expelled on a normal breath. This is shown in the following table:

	Inhaled Air	Exhaled Air
Nitrogen and other inert gases	79.02%	79.07%
Oxygen	20.94%	16.03%
Carbon dioxide	0.04%	4.40%

Control of Breathing

The rate at which we breathe is controlled by two different factors. One of these is *neural* (nervous) control, while the other is *chemical* in nature. Both of these may operate simultaneously, but each is independent of the other. The neural control of the respiratory movements is the result of stimuli affecting the *respiratory center* in the medulla of the brain, while the chemical control depends on a change in the pH of the blood which also stimulates the neurons in the respiratory center.

Neural control was first demonstrated by Galen in the 1st Century A.D. He showed that a center in the lower brain was responsible for this activity. More modern experiments carried out under very rigorously controlled conditions have shown that Galen was essentially correct in his conclusions.

Located near the upper part of the medulla is an island of cells that constitutes the respiratory center (Fig. 19-1). The exact location of this region has been well established by experiments using electrical and thermal stimuli. A slight stimulus of any nature applied to the center causes the rate of respiration to increase or decrease. It is thought that the center is actually composed of two distinct regions, one that controls inspiration and one that controls exhalation.

Two nerve pathways are concerned in control of the breathing process. One follows the *phrenic nerves* to the diaphragm and the intercostal muscles. This pathway consists largely of motor fibers, although there are sensory fibers that carry impulses to the respiratory center. The other pathway transmits afferent impulses along the vagus nerves from the lungs, skin, nose and larynx, and abdominal organs. Thus, many different stimuli from various regions of the body help to regulate the breathing process.

Stimuli that arise in the surface membranes of the body can change the breathing rhythm in several familiar ways. The prick of a pin or a sudden drenching with cold water makes a person gasp. Irritation of the nasal membranes or the larynx results in a sneeze or a cough. Likewise, a thought, sight, or sound may bring about a change in the respiratory rate. We have all gasped in surprise. It is also possible to alter voluntarily the rate of breathing by holding the breath. All of these stimuli arise either in body surfaces or in the cortex of the cerebrum.

Fig. 19-1   The nerve pathways involved in the control of respiration.

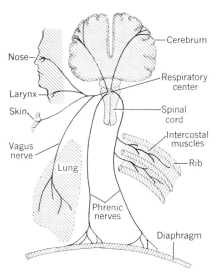

Impulses that normally regulate the rhythm of breathing travel along branches of the vagi to the lungs. As the lungs expand, nerve endings in their tissues are stimulated and set up impulses that pass to the respiratory center. Here these impulses initiate a series of stimuli that eventually pass along the phrenic nerves and their branches and cause the diaphragm and external intercostals to relax. At the same time, the internal intercostals contract and the ribs are drawn back to their resting position. When complete exhalation has occurred, the process is reversed and inspiratory impulses travel to the diaphragm and external intercostals, causing them to contract. In the action of these muscles during breathing, we have another example of reciprocal innervation such as that illustrated by the biceps and triceps.

Respiration is chemically controlled through the carbon dioxide in the blood. When the blood passes through an active tissue it picks up carbon dioxide and other products of the oxidations occurring in the tissues. The increase in carbon dioxide stimulates the respiratory center, which is richly supplied with blood vessels. Thus a person who is doing strenuous work breathes deeply and rapidly.

It is easy to demonstrate the effect of carbon dioxide on the rate of respiration. After you have determined the number of breaths (inhalations) you take per minute, inhale deeply eight or ten times, or until your fingertips tingle. Then observe the second-hand on a watch and note how much time elapses between your last deep breath and your next normal inhalation. If you do this carefully, you will find that the neces-sity to breathe has been delayed. Thus if you find that you normally breathe 15 times per minute (four seconds between each inhalation), you may not require another breath for 20 seconds following your last forced inhalation. In an experiment, a person who inhaled pure oxygen to rid his blood of carbon dioxide was able to hold his breath for 21 minutes before the accumulation of carbon dioxide in his blood forced him to exale. (Do not experiment with deep breathing more than a few seconds at a time, as dizziness and cramps may result.)

Atmospheric Changes

Some of the effects of changes in atmospheric pressure have been mentioned previously. It must be remembered that man is best adapted to air pressure at sea level. If he ventures far above or below this level, he must provide his own atmosphere to approximate that to which he is accustomed.

On the surface of the earth at sea level, the body is subjected to an air pressure of approximately 10,330 kilograms per square meter. Since a man of average size has approximately 1.85 square meters of body surface, the force exerted by the atmosphere reaches the total of 19,110.5 kilograms. Under ordinary circumstances, this is simply a mathematical curiosity because the external force is equalized by a corresponding one within the body cells. We are, therefore, not conscious of this tremendous weight. But for some people who may be subjected to rapidly decreasing pressure, these figures have a very real significance. If the pilot of an airplane is not provided with a pressurized cabin or pressure suit, he may experience considerable discomfort and danger if he ascends too rapidly, because the internal pressure cannot decrease as fast as the external pressure. Gases in the digestive tract and other body cavities expand (Boyle's Law) to a degree that results in acute pain and even disability.

As we ascend into the atmosphere, the weight of the column of air above us gradually decreases (Fig. 19-2). This not only has an effect on the internal pressures, but it has a drastic effect on our breathing. We frequently hear that a person at a high altitude has difficulty in breathing because there is not enough oxygen. To a certain extent, this is correct. It would be necessary, however, to ascend to an altitude of over about 19 kilometers before the oxygen content of the air dropped much more than two percent; this would not be more than the decrease normally found in a stuffy room occupied by several people. Unless proper precautions against pressure changes are taken during an ascent to the height of 19 kilometers a person will collapse regardless of the oxygen content of the atmosphere.

The primary reason for such a collapse is that a point is reached during the ascent when negative pressure cannot be established in the thoracic cavity. The atmospheric pressure decreases until it equals that of the thoracic cavity when the lungs are expanded; hence, no air can be inhaled. This leads to decreased oxygenation of the blood, known as *anoxia*. The decrease in pressure produces anoxia in another way as well. At any altitude,

the partial pressure (Dalton's Law) of oxygen is 20.94 percent of the barometric pressure. The result is that as we ascend above the surface of the earth, the molecules of oxygen exert less pressure on the walls of the alveoli and, therefore, do not enter the blood stream as effectively.

People who live at high altitudes have become adapted to their environment in two ways. In the first place, their thoracic cavities are larger than those of persons living at lower levels. Secondly, the number of red blood corpuscles is greater insuring adequate storage and transportation of oxygen around the body. The following table shows the effect of altitude on the number of red cells and indicates the increased efficiency of the circulatory system under unusual conditions.

Elevation in Meters	Millions of Erythrocytes per Milliliter
Sea level	4.93
1525	5.02
3270	5.82
4510	6.46
5335	7.37

The above data apply to residents at these various elevations. People who are visitors at high elevations may increase their red cell count by as much as 500,000 per milliliter of blood during a stay of less than two weeks. A return to sea level results in a reduction in the number of red corpuscles.

Anoxia is a condition that may occur in circumstances other than those associated with altitude. Drowning results in the failure of the blood to absorb a sufficient amount of gaseous oxygen. Electric shock and certain drugs depress the rate of breathing to a point where the body suffers from a lack of oxygen. Certain gases, such as carbon monoxide (CO), combine easily with hemoglobin and deprive it of its ability to carry an adequate amount of oxygen to the tissues. If the normal breathing mechanism fails, various manual and mechanical methods may be used to insure receiving the minimum amount of oxygen necessary to maintain life. Some of these methods will be discussed in the next chapter under the heading *artificial respiration.*

Rapid changes in pressure may have other dangerous effects besides anoxia. When the body is exposed to pressures that are higher than normal and is then decompressed rapidly, a condition may arise that is known as *caisson disease* or the bends.

Divers are the people most liable to suffer from rapid decompression. They work in rubber suits in which the internal pressure balances that of the surrounding water. For each 10.06 meters (33 feet) the diver descends below the surface of the sea—10.36 meters (34 feet) in fresh water—his body is subjected to an increased pressure of 1 atmosphere (the pressure of 760 millimeters of mercury). Therefore, as he descends, the pressure of air within his suit is increased to balance the external pressure. This will allow him freedom of movement as well as the ability to breathe. Under these conditions, the body is most affected by the nitrogen in the air; the oxygen is being constantly used up, the carbon dioxide

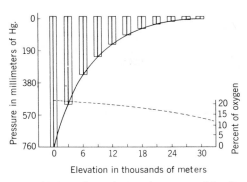

Fig. 19-2   The relation between altitude, air pressure, and the amount of oxygen in the atmosphere.

is being exhaled, but the nitrogen is neither used nor expelled. The blood of a person exposed to increased air pressures takes up the gases in direct proportion to their partial pressures (Henry's Law). The result is that the blood and tissues both become saturated with nitrogen. The dangers in this situation are, first, the nitrogen diffuses very slowly through tissues and does not reach an equilibrium rapidly. A second danger lies in the ease with which nitrogen dissolves in fats. In these materials, its rate of solution is approximately five times faster than in water. Since the nervous system contains large amounts of fats and fat-like compounds, the brain rapidly becomes affected and what is known as *nitrogen narcosis* develops. In this condition there may be a feeling of extreme well-being, accompanied by impaired mental activity and the inability to concentrate on alloted tasks.

If a diver with tissues saturated with nitrogen is brought to the surface too rapidly, the nitrogen forms bubbles in his tissues in much the same manner as bubbles of carbon dioxide appear in a bottle of a carbonated drink when the cap is removed. In both cases, the bubbles are the result of a sudden decrease in pressure on the liquids. X-ray photographs of victims of caisson disease show that the gas collects first in the spaces surrounding the joints. In severe cases (sometimes fatal) extensive bleeding occurs in the brain and spinal cord. The treatment for persons suffering from this disease is to put them into a chamber that can be pressurized to a level equal to that under which they had been working. Very slow decompression will then return the internal pressure to normal levels. Skin divers who descend to great depths under water must also allow time for decompression as they rise to the surface in order to prevent the bends.

Recently helium has been substituted for nitrogen in the gas that is given to divers. The reason for this is that helium is about one-half as soluble in body fluids as nitrogen is. Also, it diffuses through cell membranes about twice as rapidly. This means that at a given pressure, the amount of helium taken up by the body would be considerably less than the amount of nitrogen absorbed under the same conditions. Also, the rate at which the helium is liberated during decompression is considerably more rapid than is the case with a nitrogen-oxygen mixture.

It should be pointed out that an aviator in an unpressurized plane in ascending to a height of 10,060 meters (33,000 feet) would undergo the same amount of decompression that a diver who had been working at a depth of 30.5 meters (100 feet) would in coming to the surface. This effect of altitude is the result of decreasing the atmospheric pressure from one atmosphere to about one-quarter of an atmosphere.

VOCABULARY REVIEW

Match the phrase in the left column with the correct word in the right column. *Do not write in this book.*

1 Refers to the maximum total amount of air contained by the lungs.
2 The volume of air remaining in the lungs following maximum respiratory effort.
3 Results from nitrogen taken up by the blood when the body is under high atmospheric pressure.

a anoxia
b Boyle's Law
c caisson disease
d Charles' Law
e Dalton's Law
f expiratory reserve volume
g Henry's Law

4 A device for measuring lung capacity.
5 The volume of air entering and leaving the body with each breath.
6 Stimulation of this by a change in pH of the blood results in a change in the rate of breathing.
7 Stimuli passing along these afferent nerves control the rate of breathing.
8 The intercostal muscles and diaphragm are stimulated by impulses that enter them through these nerves.
9 The gas law concerned with the effect of temperature on the volume of a gas.
10 The gas law stating the relation between the partial pressure of a gas and its solubility.

h inspiratory reserve volume
i phrenic nerves
j residual volume
k respiratory center
l spirometer
m tidal volume
n vagus nerves
o vital capacity

TEST YOUR KNOWLEDGE

Group A

Select the correct statement. *Do not write in this book.*

1 The gas in the atmosphere that is most closely concerned with respiration is (a) nitrogen (b) water vapor (c) oxygen (d) hydrogen.
2 The brain center for the control of breathing is in the (a) cerebrum (b) cerebellum (c) medulla (d) pons.
3 The gas in the blood that exerts chemical control over the rate of breathing is (a) oxygen (b) nitrogen (c) carbon dioxide (d) hydrogen.
4 The amount of carbon dioxide in inhaled air is 0.04% and the amount in exhaled air is about (a) 0.04% (b) 0.004% (c) 0.4% (d) 4.4%.
5 Boyle's Law pertains to the (a) number of molecules in equal volumes of gases (b) effect of temperature on gas volume (c) effect of pressure on gas volume (d) effect of pressure on the solubility of gases.
6 Gases are more easily absorbed into the blood if they have (a) high molecular weight (b) low molecular weight (c) high density (d) low pressure.
7 The volume of air taken in with maximal inhalation is called the (a) inspiratory reserve volume (b) tidal volume (c) residual volume (d) expiratory reserve volume.

8   Decreased oxygenation of the blood is known as (a) anoxia (b) the bends (c) nitrogen narcosis (d) decompression.

9   The amount of oxygen in inhaled air is about 20.9%, and in exhaled air is about (a) 4.4% (b) 79% (c) 21% (d) 16%.

10   The gas law concerning partial pressures is attributed to (a) Avogadro (b) Charles (c) Dalton (d) Henry.

Group B

1   What is the role of each of the following in the process of respiration (a) nitrogen (b) oxygen (c) carbon dioxide (d) water vapor?

2   Which of the gas laws determines the entrance of gases into the blood under the following conditions? Explain. (a) Climbing to an altitude of over 12,000 feet. (b) The need to warm air before its gases can be absorbed by the blood. (c) The absorption of carbon dioxide by the plasma (fluid) of the blood.

3   Explain the meaning of each of the following (a) tidal air (b) inspiratory reserve volume (c) expiratory reserve volume (d) residual air (e) minimal air.

4   What correlation exists between the vital capacity of the lungs and the total surface area of the body?

5   How do changes in the pH of the blood affect the rate of breathing?

Chapter 20 Gas Exchange. Phonation

Transport of Gases

In the preceding chapter we spoke of the close relation between the respiratory and circulatory systems. We noted that the blood was responsible for the transportation of oxygen and carbon dioxide around the body. However, internal respiration is not a process involving simply the physical solution of gases in the blood and other body fluids. It includes, in addition, a complex series of chemical reactions by which the cells receive an adequate supply of oxygen for all the reactions involved in cell metabolism, and dispose of waste carbon dioxide.

The underlying physical process by which both oxygen and carbon dioxide enter or leave the blood is diffusion. These gases must pass through several layers of tissues or fluids as indicated in Fig. 20-1. These layers include the capillary walls and epithelium of the lungs, the plasma of the blood, the tissue fluid surrounding the cells, and the membrane of the red corpuscles. During the passage of the gases into or out of the blood, they are at some times dissolved in the fluids, but this is not their principal method of transport around the body.

The red blood corpuscle is a minute biconcave disk. This shape is of distinct advantage in its functions, as will be pointed out in a later chapter. A red corpuscle is surrounded by a membrane composed of layers of proteins and fats. The membrane encloses a small quantity of *hemoglobin,* a reddish colored protein. One symbol for hemoglobin is HHb, indicating that it is slightly acid in nature. Hemoglobin combines readily with oxygen to form a compound known as *oxyhemoglobin* ($HHbO_2$), and this oxygen can be freed from the HHb with ease. Hemoglobin also forms *carbamino hemoglobin* ($HHbCO_2$) in contact with carbon dioxide. The name of this compound indicates that the CO_2 is attached to one of the amine groups in the hemoglobin molecule.

The method by which the gases are transported around the body by the blood can be summarized in the following manner. (The italic *numerals* in the description refer to the numbers of the chemical equations shown in Fig. 20-1 on page 278).

277

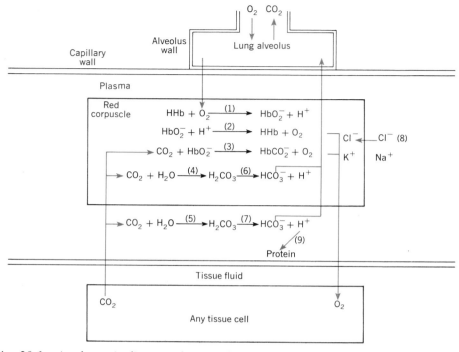

Fig. 20-1   *A schematic diagram showing the chemical reactions taking place in the exchange of gases between the blood and the tissues. Numbers in parentheses are keyed to the italic numerals in the text on pages 278–279.*

Oxygen. Oxygen diffuses into the plasma of the blood as it passes through the walls of the alveoli of the lungs. It then enters a red blood corpuscle and combines with the HHb to form HbO_2^-, freeing an ion of hydrogen (H^+) *(1)*. The hydrogen ions that are produced in this manner would have a tendency to make the blood acid were it not for one of two factors operating in the blood. In the first place, some of these free hydrogen ions may recombine with HbO_2^- and thereby liberate more O_2 *(2)*. Others may be absorbed by some of the other proteins of the

blood in an action known as *buffering* *(9)*. Thus the acidity of the blood is kept slightly alkaline at a pH of 7.40.

Carbon dioxide. Approximately 10 percent of the carbon dioxide produced by cellular oxidations actually dissolves in the plasma of the blood; most of it is carried as chemical compounds *(3)*. Approximately 20 percent is carbamino hemoglobin ($HHbCO_2$), about 25 percent is carbonic acid (H_2CO_3) in the corpuscle itself, and 45 percent H_2CO_3 in the plasma.

The formation of carbonic acid in the blood can occur either quite slowly in the plasma *(5)* or rapidly in the corpuscle *(4)*. In the slower process the gas combines with water in much the same way that carbonic acid is formed in a carbonated drink, but this requires an appreciable period of time. In the hemoglobin, however, the process takes only about 0.7 second as the result of the presence of an enzyme known as *carbonic anhydrase.*

Several of the reactions involving oxygen and carbon dioxide result in the liberation of hydrogen ions *(6)* and *(7)*. These are positively charged particles to which the wall of the corpuscle is not very permeable, because the cell membrane is itself positively charged over most of its surface and will repel positively charged ions. On the other hand, negatively charged ions such as (HbO_2^-) and carbonate (CO_3^{--}), will pass through the membrane easily.

As a result of this ionization, oxyhemoglobin and the carbonate compounds are in a highly unstable state and give up O_2 or CO_2 easily when they pass through a tissue that has a low concentration (tension) of these substances. Thus, if the oxygen tension is low in a muscle tissue, oxygen will leave the blood; carbon dioxide leaves the blood in the alveoli because of the low CO_2 tension there. The reverse processes occur whenever the blood passes through a tissue or region where either of these gases is in a higher concentration than it is in the blood. The $HHbO_2$ can then have its oxygen replaced by carbon dioxide, if the tissue has an especially high concentration of this substance *(3)*.

Another result of the process of ionization that occurs within the blood is the possibility of an imbalance between the positive and negative ions. Since the positive ions do not diffuse through the cell membrane easily, any retention of an excess of these ions on one side of the membrane will affect its electrical properties. To prevent this, there is a diffusion of chloride ions (Cl^-) from the plasma into the cell. These negative ions balance the positive ions in a process known as the *chloride shift (8).*

As a result of these chemical reactions between the blood gases and the blood, there is much more carbon dioxide and oxygen carried than would be true if they were simply held in a saline solution or water. Fig. 20-2 illustrates this difference. At the bottom of the figure there are two dashed lines (Free CO_2 and Free O_2). These represent the amount of the two gases that can be carried by the fluid part of the blood when no chemical reactions are involved. The concentrations of these free gases in the blood then follow Henry's Law. The two upper curves represent the actual carrying capacity of the blood, the difference between the two sets being the result of chemical reactions.

Respiratory Failure

Stoppage of breathing most frequently is considered in terms of drowning, but there are other physical and chemical causes for failure in the normal breathing rhythm. Electrocution, the use of certain drugs which depress the nervous impulses passing to the respiratory muscles, poisonous gases such as carbon monoxide, or mechanical obstruction of breathing by

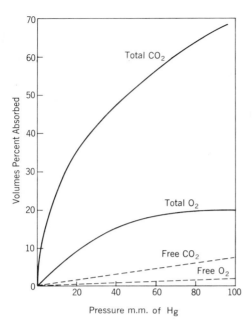

Fig. 20-2   The carrying capacity of the blood. Dashed lines show uncombined gas carried in solution. Continuous lines show total (combined and uncombined) gas carried by the blood.

the clogging of the respiratory passages —all call for artificial respiration.

One of the most common causes for the failure of respiratory movements is carbon monoxide poisoning. This gas is produced whenever a carbon-containing substance is incompletely oxidized. Under this condition, each atom of carbon combines with only one atom of oxygen. Carbon monoxide has the ability to combine with the hemoglobin of the blood 210 times more easily than oxygen. This means that the hemoglobin molecules that have combined with carbon monoxide are unable to accept oxygen and the cells of the body are therefore deprived of this substance. The carbon monoxide-hemoglobin compound (COHb) is also quite stable chemically so that the carbon monoxide is not readily given off in the lungs.

Carbon monoxide is odorless, tasteless, and slightly heavier than air. The exhaust from an efficiently operating automobile engine contains about 6 percent of this gas, but while the engine is warming up, the amount is much higher. Some fuel gases contain between 30 and 40 percent carbon monoxide. To avoid accidental gas poisoning by the odorless carbon monoxide, producers of such fuel gases frequently add to them an odorous substance so that consumers can more easily detect gas leaks. There is no appreciable percentage of carbon monoxide in natural gas, so it is much safer to use for heating and cooking purposes in the home, and in industry.

The early symptoms of poisoning from carbon monoxide are nausea, disturbed vision, and headaches. Later, behavior may become irrational. These symptoms may not be readily recognized since they are the same as those often associated with the stuffiness in poorly ventilated rooms. The mental disturbance moreover, may appear some time after a person has been removed from the poisonous atmosphere. A classic example of this mental reaction is the case of the man who very thoughtfully thanked his rescuers for removing him from the burning building and then promptly knocked one of them unconscious.

The length of time a person is exposed to carbon monoxide is as important as the amount of gas present in the atmosphere. An exposure for two hours to air with a concentration of 0.075 percent carbon monoxide would be as dangerous as an exposure of one hour to 0.15 percent or 0.3 percent for half an hour. A person at rest can withstand a 50 percent saturation of the blood by the gas before collapsing, but if he is exercising, the limit is about 30 percent. This difference is the result of the increased consumption of oxygen during exercise and the more rapid rate of absorption of carbon monoxide. A person may sit quietly in a room containing a small percentage of carbon monoxide, and then collapse suddenly when he gets up to try to open a window.

Artificial Respiration

In Chapter 18 we discussed the relationship of air pressure to breathing. We pointed out that each breath we take is the result of a change within the thoracic cavity which either reduces the air pressure below that of the surrounding atmosphere or raises it. When the pressure is reduced, air is drawn into the lungs. It is expelled when the internal air pressure rises. The motions which bring about these changes result, normally, from the contraction of muscles, but they can also be duplicated mechanically when the muscles fail to act efficiently.

In all instances of stoppage of breathing, it must be remembered that complete lack of oxygenation of the blood for more than four or five minutes may result in permanent injury or death. If artificial respiration is required, it should be started immediately and continued for as long as necessary. Oc-

casionally, victims have been saved from death by continued artificial respiration over a thirty-hour period. Thus, it is evident that once started, the process should be continued as long as there is the faintest hope for revival. In a few cases victims have been revived even after they had been pronounced dead by qualified physicians.

The best method for artificial respiration is based on the introduction of air, or a mixture of gases, directly into the lungs. This is accomplished by mechanical devices or by breathing directly into the victim's mouth.

One mechanical device that has been in use for some years is the *inspirator*. This consists of a mask that fits over the mouth and nose of the victim. Through the mask, compressed air is pumped into the lungs and then withdrawn. While the theory of this method is good, its successful operation depends on the skill of the operator. In the hands of a poorly trained person, it is highly dangerous because of the possibility of exerting too much pressure— either positive or negative—on the thin-walled air sacs.

Modern *resuscitators* are made to work on the same principle as the inspirator except that they use oxygen and are equipped with a complex series of valves that automatically regulate the pressure of the oxygen entering or leaving the lungs. These machines develop a positive pressure of about eight millimeters of mercury to force air into the lungs, and then withdraw it at about the same pressure. The resuscitator is equipped with one or more tanks of pure oxygen under a pressure of about one hundred atmospheres. It is able to apply resuscitation for periods of twenty minutes to one hour depending on the number and size of tanks available. See Fig. 20-3.

It is possible, and at times desirable, to equip a resuscitator with a tank of carbon dioxide so that this gas may be administered with the oxygen. A small amount of CO_2 (5 percent) is considered suitable because of the stimulating effect it may have on the respiratory center in the medulla oblongata. Carbon dioxide mixed with oxygen is especially valuable in reviving victims of carbon monoxide poisoning because

CO_2 competes with CO in combining with the hemoglobin.

Currently, the American Red Cross advocates *mouth-to-mouth* resuscitation as the best method of artificial respiration. As Fig. 20-4 shows, this method consists in breathing into the mouth of the victim after checking to insure that his respiratory tract is free of obstructions. This method is a safe one from the standpoint of both the victim and the operator because at no time can the pressure within the lungs of either reach a dangerous level. It

Fig. 20-3   *A demonstration of a resuscitator in use. Resuscitators are valuable in cases of carbon monoxide poisoning, drowning, electric shock, and other conditions of suffocation. (Stephenson Corporation)*

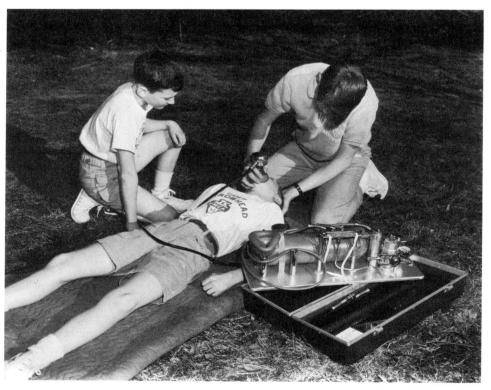

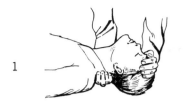

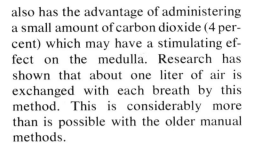

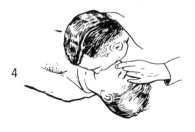

also has the advantage of administering a small amount of carbon dioxide (4 percent) which may have a stimulating effect on the medulla. Research has shown that about one liter of air is exchanged with each breath by this method. This is considerably more than is possible with the older manual methods.

Fig. 20-4 Before starting the mouth-to-mouth method of artificial respiration, look to see if there is any foreign matter in the victim's mouth. If so, wipe it out with your fingers, then proceed according to the following steps:

First, tilt the head back so the chin is pointing upward (1). Pull or push the jaw into a jutting-out position (2 and 3). These maneuvers should relieve obstruction of the airway by moving the base of the tongue away from the back of the throat.

Second, open your mouth wide and place it tightly over the victim's mouth. At the same time pinch the victim's nostrils shut (4) or close the nostrils with your cheek (5). Or close the victim's mouth and place your mouth over his nose. Blow into the victim's mouth or nose.

Third, remove your mouth, turn your head to the side, and listen for the return rush of air that indicates air exchange. Repeat the blowing effort. For an adult, blow vigorously at the rate of 12 breaths per minute. For a child, take relatively shallow breaths appropriate for the child's size at the rate of about 20 per minute.

Fourth, if you are not getting air exchange, recheck the head and jaw position. If you still do not get air exchange, quickly turn the victim on his side and administer several sharp blows between the shoulder blades to dislodge any foreign matter in the airway. (American National Red Cross)

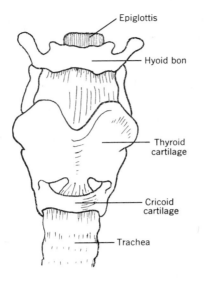

Fig. 20-5 Front view of the larynx show-ing the positions of the various cartilages.

as the muscles that are attached to it bring it upward ·to aid in preventing food from entering the trachea. These two cartilages form the so-called Adam's apple which is so much more prominent in men than in women. In males, the larger size of the larynx results in a deeper pitch of voice. As boys mature, the larynx grows larger, producing a lowering of the pitch of the voice. In girls, the enlargement of the larynx does not occur, and their voices remain high pitched.

The cricoid cartilage is shaped some-what like a signet ring with the thick-ened area pointing toward the back of the trachea. It lies below the thyroid cartilage, and is connected with it and the upper part of the trachea by means of membranes.

PHONATION

Organs of Phonation

Man and other mammals can produce sounds because they have a special organ of *phonation* (sound production), the larynx. Among all of the mammals, the giraffe alone lacks a larynx. We might consider that the larynx of this animal has been sacrificed in favor of a tremendously long neck.

As we have said previously, the walls of the larynx are composed of cartilages derived from the arches that supported the pharyngeal pouches of the early em-bryo. The largest of these cartilages are the *thyroid* and *cricoid cartilages*. The thyroid cartilage is formed by two sepa-rate plates to form a prominent struc-ture somewhat like the prow of a ship (Fig. 20-5). Its movement can be seen and felt during the act of swallowing,

Fig. 20-6 Side view of the larynx with the thyroid cartilage shown as though it were transparent.

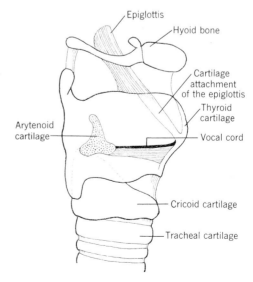

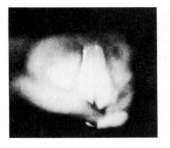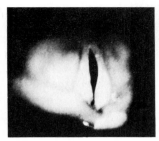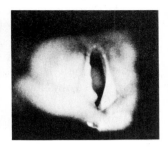
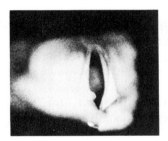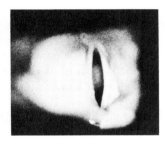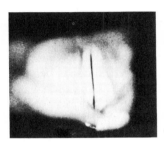

Fig. 20-7 The action of the vocal cords. Successive photographs show one cycle of cord motion beginning top left and ending bottom right. At this pitch, 120 cycles occur each second, and the vocal cords are very relaxed. (From the film High-Speed Motion Pictures of the Human Vocal Cords, *Bell Telephone Laboratories)*

There are seven other cartilages connected with the larynx. Two of these are the *arytenoid cartilages* that are attached to the *vocal cords.* Another is the attachment of the epiglottis to the front wall of the thyroid. These cartilages are shown in Fig. 20-6.

Attached to the inner walls of the larynx, at one end by the arytenoid cartilages and at the other by membranes, are the vocal cords (Fig. 20-7). The space between them is the opening of the glottis. When these cords are in a neutral position, as they are during quiet breathing, the glottis is somewhat triangular in shape with its apex pointing forward. The shape of the glottis can be changed, however, and these changes result in the different qualities of the voice.

Characteristics of Human Sounds

Three characteristics of sound (Chapter 12) are important in speech. *Loudness* (volume, or amplitude), is the result of changing the force with which air is expelled from the lungs when the vocal cords are held in the proper position to produce sound. You need only try to speak on an inhalation to realize that the sounds are quite indistinct; clear speech is possible only on an outgoing column of air. If, then, the exhalation is very forceful, the sounds produced will be much louder than when the vocal cords are not subjected to strong and violent passage of air.

A second property of sound is its *pitch*. The pitch of the voice changes as the vocal cords increase or decrease in length because of the tension exerted on them by the arytenoid cartilages. The vocal cords behave very much like a violin string, which emits a low pitch when it is relatively relaxed, but a high pitch when drawn taut. In males, the vocal cords behave very much like the G string on a violin. Even though they are drawn as tight as possible, their upper range of pitch cannot equal the high notes produced by women, which might be likened to the tone of the E string of a violin. In either the violin or the vocal cords, the smaller and shorter the vibrating string, the higher the pitch of the note it will produce.

A third characteristic of the voice is its *timbre*, or *quality*. This is a complex aspect of speech in which harmonics (overtones) are formed as a result of vibrations occurring within the cavities of the nose, throat, and thorax. A violin and a cello differ in tone because of their variation in size and in the resonating qualities of the woods used in making them. Just so, the various cavities of the body give a quality to the human voice that is quite separate from that produced by the vocal cords. If the larynx, chest cavities, and sinuses are large, the pleasing quality of the voice is increased. In trained singers, the harmonics that develop in these cavities give particular qualities to their voices that distinguish them from untrained singers. It is only by rigorous training that the full value of the resonating chambers of the body can be attained.

The sounds of the human voice are the result of vibrations of the vocal cords controlled so as to produce differences in pitch and volume. Variation in pitch is limited by the size of the vocal cords. Those of a man are between 2 and 2.54 centimeters long, while those of a woman average about 1.78 centimeters in length. Since this fundamental difference exists, the pitch of the male voice is somewhat lower than that of a woman.

It is possible, however, for both sexes to alter the length of the cords by voluntarily controlling the movement of the arytenoid cartilages. This conscious contraction and relaxation of the vocal cords results in the range of a person's voice. In a few individuals this may extend over three or four octaves on the musical scale. Only with special training can these limits be exceeded. Based on the rate of vibration of the cords, the ranges of human voices may be classified as shown.

	Vibrations per Second
Bass	80–250
Baritone	100–350
Tenor	125–425
Contralto	150–500
Mezzo-soprano	200–650
Soprano	250–750, or higher

The individual speech sounds are the result of the action of the tongue, teeth, and lips, and the partial closing of the glottis. In making the sounds of the various vowels, the tongue and the lips play the major roles. The back of the tongue may be elevated in forming some sounds, while the front may be active in making others. Thus, in making the sound of ee (as in cheese), the front part of the tongue is slightly elevated and the lips drawn back. To sound u (as in rule), the lips are pursed and the back of the tongue raised.

In the formation of consonants, the lips, teeth, and tongue are important. As you say the letters p or b, you notice that the sound is made by the sudden opening of both lips. The difference between the p and the b depends on the

Fig. 20-8   Voiceprints are "pictures" of one word of a person's speech. They reveal patterns of voice energy at various levels of pitch—patterns which are distinctive and, like fingerprints, can be used for identification. In the six voiceprints below: time is plotted from left to right; low pitch is at the bottom with higher pitches toward the top; intensity is represented by the contours. These six voiceprints were made of five different speakers' utterances of the word "you." One speaker said it twice. Which two voiceprints are from the same speaker? (Bell Telephone Laboratories)

position of the vocal cords. The *b* is *voiced;* that is, it has tone produced by vibration of the vocal cords. The *p* is *voiceless.*

Other consonants in which the lips and teeth are used are *f* and *n.* Various combinations of sounds result from different positions of the organs in the mouth region, or from the regulation of the rate at which air is expelled. In some instances a nasal tone is produced by the passage of the column of air through the nose. Thus a word like *sing* has a nasal quality that can be easily detected if the nose is held while the word is being pronounced.

VOCABULARY REVIEW

Match the phrase in the left column with the correct word in the right column. *Do not write in this book.*

1 A compound formed when hemoglobin unites with oxygen.
2 A chemical compound that unites with hemoglobin more easily than oxygen does.
3 A chemical substance that helps to maintain the pH of the blood.
4 A mechanical device used to restore breathing after it has failed.
5 A term used to refer to the loudness of a sound.
6 The quality of a voice is due to this characteristic.
7 The tautness of the vocal cords is due to the action of these.
8 An enzyme concerned with the absorption of carbon dioxide.
9 Membranes that respond to currents of air and produce sounds.
10 Process maintaining the electrical balance of the red blood cell membrane.

a artificial respiration
b arytenoid cartilages
c buffer
d carbamino hemoglobin
e carbonic anhydride
f carbon monoxide
g chloride shift
h oxyhemoglobin
i phonation
j pitch of sound
k resuscitator
l timbre of sound
m vocal cords
n volume of sound

TEST YOUR KNOWLEDGE

Group A

Select the correct statement. *Do not write in this book.*

1 The process by which cells receive oxygen and dispose of carbon dioxide is called (a) internal respiration (b) external respiration (c) inhalation (d) exhalation.

2 The physical process by which gases enter and leave the blood is (a) dispersion (b) absorption (c) diffusion (d) osmosis.

3 Hemoglobin in contact with carbon dioxide forms (a) oxyhemoglobin (b) carbamino hemoglobin (c) carbonic anhydrase (d) carbonic acid.

4 The pH of the blood is normally about (a) 3 (b) 6.6 (c) 7.4 (d) 10.

5 The acidity of the blood is held within a narrow range by a chemical reaction called (a) chloride shift (b) oxidation (c) buffering (d) neutralization.

6 Human sounds are produced by vibration of the (a) epiglottis (b) vocal cords (c) arytenoid cartilages (d) cricoid cartilage.

7 The so-called Adam's apple is properly known as the (a) pharynx (b) trachea (c) larynx (d) bronchus.

8 The characteristic of the human voice that depends on the rate of vibrations is (a) volume (b) amplitude (c) pitch (d) timbre.

Group B

1 Briefly describe the structure of the larynx and the relation of the vocal cords to it.

2 Distinguish between external and internal respiration.

3 Briefly outline the chemical changes that occur in the blood when it is in contact with oxygen, and carbon dioxide.

4 Outline the procedure for carrying out resuscitation by the mouth-to-mouth method.

Chapter 21 Internal Transport

Body Fluids

Much of the human body is composed of water in which are dissolved or suspended those essential materials which supply the cells with nourishment, or are the products of the cells. This fluid, as shown in Fig. 21-1, may be divided into the fluid that is within the cells, the *intracellular fluid,* and that which is outside of the cells, the *extracellular fluid.* If one excludes the weight that is represented by fat in the body, and this is essentially a water-free substance, the amount of intracellular and extracellular fluid accounts for about 72 percent of the total weight of an average adult. Of this, approximately one-third of the body water is extracellular in the adult. In infants and young children, about one-half lies outside of the cells.

The extracellular fluid can be further designated by its location. If it lies within blood vessels, it is called *intravascular fluid,* but if it surrounds the cells, it is called *intercellular fluid.* It is this fluid that lies in the minute spaces between cells that serves as a "middle man" in the diffusion of materials between the blood and the cells.

It is the intravascular fluid that is responsible for the transport of materials around the body. Human beings, in common with all vertebrates, have a circulatory system that is composed of blood vessels that form a continuous pathway from which the whole blood never escapes. This is called a *closed* system to distinguish it from the type of circulatory system found in certain invertebrates where the blood is able to leave the vessels and wander among the tissues.

Functions of Blood

The blood plays a major role in maintaining optimum conditions in the internal environment of the body for the efficient function of all tissues. It regulates the distribution of heat, food, oxygen, the removal of the waste products of metabolism, the water content and chemical composition of the body fluids, and helps keep all these in a state of dynamic equilibrium. This self-regulation and maintenance of a relatively stable internal environment is called *homeostasis (homeo,* similar).

The blood is a circulating tissue. The blood cells are the solid parts of blood, while the fluid in which they float, the *plasma,* serves to convey the cells around the body. Both the cells and the fluid play very distinct and individual roles in the distribution of materials. The solid parts of blood are primarily responsible for the transport of gases, protection against invasion by foreign bodies, and prevention of excessive loss of blood when blood vessels are broken. The fluid part carries dissolved materials which nourish the cells, or removes the waste products of cell metabolism.

The functions of the blood are closely tied to the functions of all the organ systems. The blood *transports* all materials that are required by the cells or discarded by them. As we have seen (Chapter 19) the gases concerned in respiration are carried by the blood. Oxygen is carried from the lungs to the cells in chemical combination with hemoglobin. Some carbon dioxide is also transported in this manner, although most of this gas is carried dissolved in the plasma.

Foods that have been rendered soluble by digestion are absorbed into the plasma and are thus transported around the body. In a similar manner, soluble wastes are removed from the cells and carried by the plasma to those organs where they can either be eliminated or changed into compounds that are useful for other purposes. The glands that manufacture chemical regulators of bodily functions, hormones, empty their secretions directly into the blood for transportation around the body in the plasma. (Chapters 30, 31)

Maintenance of water content of the tissues is one of the principal functions of the circulatory system. If the water content of a particular region of the

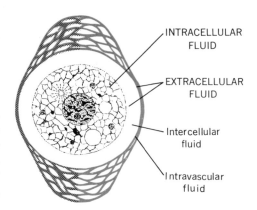

Fig. 21-1   *The relations of the various body fluids to each other shown diagrammatically. (Adapted from* Fluid and Electrolytes, *the Abbott Laboratories, 1957)*

body is lowered as a result of increased chemical activity, the blood makes up this deficit. It then draws on reserve supplies of water to maintain its own fluid nature.

The blood, as a whole, helps to *maintain body temperature.* If some tissue, such as muscle, is very active, the increased rate of oxidation occurring in it raises its temperature. Continued exposure to high temperature has a serious effect on enzymatic reactions in the cells. It is therefore essential to bring the temperature of the tissue to a normal point as rapidly as possible. Thus any increase in the activity of a tissue is also accompanied by an increase in the circulatory rate so that the blood passing through it is heated. This excess heat is then given off over the surface of the body, through the lungs, or by excretions.

Protection against disease is another important function of the blood. This is accomplished by the white cells, or *leukocytes,* in one of two ways. First, some leukocytes are able to engulf bacteria and other foreign bodies that have succeeded in penetrating the surface covering. Second, still other types of white corpuscles have the ability to manufacture chemical substances to destroy bacteria or to neutralize the toxins (poisons) produced by invading organisms.

The *acid-base equilibrium* of the blood is maintained by the buffering action of bicarbonates and phosphates which neutralize the usually small amounts of acids or alkalis that the blood absorbs.

Composition of Blood

A person weighing 68 kilograms (150 pounds) has approximately five liters of blood. The exact amount of blood in the body cannot be determined with accuracy, since the blood is normally contained within vessels that have a total length of almost 161,000 kilometers (100,000 miles). Many of these vessels are of microscopic diameter, so it is obvious that there will always be some blood adhering to their walls. Measurements of the quantity of blood in the circulatory system must be carried out by indirect methods. In health, the quantity of blood in any individual varies only slightly from time to time, the total volume making up about nine percent of the total body weight.

Blood is approximately 55 percent fluid and 45 percent solid matter. Although both the parts of the blood will be discussed in detail in later chapters, a brief statement of the function of each is helpful here.

Plasma

The plasma contains the following materials in solution:

Water makes up about 92 percent of the total volume of plasma.

Proteins. Approximately 60 different types of proteins have been identified in the plasma. Collectively, these are known as *plasma proteins.* Some of them are highly specific in their action, such as fibrinogen which aids in the clotting of blood, or those which protect against specific diseases. The very large molecules of these, regardless of what other roles they may play, are of extreme importance in that they cannot leave even the thin-walled vessels, the capillaries. As a result, they exert an osmotic force across the walls of the capillaries and thus play an important part in regulating the distribution of water. Some of these plasma proteins will be discussed in other chapters.

Inorganic salts. These relatively simple compounds are obtained from foods. Some of their ions act as buffers in helping to maintain the acid-base equilibrium of the blood.

Nutritive materials are absorbed from the digestive system. The monosaccharides, the neutral fats, and the amino acids are all dissolved in the plasma.

Hormones, vitamins, and *enzymes* are present in minute quantities. Some of these enter the blood in specific organs, while others originate in various tissues scattered in many places throughout the body.

Waste products. Each cell of the body may be considered a complex chemical factory that manufactures substances necessary for its own existence. The blood carries the waste products of cell activity.

Solid Parts of Blood

Floating in the plasma are the following solid cellular structures:

Red blood corpuscles (erythrocytes). When fully developed, these can hardly be considered true cells because they lack nuclei and therefore are unable to reproduce. Their function is to transport gases around the body. Each contains a minute amount of hemoglobin which combines chemically with oxygen and carbon dioxide, and which gives the blood its characteristic red color.

White corpuscles (leukocytes). These corpuscles do not contain hemoglobin and are therefore almost colorless. They engulf and destroy invading bacteria and other foreign particles. Some, also, are responsible for the destruction of worn-out red corpuscles.

Blood platelets (thrombocytes). These are the smallest of the blood solids. Although we know little about their structure and origin, we know they play an important role in the formation of a blood colt.

The average composition of the blood is summarized in the table (right).

Circulation

Among animals, the general pattern of the circulatory system falls into one of two categories. It is either an open system or a closed system. The *open circulatory system* is present only in those animals that lack a backbone (invertebrates). The *closed system* is found in some invertebrates, but it is the only type present in vertebrates.

Blood Plasma
Males: 54% by volume
Females: 58% by volume
pH: 7.3 to 7.4 at 38°C.

Red Blood Corpuscles
Males: 46% by volume, 5,400,000 per cu. mm.
Hemoglobin: 15.6 gm. per 100 ml. of blood
Oxygen capacity: 20.9 ml. per 100 ml. of blood.
Females: 42% by volume, 4,900,000 per cu. mm.
Hemoglobin: 14.2 gm. per 100 ml. of blood
Oxygen capacity: 19.0 ml. per 100 ml. of blood

White Blood Corpuscles
6,000 per cu. mm. of blood

Platelets
250,000 per cu. mm. of blood

In the open circulatory system the blood is not always confined within blood vessels. After having been forced out from the heart through vessels, the blood leaves them and passes among the tissues, returning to the heart as a result of the contraction of muscles of the body. In the closed type of circulation, blood never normally leaves vessels to flow among the tissues. In humans, as in other vertebrates, blood is forced *outward* from the heart through *arteries* which become increasingly smaller as they branch out among the tissues. These finer branches which retain the principal characteristics of the arteries are known as *arterioles*. Eventually, these branch further to become tubes whose walls are composed of only a single thickness of endothelial cells, called *capillaries*. These extremely small vessels pass between cells, or layers of cells, bringing blood into intimate contact with the cells and allowing materials to be easily exchanged. The capillaries then join together to form *venules,* larger tubes with thicker walls. These become increasingly larger as they are joined by others. Finally, *veins* are formed, which return the blood to the heart.

The *heart* is a muscular pump that is responsible for forcing the blood around the body. It is four-chambered in man, and divided into right and left halves. Each half is composed of an *atrium* (auricle) and a *ventricle.* The former is the receiving chamber for the blood as it enters the heart. The latter is a highly muscular chamber from which blood is forced out through the arteries. Blood in the right side is *deoxygenated* and is pumped only to the lungs. The left side contains *oxygenated* blood that is pumped throughout the body.

The blood must pass through the heart twice in making a complete circulation of the body. For example, if we follow a drop of blood from the time it leaves the wrist in a vein until it returns to the same spot, we will find that it has passed into the right side of the heart and from there to the lungs.

In the lungs it gives off carbon dioxide, takes on a supply of oxygen, then returns to the left atrium, and thence to the left ventricle. From here it is forced out through the aorta to arteries that eventually bring it to the capillaries of the hand. Blood from the capillaries then flows into veins and returns to the point from which it started. As Fig. 21-2 shows in a diagrammatic fashion, the blood goes through a long and twisted path in its passage around the body.

A diagrammatic representation of some of the vessels of the lymphatic system is also shown in Fig. 21-2. The vessels of this system will be described in Chapter 25. They drain waste materials from the tissues and return them, with the fluid in which they are dissolved, to the blood stream by way of right and left subclavian veins. As we will explain later, there is a constant passage of fluid from the blood to the tissues. Unless this fluid is returned to the blood, the body suffers seriously from the loss of an important transporting medium.

The table below summarizes the principal materials that are exchanged between the blood and the tissues.

Material	Enters the Blood at the	Leaves the Blood at the
Oxygen	Lungs	Cells
Carbon dioxide	Cells	Lungs
Food	Digestive organs	Cells, for use or storage
Inorganic salts	Digestive organs	Cells or as excretory products
Hormones	Endocrine glands	Cells
Water	Digestive organs	Cells
	Cells	Skin, lungs, kidneys
Organic wastes	Cells	Excretory organs
Erythrocytes	Bone marrow	———
Leukocytes	Bone marrow, lymph tissue	———
Platelets	Bone marrow	———

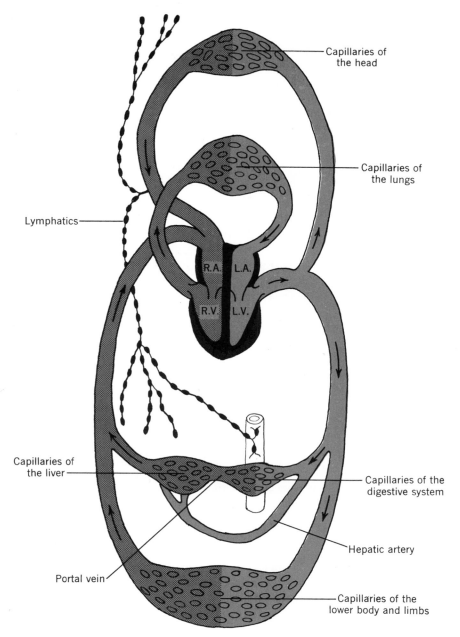

Fig. 21-2 Diagram of the circulatory system showing the path of the blood through the principal regions of the body. Note that the blood passes through the heart twice during each complete circuit of the body. The vessels in the dark red portions of the diagram carry deoxygenated blood: the bright red vessels carry oxygenated blood.

VOCABULARY REVIEW

Match the phrase in the left column with the correct word in the right column. *Do not write in this book.*

1	A fluid which is found within the cells.	**a** arteriole
2	The type of vessel which returns blood to the heart.	**b** artery
3	Solid parts of the blood which aid in the formation of a clot.	**c** atrium
		d erythrocyte
4	A part of the heart which receives blood.	**e** extracellular
5	Fine branches of arteries that have the characteristics of arteries.	**f** intercellular
		g interstitial
6	Blood solids that can combat bacterial invasion.	**h** intracellular
7	Another name for intercellular fluid.	**i** intravascular
8	A type of fluid that does not normally leave vessels.	**j** leukocytes
9	Transports the majority of the oxygen and carbon dioxide around the body.	**k** plasma
		l thrombocytes
10	A type of vessel that carries blood away from the heart.	**m** vein

TEST YOUR KNOWLEDGE

Group A

Select the correct statement. *Do not write in this book.*

1 In making a complete circuit through the body, the blood passes through the heart (a) once (b) twice (c) three times (d) four times.
2 Materials in blood that are absorbed from the digestive organs are (a) gases (b) antitoxins (c) vitamins (d) antibodies.
3 Blood returning from the arms and legs enters the heart through the (a) left atrium (b) right atrium (c) left ventricle (d) right ventricle.
4 The quantity of blood in a 68-kilogram person is about (a) one liter (b) three liters (c) five liters (d) ten liters.
5 The most abundant material in blood is (a) protein (b) red corpuscles (c) white corpuscles (d) water.
6 The fluid within the cells in said to be (a) extracellular (b) intracellular (c) intravascular (d) intercellular.
7 Of the following, select one that is *not* a solid part of the blood: (a) erythrocytes (b) leukocytes (c) thrombocytes (d) globulins.

8 Of the following, select one that is *not* a blood vessel: (a) arteriole (b) bronchiole (c) venule (d) aorta.

9 A plasma protein that aids in the clotting of blood is (a) gamma globulin (b) fibrinogen (c) serum (d) vitamin K.

Group B

1 List the general functions of the blood and mention what part of the blood is concerned in the performance of each function.

2 (a) What are the general types of materials present in the plasma? (b) What materials are the most common and what are their functions?

3 (a) Why does the blood flow through the heart twice in making a complete circulation of the body? (b) What roles do the arteries, veins, and capillaries play in the circulation of the blood?

4 What are the solid parts of the blood, and what are the function, or functions, of each?

5 Distinguish between the members of the following pairs of terms: (a) intravascular fluid—intracellular fluid (b) artery—vein (c) atrium—ventricle (d) plasma—corpuscle.

Chapter 22 Blood Solids

The solid parts of the blood, the red blood cells (erythrocytes), the majority of the white blood cells (leukocytes), and the platelets have a common origin. They are all formed in the red marrow of many bones. Two types of white blood cell, however, the lymphocytes and the monocytes, are formed in lymphatic tissue. As these blood solids are being constantly lost as a result of their activities or injury, it is necessary for the body to replace them at a rate corresponding to their loss. It is one of the primary functions of the red marrow of bones to meet this demand.

Functions of the Marrow

The marrow found in bones is of two different types: *red marrow* which owes its name to the presence of a large amount of blood, and *yellow marrow*. The red marrow is the tissue responsible for the formation of most corpuscles and platelets. The yellow marrow is evidently a storage depot for fats, although it may have some other functions not clearly understood at present. In early embryonic stages, only red marrow is present in the bone cavities but is gradually replaced with yellow marrow. In cases of severe damage or loss of red marrow the yellow marrow may be replaced to some extent by red marrow and assume the function of forming blood cells. Normally, however, in adults red marrow is confined to certain bones: the vertebrae, thorax, base of the cranium, and upper ends of the femur and humerus. All other bones contain yellow marrow.

Within the blood-forming tissues of the red marrow appear large, primitive cells that contain distinct nuclei. These are known as the *reticular cells*. The process of division and differentiation by which these cells develop is shown in Fig. 22-1. Note that the reticular cells give rise to an intermediate form, the *hemocytoblasts*. The hemocytoblasts eventually form the red cells, most of the white corpuscles, and the platelets in red marrow. In lymphatic tissue they form mainly lymphocytes and monocytes; two types of leukocytes. The platelets are formed from a special type of cell that arises from the hemocytoblast and is known as a *megakaryocyte*.

During the changes that occur in these cells of the red marrow, the future erythrocytes lose their nuclei, and

298

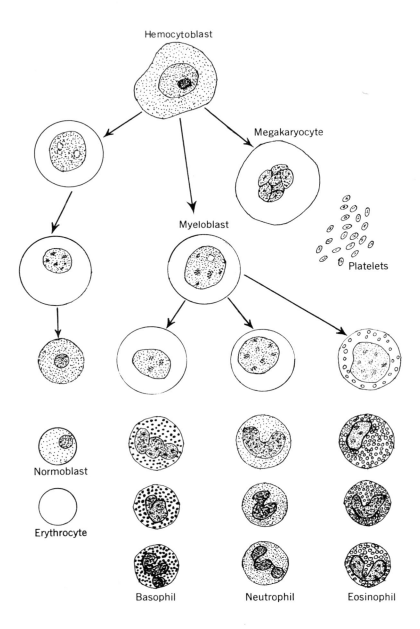

Fig. 22-1 The development of the various solid structures of the blood from a hemocyto-blast in the red bone marrow, shown schematically.

the leukocytes develop cytoplasmic structures that will later differentiate them into their various types. When blood cell formation is occurring in the marrow, the blood vessels in which the cells form are temporarily closed off from the general circulation. Within these tubes the cells pass through their various stages. When the blood cells have become mature, the vessels open and discharge their contents into the general circulation.

Erythrocytes

Mature red corpuscles normally lack nuclei, the chromatin material having been lost before the cells entered the blood stream. The *reticulocytes,* which make up from one to three percent of the erythrocytes in the blood, contain a material that may represent the remnants of the former nucleus. A normal red blood corpuscle is round in outline when viewed from above. If taken from an artery and examined in a fresh film, it may appear yellowish red, but if it has come from a vein, the color will tend toward a yellowish green. Blood owes its typical red color to the presence of large numbers of these cells; the individual corpuscles in fresh blood do not have a distinctly red appearance.

A red blood corpuscle (Fig. 22-2) has an average diameter of 7.2 microns (1 micron = 0.001 millimeter), and it is usually described as a biconcave disk, because it is thicker along the margins than in the middle. At the margin, the cell measures about 2.2 microns in thickness, but the center is only approximately half as thick. This inequality in thickness causes red blood cor-

puscles to appear doughnut shaped when viewed from above. The shape of these cells is of great importance from the standpoint of their function. It has been calculated that if the red corpuscle were a spherical body with a diameter of 7.2 microns, its surface area would be 163 square microns and its volume 197 cubic microns. The biconcave disk form, however, gives it a surface area of 302 square microns while still retaining its volume of 197 cubic microns.

The membrane that surrounds the corpuscle is complex; it is made up of three layers. On the outside is a protein layer three or four molecules in thickness. Within this is a layer of lipid molecules. The innermost layer is composed again of protein molecules. Within this envelope is a mixture of hemoglobin, salts, and other materials that play roles in the transport of the gases. There also appears to be a supply of ATP which furnishes the corpuscle with energy. Part of the energy that is released is used in maintaining the characteristic shape of the corpuscle. It has been found that if the ATP is inhibited in its action, the corpuscle assumes the form of a sphere. When ATP is added to these spherical forms, they assume

Fig. 22-2 Outline and dimensions of a typical red blood corpuscle. (From Marsland, Principles of Modern Biology, *Holt, 1957)*

their biconcave shape. Thus, maintaining their typical form requires the expenditure of energy.

From high-speed motion picture studies made of red blood corpuscles flowing through capillaries it has been found that the envelope of the corpuscle is highly flexible. Many of the capillaries in the tissues are smaller in diameter than the corpuscles themselves. The result is that when they pass through these minute tubes they assume a cone-like shape with the apex of the cone pointing forward; the trailing side is then concave. Just what effect this change in shape has on the exchange of gases between the tissues outside of the capillary and the corpuscle is still a question. It may, however, hasten the exchange by increasing the surface area that is in direct contact with the capillary wall.

Hemoglobin is composed of a protein, *globin,* and an iron compound, *heme.* A complex compound, *hemin,* is easily obtained from blood. If a drop of blood is heated with a drop of glacial acetic acid, long brown needles of hemin are formed. This is considered a reliable test for the presence of blood. Globin and heme combine in the bone marrow before cells are released into circulation.

Hemoglobin is a remarkable substance in two respects. In the first place, it is able to change from the deoxygenated to the oxygenated state in a remarkably short period of time—about 0.01 second. Secondly, it is able to retain its oxygen content until such time as it comes in contact with a tissue that is suffering from oxygen want. Only under this condition does the hemoglobin release its supply of oxygen. It has been calculated that if the blood lacked hemoglobin, a person of average size would require approximately 265 liters of circulating fluid to dissolve sufficient oxygen to supply the needs of the cells. The reason for this is that oxygen dissolves in plasma at sea level only to the extent of 1.23 milliliters per 100 milliliters of fluid. In the presence of hemoglobin, full saturation of the blood with oxygen results when 20.5 milliliters of oxygen is present in each 100 milliliters of blood, since each gram of hemoglobin will combine with 1.34 milliliters of oxygen. This gives a ratio of one part of oxygen dissolved in the plasma, as compared with 192 parts combined with hemoglobin.

Several different types, about 20, of hemoglobin have been identified in human beings. Some of these will be discussed later, but mention should be made now of a change that normally occurs during a person's life. Before birth, the fetus has a type of hemoglobin that is different from that of the adult. This fetal hemoglobin (HbF) differs from the adult type in that it can be saturated with oxygen at a lower oxygen tension and give up carbon dioxide at high CO_2 tensions. This is a special adaptation of the fetus who must receive all of its oxygen and release its carbon dioxide through the maternal circulation. The HbF is replaced by adult hemoglobin during the first few months after birth.

Leukocytes

The white blood cells range from one and one-quarter to two times the diameter of the red cells, and can move independently. Whereas red corpuscles must be carried by the plasma and cannot change their shape except as they are squeezed in passing through very small vessels, the white corpuscles are capable of changing shape when the proper chemical stimulus is applied. They can move out from the blood stream by passing between the endothelial cells that comprise the walls of the capillaries. In this respect, they resemble the single-celled animal, *Ameba*. In both, the cytoplasm flows toward a stimulus. An ameba, for example, will move toward a source of food or oxygen by an oozing, flowing movement that is generally believed to be caused by a change in the surface tension of the cell membrane on the side toward the stimulus. The same general movement is found in white corpuscles. These cells are able to change their shape in such a manner

that they can escape from the capillaries. It is thought that they are able to penetrate the capillary walls by dissolving the cement substance that holds the endothelial cells together.

The diagram in Fig. 22-3 shows how leukocytes pass through a capillary wall toward a stimulus, represented by a small wound that has allowed bacteria to gain entrance through the skin. The bacteria produce various chemical substances (toxins) which are the products of their metabolic activities, and which act as stimuli for the leukocytes.

One of the first results of the presence of foreign materials in tissue is the dilation of the capillaries in the area. As a result, more blood enters the tissue and there is a corresponding rise in the local temperature. Also, with the increased blood flow, more white corpuscles are carried to the area. We do not yet know what mechanism stimu-

Fig. 22-3   Diagram showing how neutrophils move from the blood vessels to attack invading bacteria.

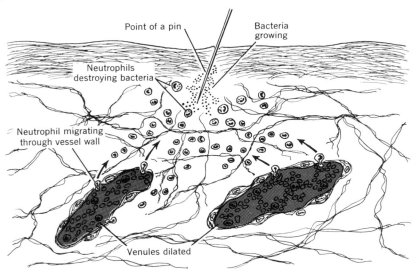

lates the production of a larger number of leukocytes or attracts them to a specific region of the body. On arrival in the infected area, the white corpuscles pass between the cells of the capillary walls and move toward the invading bacteria. As they approach the bacteria, the cells send out projections that surround and engulf the bacteria. Once inside the leukocyte, the bacterial cell is digested and rendered harmless. This process is known as *phagocytosis* (Greek, *phagein,* to eat), and any cells that perform this function are called *phagocytes*. The number of bacterial cells that a single leukocyte ingests may be greater than its digestive capacity. When this occurs, the leukocyte dies. Its remains, plus the active white cells and invading bacteria, are the material commonly known as *pus*. As pus collects in the tissue, materials produced by the bacteria cause a breakdown of the surrounding tissue. An abscess then develops as a wall of tissue is formed around the area by specialized white cells known as *fibroblasts*. Meanwhile, more blood is carried to the area with resulting redness. The nerve endings of pain are also stimulated. Inflammation is recognized by swelling due to the accumulation of fluids, redness and heat as a result of the increased blood supply and pain caused by stimulation of the nearby receptors.

In addition to engulfing invading organisms, another phagocytic activity of the white cells is their ability to destroy red cells that cannot function normally. Erythrocytes become damaged as they pass through the smallest blood vessels, especially if these are in active muscle tissue. Indeed, there is evidence that red corpuscles may be broken up into small fragments as a result of the activity of a muscle. These damaged cells, or their parts, are engulfed by some of the white blood cells. The regions where this destruction of red cells occurs most actively are the spleen, red bone marrow, liver, and certain lymph glands. The digestive activities of the leukocytes break down hemoglobin into globulin and heme. The latter is further reduced to *hemotoidin,* which is then transformed into bilirubin by the liver. Another product of this decomposition is *hemosiderin,* which may be stored as an iron compound for later conversion into hemoglobin by steps that are not thoroughly understood at present. The rate at which the red cells are destroyed is almost unbelievable; it is estimated that there are an average of 3,500,000,000 red corpuscles consumed per kilogram of body weight each 24 hours. Of course, there is a corresponding production of cells by the bone marrow. If an excess of erythrocytes is injected into the blood stream, they are rapidly removed until the count returns to a normal level. This indicates that the number of erythrocytes is kept at a fairly constant level, but the mechanism which accomplishes this is not known.

The types of white corpuscles can be identified by different methods of staining. If a mixed dye is used, certain cellular structures take up the different components of the dye in a definite manner. The most commonly used stain for blood smears is *Wright's stain*. It is a mixture of *methylene blue,* a dye that stains alkaline materials, and *eosin,* a bright red stain that has an affinity for acidic materials. When a blood smear is stained with Wright's stain, the nuclear materials of the cells are colored blue and any acidic cytoplasmic granules are stained red. The erythrocytes, since they lack nuclei, take up the red stain lightly and appear as pink objects. On the other hand, white corpuscles show a nucleus that is stained deep blue. Some of these cells may contain relatively large cytoplasmic granules that stain red, others may have granules that are blue, while other cells may not show any large distinct blue or red inclusions. In some of the leukocytes, the cytoplasm shows no staining reaction whatever. Thus these cells may appear red, blue, an intermediate shade or almost transparent when stained.

On the basis of their staining reactions, white blood cells are classified as *neutrophils, eosinophils, basophils, monocytes,* and *lymphocytes.* The percentages of each in one cubic milliliter of the blood varies widely even in healthy people as the following table shows.

Neutrophils	33–75
Lymphocytes	15–60
Monocytes	0–9
Eosinophils	0–6
Basophils	0–2

Wide deviations from these percentages can be useful in diagnosing certain diseases.

Blood Platelets

These are the smallest of the solid parts of blood. In a fresh preparation of blood, they have an oval outline and appear as individual structures. In a dried smear, however, the platelets appear round and are slumped together in small groups. As Fig. 22-1 shows, they arise from a fragmentation of the megakaryocytes that are formed in the red bone marrow.

The number of blood platelets is difficult to determine with accuracy, since it varies in different regions of the body, and under varying physiological conditions. In the next chapter we will discuss the role of the platelets in the clotting of blood, since they are most concerned with this process.

Regulation of Blood Cell Numbers

The formation of new platelets, and red or white corpuscles, is thought to be controlled by definite chemical substances which experimental evidence suggests develop in the yellow bone marrow in response to a stimulus. Red blood corpuscle formation, for example, may be stimulated by continued low barometric pressure such as exists at high altitudes, which results in a marked increase in the number of red corpuscles in the blood stream. White blood cells will also respond to a group of stimuli which increase their numbers. This, however, is quite different from the condition resulting from the release of leukocytes following an infection. In all of these cases so long as there is a normal number of each type of blood solid present, there is a mechanism which prevents the formation of the growth-stimulating factor.

Anemia

Deficiency in the number of red cells is called *anemia.* Studies of this abnormal condition have contributed greatly to our knowledge of blood formation. From a physiological standpoint, anemia results from one of the following principal causes: an acute or chronic loss of blood (hemorrhage) which reduces the number of red corpuscles; the excessive destruction of red corpuscles, as for example in malaria; or the improper formation of hemoglobin or red cells by the body. Regardless of the immediate cause of anemia, there is always a lack of hemoglobin with a corresponding deficiency in the oxygen supply of the body. Immediately following loss of blood, the arterial vessels constrict and thus reduce the flow of blood through them. Fluid from the tissues then passes into the blood vessels, restoring the normal blood volume. Approximately forty-eight hours after the loss, many immature red cells appear in the blood stream along with an excess of leukocytes. This indicates that the red marrow has been stimulated to increase its production of blood cells.

A chronic loss of blood caused by hookworm infection may result in a prolonged recovery period after the worms have been killed. The total loss may be considerable, since there are usually a great many hookworms embedded in the mucous lining of the intestinal wall and using their host's blood as food. Not only do these parasites cause an actual decrease in the blood supply, but they also produce toxins which impair the body's ability to regenerate new cells.

A second type of anemia results from the destruction of red corpuscles caused by the ingestion or absorption of poisonous chemicals, or of toxins produced by bacteria that have penetrated beyond the first lines of defense. Lead poisoning, an occupational disease, often results from carelessness in the use of substances containing lead. A house painter, for example, may not wash his hands thoroughly before eating. In this case, minute traces of lead may be ingested. Spray painters may also inhale lead particles. Over a long period of time, these small quantities of lead accumulate in the red marrow of the bones and destroy some of the blood-forming cells. The hemoglobin is thus reduced. In severe cases of lead poisoning, enough of the metal may accumulate in the bones to be detected by means of X rays. An early symptom of lead poisoning is the appearance of a dark blue line in the gums just above the teeth. This *lead line* is caused by the deposition of lead sulfate in the walls of the capillaries. In cases of lead poisoning, the red corpuscles show a blue mottling when they are stained because lead salts in their cytoplasm are able to absorb the methylene blue of the stain.

Certain bacteria may invade the blood stream and cause decomposition of the red corpuscles, with resulting anemia. Such an invasion is usually termed *septicemia*. Once bacteria or other invading organisms have succeeded in passing the defensive lines of the skin, the white cells, and the antitoxins, their spread throughout the body is relatively easy. Bacteria and other organisms (popularly known as germs) may occasionally destroy red cells. A good example of this is seen in the malarial organisms, which kill large numbers of erythrocytes in the course of their reproductive cycles. Bacteria may also produce toxins, which have a much more rapid and drastic action because they chemically affect the cellular envelopes of the corpuscles and destroy the hemoglobin.

A third type of anemia results from improper formation of red corpuscles. This is the result of either an inherited defect in the blood-forming tissues or an acquired inability to form blood in a normal manner. An example of the former is the condition called *sickle cell anemia*. The replacement of one single amino acid by another, different amino acid in a unit part (peptide) of the hemoglobin molecule is responsible for this condition. The mutant molecules tend to link and form strange structural configurations which distort the shape of the red blood cell. This occurs almost exclusively among Negroes and is characterized by red cells that are curved so that they resemble sickles. If the number of affected cells is relatively low, there are no outward symptoms evident, although the cells can be identified in blood smears. In severe cases, however, the red cells are rapidly destroyed and jaundice develops as a result of the accumulation of decomposed hemoglobin in the blood stream. There is an accompanying loss of the oxygen-carrying capacity of the blood. A somewhat similar condition, known as *Cooley's anemia,* occurs among members of the Mediterranean peoples. *Familial hemolytic jaundice* results from an abnormal activity of the spleen which makes the red cells fragile, so that they break up easily and release their hemoglobin into the blood. Removal of the enlarged spleen usually cures this.

Pernicious anemia is the result of a vitamin B_{12} deficiency. In this disease, proper cell division and development is affected so that fewer than normal, and abnormally large and misshapen, erythrocytes are formed. Each of these, however, contains an amount of hemoglobin that is consistent with its size, so the onset of this disease is very difficult to diagnose. If the disease progresses without being identified, death usually results. Since the discovery of the roles of vitamin B_{12} and the intrinsic factor, an unidentified substance secreted in gastric juice which facilitates the absorption of B_{12} by the stomach, the disease should no longer be considered "pernicious." If the treatment is started early, before there is serious damage to the central nervous system, the outlook is excellent. The use of the vitamin in combination with the gastric secretion restores the normal blood-forming functions of the red bone marrow. If used alone B_{12} is injected.

Iron-deficiency anemia is of rather frequent occurrence in growing children. This is caused by the absence of an adequate amount of iron in the diet, with the result that there is insufficient hemoglobin in the red corpuscles. Since iron is a common element found in the majority of plant foods, a well-balanced diet should prevent this condition. It is rather interesting to note that the minute amount of copper found in most plant foods aids in the absorption and utilization of iron by the body. Copper is usually considered a violent protoplasmic poison in high concentrations, but its presence in concentrations such as those found normally in plants is evidently essential for the formation of hemoglobin to take place effectively in animals.

Leukemia

This is a blood disease that is usually fatal. It is so named because the blood may show a whitish color caused by the presence of an excessively large number of leukocytes. In extreme cases of this disease, the number of white cells may equal or exceed the number of erythrocytes. There are several well-recognized types of leukemia: some are acute in the sense that they develop rapidly and run their course quickly; others are said to be chronic because their action is slower and may have little effect on the victim over a period of many years. Acute types of leukemia are more frequently found among children and young adults.

An excess of leukocytes in the blood may replace the erythrocytes to the extent that the blood can no longer carry the needed oxygen to the tissues. It may also impare the ability of the bone marrow to produce more red cells.

Today there is increased interest in leukemia because it has been discovered that the disease can be caused by excessive exposure to X rays and the radiations from radioactive elements. At present research workers cannot agree on the exact relationship between the amount of radiation exposure and the occurrence of the disease. However, it has been proved beyond any doubt that exposure to these penetrating radiations can cause increased production of leukocytes in the body.

VOCABULARY REVIEW

Match the phrase in the left column with the correct word in the right column. *Do not write in this book.*

1 An abnormal condition resulting from a reduction in the amount of hemoglobin.	**a** anemia
2 The process in which leukocytes engulf bacteria or other foreign bodies.	**b** eosinophils
3 An erythrocyte which has been recently released into the blood stream and has not lost all of its nuclear materials.	**c** globin
	d heme
	e leukemia
	f leukocytosis
	g erythrocytes
4 An abnormal condition resulting from the overproduction of white blood cells.	**h** phagocytes
5 The principal blood-forming tissue.	**i** phagocytosis
6 The protein part of the hemoglobin molecule.	**j** red marrow
7 The smallest solid part of the blood.	**k** reticulocytes
8 A type of blood cell especially active in the destruction of bacteria.	**l** septicemia
9 An iron-containing compound.	**m** thrombocytes
	n Wright's stain
	o yellow marrow

TEST YOUR KNOWLEDGE

Group A

Select the correct statement. *Do not write in this book.*

1   Blood cells that have no nucleus are the (a) leukocytes (b) monocytes (c) erythrocytes (d) white corpuscles.
2   Red blood corpuscles are formed mainly in the (a) yellow marrow (b) liver (c) red marrow (d) lymph nodes.
3   Anemia may result from the lack of (a) oxygen (b) nitrogen (c) sulfur (d) iron.
4   The blood cells that engulf bacteria are (a) erythrocytes (b) red corpuscles (c) thrombocytes (d) leukocytes.
5   The most abundant white cells in the blood are (a) basophils (b) monocytes (c) lymphocytes (d) neutrophils.
6   The only blood cells that are normally found outside the blood vessels are (a) erythrocytes (b) reticulocytes (c) leukocytes (d) platelets.
7   A pathological excess of white corpuscles in the blood is called (a) anoxia (b) anemia (c) leukemia (d) septicemia.
8   The parts of the blood that are mainly concerned with clotting are the (a) basophils (b) erythrocytes (c) thrombocytes (d) monocytes.
9   Of the following, the only cells that are *not* concerned with the blood-forming functions are (a) reticular cells (b) megakaryocytes (c) normoblasts (d) eosinophils.
10   Leukocytes pass from the blood into the tissues chiefly through the walls of the (a) bronchioles (b) venules (c) capillaries (d) arterioles.

Group B

1   Briefly describe the development of the various blood solids and point out how they differ in structure from each other when they become mature.
2   (a) Describe the structure of an erythrocyte. (b) What is the function of hemoglobin? (c) What is the fate of worn-out hemoglobin?
3   Briefly describe three different types of anemia and relate each to the structure of the blood corpuscles.
4   Identify each of the following: (a) leukopenia (b) pus (c) leukemia (d) yellow bone marrow (e) monocytes.

Chapter 23 Blood Plasma

Plasma is the highly complex part of blood which carries on most of the functions of the circulatory system. Basically, it is a water solution of various substances which also contains a suspension of other materials. Since some of these materials are colloids, the plasma acts in many ways like a colloidal suspension. When separated from the corpuscles, plasma is a yellowish straw-colored fluid.

The composition of plasma is not constant. It varies from person to person and also within any one person from time to time, depending on the materials that are added to it or withdrawn from it by the various activities of the body. Following a meal, for example, it contains an increased amount of nutrient materials. On the other hand, after a period of strenuous exercise, plasma contains wastes that the blood has removed from active tissues. Thus from one moment to the next, plasma reflects the different physiological states of the body. This makes the exact analysis of its composition impossible. The table shows its average composition.

Water	92%
Blood proteins	7%
Fibrinogen	0.3%
Serum albumin	4.4%
Serum globulin	2.3%
Other organic substances	0.136%
Nonprotein nitrogen (urea)	
Organic nutrients	
Inorganic ions	0.931%
Calcium, Chlorine,	
Magnesium, Potassium,	
Sodium, Bicarbonates,	
Carbonates, Phosphates	

Formation of Blood Clots

The plasma protein, fibrinogen, responsible for clotting blood is probably formed in the liver, but the exact steps in its production are not clearly understood. It is a substance of uniform chemical composition with a molecular weight of approximately 500,000, which indicates its complex nature.

Several chemical reactions occur during *coagulation,* the formation of a blood clot. The exact steps of the process are not definitely known, and several theories have been advanced. An explanation of each of these may be found in advanced texts in medical physiology. While there is some difference of opinion on the reactions involved, there is also agreement on the fundamental processes that occur.

All the tissues of the body contain a substance that hastens the coagulation of blood. This material, *thromboplastin*, is produced whenever a tissue is injured. Therefore, when blood comes in contact with cells that have been cut or otherwise injured, thromboplastin initiates the formation of a clot.

Thromboplastin can only cause clot formation if *calcium ions* and the plasma protein *prothrombin* are also present in the plasma. The prothrombin is formed in the liver as is shown by the decreased amount in plasma in cases of injury or disease of the liver. This first step in the formation of a clot is enzymatic in nature and can be represented as follows:

$$\text{prothrombin} \xrightarrow[\text{+ calcium ions}]{\text{thromboplastin}} \text{thrombin}$$

This process occurs only when blood is shed, there being no thrombin present in the circulating fluid. *Vitamin K* must also be present for it to occur.

The fact that calcium ions are required for the completion of this process is well illustrated by the effect of their removal from the blood. When a physician takes a sample of blood from a patient for examination at a later time, he adds a solution of sodium or potassium oxalate to it. This prevents the blood from clotting because the calcium ions unite very easily with the oxalate and are therefore removed from the reaction.

The second step in the formation of a clot involves fibrinogen, one of the plasma proteins, and thrombin. Thrombin acts as an enzyme catalyzing the formation of fibrin as follows:

$$\text{fibrinogen} \xrightarrow{\text{thrombin}} \text{fibrin}$$

The fibrin threads produced by this reaction form the basis of the clot. They are actually minute white strands of material which trap the corpuscles until an efficient block has formed across the opening to stop the flow of blood.

The platelets were formerly thought to be the principal source of thromboplastin. Recent research has thrown some doubt on this supposed function of platelets, since they appear to contain too little thromboplastin to account for the rapid coagulation of blood. Since they are extremely fragile objects, they break up on slight contact with a for-

Fig. 23-1 The clotting of blood. Freshly drawn blood is shown in tube A. In B some of the cells have settled and a clot is forming along the side of the tube. In C most of the cells have settled and the clot is fully developed, leaving the clear serum. A few cells are trapped in the clot.

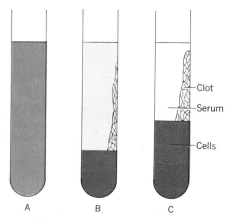

eign body. In breaking up, they release a substance, *cephalin,* which apparently aids in the formation of thrombin, but its exact action is not known.

Clotting of blood is affected by physical factors. A surface that can be wetted by water hastens the process, while one that prevents molecules of water from clinging to it retards clotting. For example, if blood is carefully removed from a vessel, so that it does not come in contact with injured tissues, and then placed in a clean glass tube, a clot will appear in a short time. If, however, the inside of the tube is coated with a layer of paraffin or wax (which the blood cannot wet), the clotting time will be greatly prolonged. Temperature also affects the rate of clotting. At a relatively high temperature, blood will clot easily, while under carefully controlled refrigeration, blood will remain fluid for a long time. There appears to be a rapid destruction of the blood platelets on contact with a rough surface.

The importance of a rough surface and its physical effect on the clotting of blood is seen whenever a tissue is cut. Since all tissues are wettable by blood, the cut surface itself aids in the formation of a clot. The process can be hastened by placing a piece of gauze or other cloth with a relatively rough surface over a wound. This further stimulates the formation of a clot.

A study of the blood defect known as *hemophilia* has shed some light on the structure of plasma and the possible function of some of the blood proteins in the process of coagulation. Hemophilia is an inherited condition in which the power to produce a sufficient amount of thromboplastin is greatly reduced. The result is a failure of the blood to clot within a normal period of time. Consequently, people suffering from this condition may bleed to death from wounds that would be minor for a normal individual. Hemophilia may occur even when blood contains a normal amount of prothrombin, calcium, and platelets, so it appears that none of these is implicated in this disease. Recent work has indicated that there is present in the normal serum globulin of the blood an anti-hemophilic factor that is necessary for the proper coagulation of the blood. Experimental evidence of this has been obtained by injecting small amounts of normal serum globulin into persons suffering from hemophilia. This temporary measure causes the blood of the victim to clot faster, but it does not change his hereditary characteristics, with the result that the person can still transmit the defect to his or her offspring.

The fact that blood does not normally clot in the blood vessels has been attributed to a substance known as *heparin.* This was first isolated from the liver but was later found to be present in other organs. At present it is obtained from the lungs of slaughtered animals in sufficient quantity to supply commercial needs. Under normal conditions, heparin is not present in the blood in sufficient quantities to be detectable as an anticoagulant.

Blood clots called *thrombi* (sing., *thrombus*) occur occasionally in blood vessels. If they become too large, they seriously hamper the flow of blood through a vessel. Thus the passage of blood to a region of the brain or to the muscles of the heart may be affected, resulting in *thrombosis.* In such cases, doctors frequently use an anticoagulant such as heparin. Another drug that is being currently used to reduce a thrombus is *Dicumarol*. This compound was first isolated from sour clover fodder that was found to be the cause of hemorrhages in cattle who suffered minor cuts. Vitamin K, which is essential in the formation of prothrombin, counteracts the action of both heparin and Dicumarol.

Plasma Proteins

The plasma proteins (other than fibrinogen) are classified as either *globulins* or *albumins*. These terms are used to designate certain types of protein molecules, and there are many individual examples of each. Both of these types of proteins are widely distributed in plant and animal tissue. A good example of an albumin that is not a constituent of blood is the white of an egg, while a globulin is present in the yolk of an egg. The serum albumins and serum globulins can be separated from the plasma when the fibrinogen has been removed.

Of the two, serum albumin is the more plentiful. Its principal function is to control the osmotic pressure of blood. As blood flows through the capillaries into the tissues, its osmotic pressure is highest just after the capillary branches from an arteriole. From that point onward through the capillary, the osmotic pressure decreases. The result of this gradient in pressure is that food and oxygen leave the blood early in its passage through the capillary. Then waste materials pass into the blood as the capillary joins others to form a venule. Under normal conditions, the blood proteins do not pass through the capillary walls because of the relatively large size of their molecules. However, since they are colloidal materials, they can give up, or take up, water-soluble substances, and thereby control the osmotic pressure within the vessels.

If there is a decrease in serum albumin, fluids leave the blood throughout the entire length of the capillaries and the surrounding tissues fill with fluid and become enlarged. Such a condition is known as *edema*. Certain types of kidney disorder result in the removal of abnormally large amounts of protein from the blood with resulting edema.

The serum globulin also helps in the maintenance of osmotic pressure, but it has other functions to perform. Under certain laboratory conditions, it is possible to obtain several different substances that show characteristic features.

Among the substances that have been extracted from serum globulin is *gamma globulin*. Associated with this substance are the *antibodies* that protect a person against invasion by bacteria and viruses. For example, if a person has had measles, he has in his blood certain chemical materials which combat another attack of this disease. These materials are present in the serum globulin as definite antibodies; substances that neutralize the effects of a future invasion of the body by the same organism. Since the lymphocytes are the source of serum globulin, they are believed responsible for production of antibodies.

Immunity and the Thymus

Since about 1960 there has been a great deal of interest shown in the relation of the thymus gland to the immune reactions of animals. Until that date, little was known about its activity. The thymus lies in the thoracic cavity just under the upper part of the sternum. At birth it weighs about 11 grams and continues to grow until puberty is reached, when its weight approaches 35 grams. The gland then gradually decreases in size until at age 60 its weight is approximately that at birth.

In the past, considerable research work was done on the thymus, but little information gained. The recent advances are the result of a carefully correlated study of the growth of the gland and the types of cells that are present in it. It was found that the first place lymphocytes appeared was the thymus, where they were found even before birth. They were formed from cells that line the organ. It is currently thought that some of these early lymphocytes enter the blood stream and are carried to different parts of the body where they establish new centers of lymph tissue. During the early years of life, the thymus continues to grow actively and produce large quantities of lymphocytes. Since these cells produce the plasma cells that form the serum globulin, antibodies can be formed in sufficient quantities to offset, or at least modify, an attack of a disease.

It has recently been found that in order to proliferate and produce plasma cells to synthesize antibodies in the presence of disease-carrying organisms, the lymphocytes in lymph tissue must be "activated" by a hormone which the thymus gland starts producing just after birth. It has been found that if the thymus is removed from an experimental animal, for example a mouse or hamster, within 48 or 72 hours of birth, the animal loses its ability to develop immunity. Also, animals that have been reared under germ-free conditions are unable to develop immunity against quite common infections and die soon after exposure to ordinary conditions.

It is possible by modern medical techniques to separate some of the antibodies from the serum globulin and use them to give temporary immunity to a person who has been exposed to a specific disease. The injection of such antibodies either prevents the disease or renders an attack less severe than it might have been without this protection. Among the types of antibodies are the *antitoxins,* which combat the *toxins,* or harmful substances, produced by bacteria.

Physiological Shock

This condition is caused by a painful injury, extensive burns, or severe loss of blood. Regardless of its cause, it results in a decreased volume of circulating blood and a reduction in blood pressure. There are two stages in this process. The first, known as *reversible shock,* is characterized by constriction of the small vessels which restricts blood to the larger arteries and veins. The recommended first aid procedure for this stage of shock takes into consideration the decreased blood flow through the smaller vessels. The injured person is kept warm and the feet and legs are slightly elevated (20 to 30 centimeters), thus ensuring a better flow of blood to the head and chest regions. This aids in increasing the amount of blood passing to the brain and assures a better supply to the heart.

The second stage of physiological shock, *irreversible shock,* is the result of rapid dilation of the smaller vessels. This greatly increases the area that must be supplied with blood, and blood pressure within the arteries drops. Blood fluids rapidly filter out of the vessels into the surrounding tissues and the volume of the blood is reduced to dangerously low levels. This collapse of the circulation is usually fatal.

Some evidence points to the presence in blood of two substances that bring about these reactions. The constriction of small vessels that occurs in cases of reversible shock is evidently due to the presence of a material formed in the kidneys. This stimulates the smooth muscle fibers in the walls of the vessels. The irreversible condition is brought on by the presence of a second substance, formed in the liver. Its action is antagonistic to that of the kidney secretion, and results in the relaxation of the muscle fibers, so that blood can pass into the smaller vessels. In reversible shock, the skin is pale and there may be profuse sweating, but this is reversed when the second stage is reached. In this condition the skin is flushed, due to the presence of an increased amount of blood in the surface vessels.

Acid-Base Equilibrium

One of the factors in maintaining a constant internal environment in which the cells can function normally is the control of the amount of acids and alkalis present. Metabolic processes are constantly forming acid waste materials. Some of these are strong acids which, if allowed to accumulate, might kill the cells. The blood normally keeps the amounts of these substances within limits that will permit the cells to function properly. Except under abnormal conditions, such as those arising during disease, the hydrogen ion concentration of the blood remains at an almost constant level.

There are three factors which help the blood to maintain its pH value: the action of buffers in the blood, the excretion of carbon dioxide by the lungs, and the excretion of fixed acids by the kidneys.

A buffer system in a solution makes possible rapid adjustment to the addition of acid or base. If the solution contains chemical elements which enter easily into compounds to form weak acids or salts, these elements "cushion" or buffer the effects of acid or base that enters the solution.

In the table on page 309 there is a list of some of the more common inorganic substances in plasma. They include chemical elements such as sodium and potassium, which may combine with bicarbonate ions to form compounds such as sodium bicarbonate ($NaHCO_3$) or form carbonates and phosphates. These compounds act as buffers by changing their composition rapidly in response to the amounts of hydrogen (H^+) or hydroxyl (OH^-) ions in the blood. Thus, for example, an alkaline substance will be neutralized by the weak carbonic acid in the blood. Bicarbonates in the blood will react with acids such as lactic acid produced by muscular exertion to neutralize them.

$$HLa + NaHCO_3 \rightarrow NaLa + H_2CO_3$$

| lactic acid | sodium bicarbonate | sodium lactate | carbonic acid |

Carbonic acid is a weak acid and therefore remains largely undissociated. Under normal conditions, therefore, the pH of the blood is maintained at a level of 7.4 by these chemical reactions.

In addition to the inorganic materials present in the plasma, the blood proteins also act as buffers. As we have pointed out, proteins are highly complex compounds containing different components which can ionize in aqueous solutions. These components react to shifts in the blood pH in much the same way as the inorganic compounds. These reactions result in the return of the pH of the blood to its normal level.

The exchange of oxygen and carbon dioxide in the tissues involves constant changes in the acidity of hemoglobin. The compound *oxyhemoglobin* is relatively acid. When blood containing a large amount of oxygen in the form of oxyhemoglobin passes along a tissue that is deficient in oxygen, hemoglobin changes chemically to release oxygen, allowing it to diffuse toward the tissue. The hemoglobin thus becomes more alkaline and accepts the carbon dioxide that the tissue has formed. This is carried to the lungs where the conditions are reversed. Here the blood encounters a region that is rich in oxygen, and hemoglobin once more becomes more acid as it accepts the oxygen while giving up the carbon dioxide it has taken on in the tissues. These reactions are buffered principally by the presence of bicarbonates in the plasma and the red cells.

As we know, blood in active tissue absorbs an extra amount of carbon dioxide resulting from oxidation in the tissue. This compound combines with water in the plasma to form carbonic acid. When blood carrying excessive amounts of carbon dioxide passes through the respiratory center of the medulla, the center is stimulated and the rate of respiration increases. The excess carbon dioxide is thus removed, and does not affect the pH of the blood.

Hemorrhage

Any undue loss of blood from the circulatory system is a hemorrhage. This loss may be external as a result of a wound, or internal. The latter may be caused by a serious puncturing wound, an infection, or the eroding action of an ulcer, which allows blood to flow into some body cavity.

The amount of blood lost in a hemorrhage determines the effect on the body. A person may lose one half-liter (approximately 10 percent of the total volume of blood) without any ill effects beyond a general weakness for a day or two. On the other hand, if the loss amounts to about 35 percent, serious damage may result, while a 50 percent reduction in volume is usually fatal. If the loss is small, as is the case in the donation of a pint of blood, the body restores the fluid in a period of six to eight hours, and the red cells are brought to normal within about two weeks. White cells are restored more quickly. In fact, following the loss of blood there is a slight leukocytosis, but this disappears rapidly and the number of white cells returns to normal within a short time. If more than one half liter is lost, the return of the plasma and solids to a normal condition requires a proportionately longer period of time.

The physiological effects of hemorrhage fall under two headings. If the loss of blood is sudden and extensive, the quantity of fluid is reduced to a point where the heart can no longer pump it around the body efficiently. This loss is indicated by a sudden drop in blood pressure. Because of this decrease in volume of blood the tissues begin to feel the effects of the lack of oxygen. The nervous system responds to the emergency, constricting the cut or broken ends of the vessels in order to reduce the flow of blood from them. Elsewhere in the body the smaller vessels are also constricted. This results in a slower flow of blood due to friction of the blood cells against the walls of the vessels.

With the decrease in the blood pressure comes an increase in the rate of the heart beat, but since there is less blood flowing, the pulse may be weak or apparently absent. There is also an increase in the rate of respiration as the body attempts to compensate for the lowered supply of oxygen going to the tissues. Increases in heart and respiratory rates are important indications of internal hemorrhages. There may be few signs other than these of the loss of blood into a body cavity. Hemorrhage is usually accompanied by signs of physiological shock.

Various artificial methods are used to stop the flow of blood from an open wound or to compensate for the deficiencies arising from a hemorrhage. These are grouped under two headings: those that stop the flow of blood by mechanical means, and those that replace various parts of the blood.

Three methods of stopping external loss of blood are as follows:

The application of some rough material directly over the wound. The American Red Cross recommends this as the primary method to be used in stoppage of bleeding. Surgical gauze or a piece of cloth will hasten the formation of a clot, since a rough surface favors the destruction of the platelets.

The application of pressure to the principal blood vessel that leads to the wounded area. This will stop the flow of blood through the vessel. Such pressure should be applied to an artery between the wound and the heart. On the other hand, if a superficial vein has been cut, pressure may be applied to the vein on the side away from the heart, since veins are returning blood to the heart. Pressure on the gauze or cloth directly over the wound is helpful. To prevent severe loss of blood from an artery, pressure should be applied at the appropriate *pressure point,* where the artery lies close to the bone. The location of the pressure points may be found in any first aid textbook. It is obvious that the only arterial pressure points of importance to the first aider are those where the arteries lie close enough to the surface of the body to make practical the use of external pressure.

The use of a mechanical device to put pressure on the principal arteries supplying an injured region. Such a device is called a *tourniquet* and usually consists of a piece of cloth that can be drawn tightly around a limb so that little blood will pass to or from the region below the point at which it is applied. To be effective, a tourniquet must compress all of the major vessels in the area in which the hemorrhage occurs.

A tourniquet should be used with *extreme caution,* since the reduced blood supply to the tissues may cause *gangrene*. This serious condition results when large numbers of cells die because they cannot get nutrients or dispose of accumulated wastes. Leading first aid authorities now recommend that amateurs should *not* apply a tourniquet, except as a last resort in cases where medical aid is not quickly available. Furthermore, they state that a tourniquet, once applied, should only be loosened by a physician or some other medically trained operator. This recommendation is important, since the premature loosening of a tourniquet may result in additional loss of blood. It may also allow toxic wastes from injured tissues to spread to other parts of the body. Medical authorities are agreed that it is safer to risk loss of a limb through continued application of a tourniquet than to risk possible death of the victim by loosening it.

Four methods that can be used to combat both internal and external loss of blood are as follows:

Injection of a saline solution is used to replace the fluids that have escaped. Although quite popular at one time, this procedure has been found to be inefficient and is used now only in cases of extreme emergency. Sodium chloride molecules are too small to be retained within the blood vessels and rapidly pass to the surrounding tissues resulting in edema. This method, therefore, does not permanently increase the amount of blood fluid.

Solutions containing molecules of larger size have been tried with varying degrees of success. Such materials are called *plasma expanders* because they artificially replace the colloids and salts of the lost blood fluid. The larger molecules prevent the loss of fluid to the tissues that accompanies the use of a saline solution. Gum acacia, isinglass, and gelatin have been used. Current research includes a study of the effectiveness of complex polysaccharides, such as dextrin, and certain plastics, such as polyvinyl pyrrolidone (PVP).

Human serum and plasma can be dehydrated and preserved for relatively long periods of time. When sterile distilled water is added to them, they can then be used to replace lost blood. The value of these substances in caring for wounded soldiers was dramatically illustrated during World War II when they were used widely for the first time and were responsible for saving many lives. Since both of these substances contain the blood proteins, loss of water to the surrounding tissues is greatly reduced. Of the two, blood plasma is the more frequently used because of its ease of preparation and the presence of a normal amount of fibrinogen, which is essential in the normal process of coagulation.

The most successful method of combating excessive loss of blood is by transfusion of whole blood into the circulatory system of the victim. It is obvious that whole blood is, in most cases, superior to any blood substitute, or fraction of blood. Recovery is hastened to a marked degree. One essential precaution must be observed in making a transfusion of whole blood: the blood of the donor and the recipient must be the same in type.

Blood Typing

Human beings may be classified in one of four general blood groups, according to the nature of certain proteins in the plasma and red blood cells. The blood of the recipient and the donor in a transfusion must be of the correct type, or a serious reaction occurs between the protein outer layer of the red corpuscles and the plasma protein of the antagonistic type of blood. The corpuscles become "sticky," and tend to clump together in groups. This clumping is called *agglutination*. The plasma proteins concerned in agglutination are called *agglutinins,* while those in the walls of the corpuscles are known as *agglutinogens.*

The blood type of each person is designated as *A, B, AB,* or *O,* depending on the kind of agglutinogen present in the red corpuscle. This system of nomenclature was suggested by Dr. Landsteiner of the Rockefeller Institute for Medical Research in New York. He also found that there were two types of agglutinins in plasma. These he called *a* and *b* (or alpha and beta). If a person, for example, belongs to blood Group *A,* his erythrocytes contain agglutinogen *A* and his serum contains agglutinin *b*. Since *b* is a protein that is antagonistic to the erythrocyte protein agglutinogen *B,* transfusion from a *B* individual to an *A* individual would result in the agglutination of the *B* type of red cells. Likewise, a *B* individual has agglutinin *a,* the anti-*A* protein, in his plasma with the result

that he cannot receive blood from a person belonging to Group *A*. If agglutination occurs, the masses of clumped corpuscles may be sufficiently great to block the flow of blood through the capillaries. Also, there is a possibility that these agglutinated corpuscles might interfere with the functioning of the kidneys by accumulating in the kidney tubules, thereby preventing the flow of urine.

To prevent transfusion of blood that is not *compatible* (of the same group), the blood of the donor and the recipient must be properly classified. As a safeguard, the blood of the recipient and the donor may be *crossmatched* before transfusion. The blood of each person is tested with the serum of the other.

The table below summarizes the distribution of the *A* and *B* agglutinogens and the *a* and *b* agglutinins. It also shows what types of blood are compatible and may be used in transfusions.

The terms *universal donor* and *universal recipient* are frequently applied to members of Groups *O* and *AB,* respectively. In cases of emergency, Group *O* blood can be transfused into most people without ill effects. Although the plasma of this type of blood contains both *a* and *b* agglutinins, it does not cause agglutination when transfused. The reason for this is not clear. Even if one quart of blood is

Blood Group	Corpuscles Contain Agglutinogen	Serum Contains Agglutinin	Can Give Blood to	Can Receive Blood From
A	A	b	A, AB	O, A
B	B	a	B, AB	O, B
AB	AB	—	AB	A, B, AB, O
O	—	a, b	A, B, AB, O	O

	O	A	B	AB
U.S.A.—white	45.0%	41.0%	10.0%	4.0%
U.S.A.—Negro	49.3%	26.0%	21.0%	3.7%
Swedish	37.9%	46.7%	10.3%	5.1%
Japanese	31.2%	38.4%	21.8%	8.6%
Hawaiian	36.5%	60.8%	2.2%	0.5%
Chinese	30.0%	25.0%	35.0%	10.0%
Australian aborigine	53.1%	44.7%	2.1%	0.0%
North American Indian	91.3%	7.7%	1.0%	0.0%

used in the transfusion, the recipient's blood is not affected. Also, since Group O corpuscles lack agglutinogens, they will not clump in the presence of either a or b agglutinins. In a like manner, a person with Group AB is a universal recipient because his blood lacks any agglutinins which affect the corpuscles of the donor. Although the corpuscles of a member of this Group contain both A and B agglutinogens, they usually do not clump together.

The distribution of blood groups varies considerably among the races and peoples of the world. Blood group is an inherited characteristic. The table on this page shows how the different groups are distributed among a few peoples.

There are several other proteins in the blood which may bring about agglutination under certain conditions. The most important of these is the Rh factor, which, like the blood type, is an inherited characteristic. In the United States, approximately 85 percent of the people have the Rh factor, and are said to be Rh positive (Rh+). The Rh factor is not active until an Rh negative (Rh−) person has been given a transfusion of blood which is Rh+. An anti-Rh agglutinin may then form in the blood of the recipient, so that a later transfusion of Rh+ blood results in agglutination. Testing for the Rh factor is done in much the same manner as for the general blood type, as described earlier.

Occasionally, damage to a child's blood occurs before birth as the result of the presence or absence of the Rh factor. If the father of a child is Rh+ and the mother Rh−, the child may be Rh+. Normally, the blood of mother and child are not mixed, but in some cases bits of the child's blood cells enter the mother's blood stream. The mother then develops an antibody against the Rh factor. The antibody is seldom strong enough to affect the first child, but its concentration is increased in later pregnancies. This seriously affects the child, which develops erythroblastosis fetalis, a condition characterized by severe anemia and jaundice. If the child lives to be born, the condition is treated by means of numerous transfusions by which the child's blood is replaced with Rh− blood. The antibodies are thus removed. Blood from female donors is most effective. Why this sex difference in the blood types exists is an un-answered question, but experience has shown it to be the case.

Statistically, about 12.7 percent of marriages should pair an *Rh* positive man with an *Rh* negative woman. The number of women showing sensitiza-tion to the *Rh* positive factor is, how-ever, only 1/25th of the theoretically possible number. One factor that may account for the nonappearance of erythroblastosis fetalis among the chil-dren of such mating is that in most cases, the antibody does not reach a critical concentration in the mother's blood.

VOCABULARY REVIEW

Match the phrase in the left column with the correct word in the right column. *Do not write in this book.*

1 A straw-colored mass of fibers that forms the ba-sis of a blood clot.
2 The plasma protein that contains antibodies.
3 Loss of blood.
4 An abnormal clotting of blood within a blood ves-sel.
5 A plasma protein that catalyzes the formation of fibrin from fibrinogen.
6 A substance which prevents the clotting of blood within vessels.
7 The result of a severe injury or psychological dis-aster.
8 Inherited blood factors which may be racial in nature.
9 Thrombin is formed from this in the presence of thromboplastin and calcium ions.

a agglutination
b anticoagulant
c blood types
d edema
e fibrin
f fibrinogen
g gamma globulin
h hemorrhage
i physiological shock
j plastin
k prothrombin
l thrombin
m thrombus

TEST YOUR KNOWLEDGE

Group A

Select the correct statement. *Do not write in this book.*

1 The most abundant material in blood plasma is (a) fibrinogen (b) serum al-bumin (c) urea (d) serum globulin.

2 Of the following, the one *not* concerned with blood clotting is (a) calcium (b) thrombin (c) vitamin K (d) iron.

3 A disease in which blood does not clot readily is (a) anemia (b) hemophilia (c) hydrophobia (d) anoxia.

4 Of the following, the one that is *not* a plasma protein is (a) fibrinogen (b) globulin (c) pepsin (d) albumin.

5 Enlargement of body parts as the tissues become filled with fluid is called (a) anemia (b) edema (c) enema (d) nephritis.

6 Materials in blood that cannot normally pass through the capillary walls are (a) hormones (b) vitamins (c) colloids (d) simple sugars.

7 The clumping together of red corpuscles is called (a) agglutinin (b) agglutinogen (c) agglutination (d) coagulum.

8 Universal blood donors have blood type (a) A (b) B (c) AB (d) O.

9 The percentage of people in the United States that have the Rh positive factor in their blood is about (a) 15% (b) 35% (c) 85% (d) 65%.

10 The formation of a blood clot is called (a) thrombin (b) coagulation (c) cephalin (d) fibrinogen.

Group B

1 What role does each of the following play in the formation of a blood clot (a) fibrinogen (b) calcium (c) thrombin.

2 What are plasma proteins, and what part does each play in the physiology of circulation?

3 (a) Write a brief description of blood types, indicating the chemistry involved. (b) Discuss the precautions that must be taken before making a blood transfusion.

4 What effect does hydrogen ion concentration have on the respiratory function of the blood?

5 Why is vitamin K sometimes administered to patients prior to a surgical operation?

6 Explain the terms "universal donor" and "universal recipient" with respect to blood types.

Chapter 24 **The Heart**

Structure of the Heart

The human *heart* is a four-chambered, muscular pump that owes its unique ability to contract rhythmically to the presence of cardiac muscle tissue. This organ lies behind the sternum and within the mediastinal cavity between the lower halves of the lungs. (See the Trans-Vision insert between pages 212 and 213.) About two-thirds of its bulk lies to the left of the middle line of the body, with the apex located between the fifth and sixth ribs (Fig. 24-1). In an adult male of average size, the heart is about 11.9 centimeters long, 8.4 centimeters wide at its greatest breadth, and 6 centimeters in thickness. Very roughly, these dimensions correspond to the size of the person's clenched fist. In men, the heart weighs from 281 to 340 grams, while in women it is between 229 and 281 grams in weight. As people grow older, the average size and weight of the heart increases more rapidly in men than in women. Before birth, we have approximately the same number of muscle cells in the heart that we will have throughout the remainder of our lives; any increase in the size of the organ as we

grow is due to an increase in the size of the individual cells. As the heart grows, the cells add new cytoplasmic material so that their diameter increases to 2.6 times their original size.

The heart is divided into two distinct halves by a muscular wall, the *septum*. Before birth, however, and for a brief time afterwards, these halves are connected by a small opening through the septum, called the *foramen ovalis* (oval window). This allows blood to flow directly from the right side of the heart to the left side without going through the lungs, since in the fetus the lungs have no oxygenating function. Shortly after the time of birth this opening closes, thus separating the two halves of the heart. From this time on, the right half receives only blood that contains deoxygenated hemoglobin, while the left side receives only blood containing oxygenated hemoglobin.

The heart lies within a double-walled sac, the *pericardium,* whose lower margin is anchored to the diaphragm with fibers extending to the inner surface of the sternum. Above the heart, it surrounds the bases of the large blood vessels. It is composed of a tough, fibrous outer part, the fibrous pericardium, and a thin inner layer, the serous pericardium. Although a space theoretically separates these two layers, it is nonexistent because of the presence of the *pericardial fluid*. This fluid prevents the two layers from rubbing against each other. Thus the heart is separated from the rest of the thoracic cavity, enclosed in a sac that serves as a protection against mechanical injury.

Each side of the heart is divided into two chambers. Those at the top of the heart are called the *atria* (sing., *atrium*), and below them are the *ventricles*. As Fig. 24-2 shows, the openings between the atria and the ventricles are guarded by *atrioventricular valves,* which prevent blood from passing back into a chamber. The valve between the right atrium and the right ventricle is the *tricuspid valve,* this name having been

Fig. 24-1 *Front view of the human heart showing the relation of the various parts and the origin of the great vessels. The coronary circulation supplies the heart with blood.*

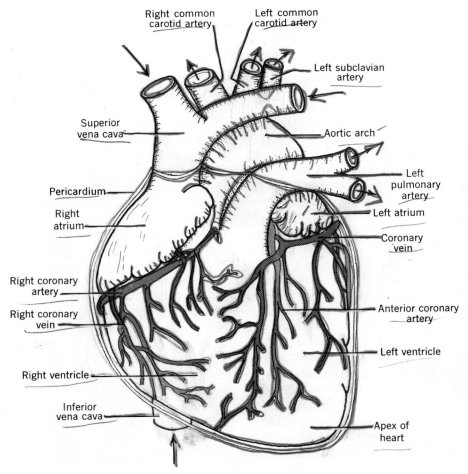

Right common carotid artery

Left common carotid artery

Left subclavian artery

Superior vena cava

Aortic arch

Pericardium

Left pulmonary artery

Right atrium

Left atrium

Coronary vein

Right coronary artery

Right coronary vein

Anterior coronary artery

Left ventricle

Right ventricle

Inferior vena cava

Apex of heart

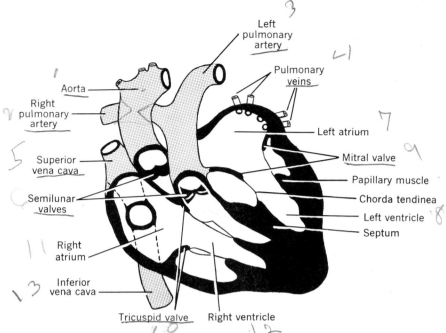

Fig. 24-2   A longitudinal section through the human heart.

given it because it has three points or cusps of attachment on the floor of the right ventricle. The opening between the left atrium and the left ventricle is controlled by the *mitral (bicuspid) valve.*

The Cardiac Cycle

The sequence of heart action, called the cardiac cycle, may be said to have three stages, as shown in Fig. 24-3. Blood enters the atria through the *pulmonary veins,* and the *superior* and *inferior venae cavae.* The former bring blood from the lungs, while the latter drain the head, neck, arms, and thorax, and the trunk and legs, respectively.

The walls of the atria are relatively thin and weak because these chambers simply receive blood from the veins and pass it into the ventricles. While the atria are filling, the valves between them and the ventricles remain open so that there is a continuous flow from the veins through the atria into the ventricles. As the ventricles fill, pressure builds up within them. When the blood in the ventricles begins to close the tricuspid and mitral valves, the muscles in the walls of the atria contract to force all the blood that remains in them into the ventricles. For the brief time of the contraction there is no blood in the atria, since the pressure within them is greater than the pressure in the veins entering them.

As the ventricles fill with blood, they become distended. Then a wave of muscular contraction sweeps over them, beginning at their lower ends and moving upward toward the atria. This not only closes the atrioventricular valves, but also forces blood out of the ventricles through either the *pulmonary artery* or the *aorta*. The blood flows from the right ventricle through the pulmonary artery and its branches to the capillaries in the lungs. Here the blood gives up its carbon dioxide and takes up oxygen. The blood containing oxyhemoglobin then returns through the pulmonary veins to the heart, entering the left atrium. Then it passes the mitral valve into the left ventricle. When the ventricle contracts, the blood is forced out through the aorta to all parts of the body except the lungs.

In order to prevent the atrioventricular valves from being forced back into the atria as the pressure builds up in the ventricles, the lower edges are anchored to the walls of the ventricles by strong elastic tendons, the *chordae tendineae*. The lower ends of the chordae are attached to the *papillary muscles* in the walls of the ventricles.

As a result, the atrioventricular valves are prevented from bulging to any great degree into the atria as the ventricles contract. Since during the contraction the ventricles shorten slightly, the contraction of the papillary muscles compensates for this decrease in length. If any injury to the valves occurs, the efficiency of the heart is lowered, since part of the blood flows back through the valves. The atrioventricular valves are surprisingly tough structures; with each normal contraction of the ventricles they must withstand a pressure of approximately 232 millimeters of mercury. During violent exercise, this pressure may exceed 360 millimeters of mercury.

When the blood leaves the right ventricle, it passes through the *semilunar valves* into the pulmonary artery. These valves are pressed against the sides of the artery by the force of the ventricle's contraction. As the ventricle relaxes, the back pressure of the blood in the pulmonary artery forces

Fig. 24-3   The sequence of the cardiac cycle. In A the atria and the ventricles are filling and the semilunar valves are closed. In B the atria are contracting. In C the ventricles are in systole with the atrioventricular valves closed and the semilunar valves open. The dotted lines show the relative size of the heart during diastole.

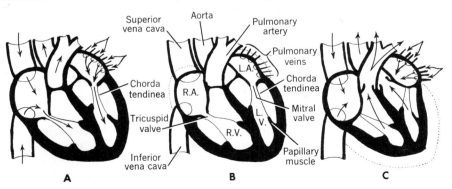

Atrial diastole	0.7 sec. (period of filling)
Atrial systole	0.1 sec. (active contraction)
Ventricular diastole	0.5 sec. (period of filling)
Ventricular systole	0.3 sec. (active contraction)
Quiescent period for entire heart	0.4 sec.

these valves closed. This prevents blood from passing back into the ventricle. A similar set of semilunar valves controls the flow of blood from the left ventricle into the aorta.

The period of the filling of the atria and the ventricles is spoken of as their *diastole*. This is a period of relaxation in which the chambers become distended. This is followed by a period of active contraction, or *systole,* which is marked by a shortening of the muscle bundles. If we take as a basis for our observations a heart beating at the rate of 70 times per minute, the duration of these phases is as shown in the table.

It should be noted that the total elapsed time for a complete cycle of both the atria and the ventricles is 0.8 second. During these cycles there is a period of 0.4 second when both the atria and ventricles are filling. This results in a quiet period when the heart muscles are not actively contracting.

The amount of blood leaving the heart with each contraction is about 160 milliliters when an adult is resting: 80 milliliters are forced out of the left ventricle into the aorta, and 80 milliliters enter the pulmonary artery from the right ventricle. This is known as the *stroke volume* of the heart. Assuming a rate of 70 beats per minute, we find that a total of approximately 5.6 liters (70 × 80) of blood is leaving each ventricle. This is known as the *minute volume* of the heart. If both ventricles are considered, we find a *cardiac output per minute* of 11.2 liters (1 liter = approximately 1.06 quarts). With an increase or decrease in stroke volume, pulse rate, or in both, the cardiac output per minute increases or decreases. When exercising, the cardiac output is greatly increased.

A study of this information shows that the heart is not a constantly moving organ. Since both atria fill with blood at the same time and contract simultaneously, they have a moderately long period of relaxation. The same is true for the ventricles. Of course, if the heart rate is increased above 70 beats per minute, the time intervals are correspondingly shortened. The differences in times required for the atria and ventricles to fill and contract are directly connected with the amounts of muscle present in the walls of these chambers.

There are three characteristics of a single cardiac cycle that must be kept in mind: the *origin* of the beat which results from the contraction of the cardiac muscle fibers in a coordinated and rhythmic manner; the *rate* at which the beat occurs and the various factors influencing it; and the *rhythm* the atria and the ventricles show during a single beat. The last two are under the control of nervous impulses and will be discussed shortly.

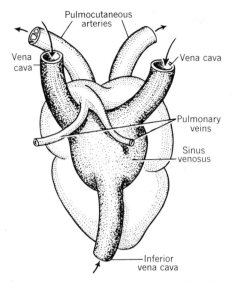

Pulmocutaneous
arteries

Vena
cava

Vena cava

Pulmonary
veins

Sinus
venosus

Inferior
vena cava

*Fig. 24-4   The dorsal surface of a frog's
heart showing the sinus venosus.*

In order to understand the fundamental processes underlying the beat of the heart, it is helpful to examine briefly the structure of the hearts of some of the lower vertebrates such as the frog and the turtle. In both of these animals there is a special chamber, separate from the atria, into which the venae cavae empty. This chamber is the *sinus venosus* (Fig. 24-4). In the course of the evolutionary development of the mammalian heart, it has become incorporated into the walls of the right atrium as a small structure known as the *atrial appendage,* which still retains one of the characteristics of the sinus venosus. If the beating heart of a frog is examined closely, it is seen that the wave of contraction that sweeps over it has its origin in the sinus venosus. In the human heart a group of nerve cells, comparable to those that

initiate the beat of the frog's heart, is located in the atrial appendage near the opening of the superior vena cava into the right atrium. These nerve cells form a very distinct area that can be isolated and stimulated artificially. When heat or an electric current is applied to them, the rate of the heart beat increases. If they are cooled, impulses flow more slowly and there is a corresponding decrease in the heart rate. This island of nerve tissue is known as the *sinoatrial node.* It is sometimes referred to as the *pacemaker* of the heart.

When the heart beats, a wave of nervous excitation from the sinoatrial node spreads out over the walls of the atria, causing their muscular layers to contract. This forces blood from the top of the chambers downward toward the atrioventricular openings. As the wave of nervous impulse approaches the junction between the atria and the ventricles, it is collected into another group of nerve fibers that make up the *atrioventricular (A-V) node.* From this, other fibers pass into the septum of the heart to form the *atrioventricular bundle.* This is also known as the *bundle of His,* in honor of the man who discovered it. It will be noted in Fig. 24-5 that the atrioventricular bundle divides into right and left branches. From each of these arise special cells which form a conducting system within the walls of the heart. These make up what is known as the *Purkinje network.* The cells in this net are somewhat similar to cardiac muscle cells, except that there is a great deal of glycogen stored in their cytoplasm and the muscle fibrillae lie lengthwise along the margins of the cells. It has been

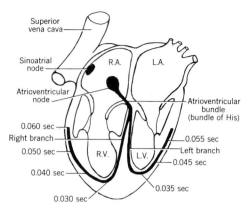

Fig. 24-5 The conducting pathways in the heart. The times indicated along the branches of the atrioventricular bundle show the rate of the passage of the nerve impulse.

found that these cells are able to conduct impulses faster than the ordinary cardiac muscle cells.

Nerve impulses pass along the right and left branches of the atrioventricular bundle to the apex of the heart and stimulate the muscle fibers in this region first. As Fig. 24-5 shows, the nervous impulse then passes upward along the walls, the time intervals for this passage being indicated in the diagram. The result of this method of the propagation of the impulse is that the bottoms of the ventricles contract first. The wave then passes upward rapidly, increasing the pressure within the chambers and forcing the blood to close the atrioventricular valves and open the semilunars. The blood then leaves the ventricles through the arteries.

Control of the Heart Beat

Nervous impulses, constantly flowing between the heart and the central nervous system, control the rate of the heart beat. These impulses are of two types: one that *inhibits,* or slows down the rate, and the other that *accelerates* it. The impulses travel over the fibers of the autonomic nervous system. By their mutually antagonistic effect, they keep the heart rate steady.

We have already mentioned the *vagus nerves* in connection with the rate of respiration, because certain branches of these nerves pass to the respiratory apparatus. They may be easily found in an anesthetized animal. One member of the pair lies on either side of the larynx. If one of these nerves is cut, there is little effect on the rate of the heart beat, but if both are severed, the rate is greatly accelerated and quickly reaches a high uniform level. Stimulation of these nerves will have an inhibitory effect on heart action. This may be shown by either a reduction in the rate, the lessening of the force of the beat without a change in rate, or the stoppage of the heart during diastole. It appears, therefore, that of the two types of impulses entering the heart by way of the vagus nerves, the inhibitory ones are the more powerful in maintaining a constant rate (Fig. 24-6).

Nerves carrying accelerating impulses also pass to the heart from one of the *cervical sympathetic ganglia.* The connection between the ganglion and the spinal cord eventually leads to the *cardioaccelerating* center in the medulla. This has connections with the *cardioinhibitory* center from which the slowing impulses arise. Thus, if the sympathetic nerve is stimulated, the rate of the heart beat increases and the flow of the inhibitory impulses is lessened.

Within the heart itself and also in the walls of some of the large vessels, there are nerve endings which help to control the rate of the heart beat. These nerves may be grouped into three categories as follows, depending on their location.

Nerve endings in the walls of the venae cavae and the right atrium are stimulated by the presence of blood in the cavities. Whenever a group of body muscles becomes very active, their pressure against the walls of the veins they contain has a milking action on these vessels and forces more blood into the general circulation. This increased flow into the veins results in increased pressure within the right atrium and the venae cavae. In order to prevent the added amount of blood from slowing down as it passes through the heart, the rate of heart beat increases. This burst of speed is known as the *Bainbridge reflex.*

Another reflex originates in the walls of the aortic arch, where there are nerve endings called the *stretch receptors (pressoreceptors).* If the pressure of blood in the aorta becomes excessively high, these receptors are stimulated by the stretching of the walls of the vessel. Impulses from them pass along the vagus nerve to the cardioinhibitory center and the rate of heart beat is reduced, with an accompanying drop in blood pressure.

If the blood pressure drops to a very low point, as it does following severe hemorrhage, these pressoreceptors are no longer stimulated and their inhibiting action is removed. This results in a rapid rise in heart rate and the small amount of blood left in the system is sent throughout the body as rapidly as possible.

The *carotid artery* passing to the right side of the head branches off from the aorta first as the *innominate artery* (Fig. 24-6). The innominate branches again to form the *right subclavian artery* that carries blood to the shoulder and arm region, and the *right common carotid artery.* From this latter arise two arteries, the *right internal carotid artery* and the *right external carotid.* The external branch supplies blood to the outer surfaces of the head, while the internal division carries blood to the brain. At the point of their separation, there is a small swelling in the internal carotid artery, known as the

Fig. 24-6 The principal nerve pathways connecting the heart and the central nervous system. The arrows indicate the directions in which the nerve impulses travel.

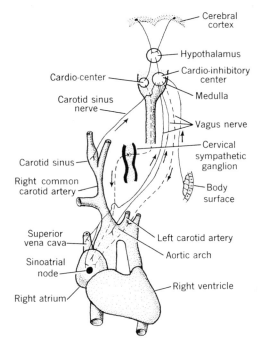

carotid sinus. Within its walls are stretch receptors from which nerve fibers pass along the ninth cranial nerves, the glossopharyngeal nerves, to the cardioinhibitory center. These form the third group of controlling nerves. If the carotid sinus becomes distended with blood, as might occur with high blood pressure, impulses stimulate the center in the medulla and the heart rate is decreased with an accompanying fall in pressure. On the other hand, if there is a decrease in blood pressure the receptors are not stimulated and the rate of the heart beat accelerates.

The name carotid comes from the Greek word, *karos,* meaning "sleep." The ancients found that certain people could be made to sleep by pressure on the carotid arteries below the region of the sinus. This effect is caused by reduction in the amount of blood in the sinus, with the accompanying drop in blood pressure in the vessels of the brain. Today we would call this loss of consciousness a fainting spell. Fainting is the result of a drop in blood pressure in the vessels supplying the brain, due to any of a number of causes, including psychological disturbances. One of the common methods of preventing a fainting spell is to lower the head below the general level of the body. This permits more blood to flow to the sinus so that the rate of heart beat is lowered, but the strength of the beat is increased. This results in a rise in the pressure within the vessels in the brain.

Exercise affects heart action. It has been observed that even the slight exercise involved in clenching the fist brings about an immediate response in the heart rate. As soon as the motion of closing the hand begins, the next dia-

stolic phase of the heart is shortened. This indicates a reflex effect, possibly started by the nerves associated with the muscles. Strenuous exercise which involves large numbers of muscles soon has a pronounced effect on the heart beat rate because the increased flow of blood toward the heart also activates the Bainbridge reflex. The reflexes that are established in muscles result in the stimulation of the cardioaccelerator center, an increase in the activity of the sympathetic accelerator nerves, and a repression of the cardioinhibitory center with a reduction in the strength of the impulses leaving it. Under the influence of these various factors, the heart rate may rise to extremes of around 190 beats per minute.

In addition to neural control of the heart rate, there is chemical control through a number of substances. If the hearts of two animals are connected in the manner shown in Fig. 24-7 and the vagus nerve stimulated electrically,

Fig. 24-7 A diagram to show how a stimulating substance is passed from one heart to another.

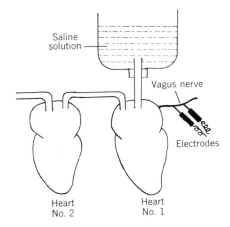

Saline solution

Vagus nerve

Electrodes

Heart No. 2

Heart No. 1

heart No. 1 will be inhibited. If this state of inhibition is continued, heart No. 2 will also become inhibited shortly. This indicates that some material formed in No. 1 passes to No. 2 through the saline solution. It is believed that the substance formed in the first heart is a minute quantity of acetylcholine. The sensitivity of the heart to this compound is so great that 0.000001 milligram of it will produce a noticeable inhibiting action.

Many chemical compounds affect the nerve endings in the heart. For example, *atropine,* the active substance obtained from deadly nightshade, has an action comparable to that obtained by cutting both vagus nerves: the heart beat is greatly accelerated. *Muscarine,* the poisonous substance found in some mushrooms, acts like acetylcholine, and will inhibit heart action to the point where the beat stops entirely. The action of both of these inhibitors can be neutralized by atropine. *Nicotine* has a double action in that it first stimulates the nerve endings and then paralyzes them. It also causes constriction of the coronary arteries that supply blood to the heart muscle. Since the effect of this drug on the heart is a pronounced one, people suffering from certain types of heart disease may be advised to stop smoking.

In a later chapter we will discuss the effects of *adrenaline* on the rate of the heart beat. This substance is a powerful heart stimulant even when present in very minute quantities. Some years ago, Dr. Cannon and his associates found that if all of the nerves to the heart except those arising in the sympathetic ganglia were cut, the rate of the heart beat was slowly accelerated.

This increase took place as much as a minute after cutting the nerves, and lasted two or three minutes. This occurred even after the adrenal glands had been removed from the body of the anesthetized animal. The cause of this acceleration was attributed to the formation of a substance called *sympathin.* Some conflicting results in later experiments indicated that the sympathin could, under certain conditions, excite the heart (sympathin E), but under other circumstances would inhibit its action (sympathin I). This led to the theory that there were actually two different substances involved. More recent work has shown that the adrenal gland does produce two different substances which have the same effects as sympathin E and sympathin I. These are *noradrenaline* and *adrenaline.* The only known difference between these hormones and sympathin is that they are produced in different organs.

Rate of Heart Beat

The rate at which the heart normally beats is determined by many factors. Age, sex, position of the body, amount of physical activity, temperature of the surroundings, thought processes are all reflected in the rate of the heart. Fig. 24-8 shows how the rate of the beat varies with age. Before birth the rate is high, but there is a steady decline after birth until a fairly constant average is reached. The vertical bars show the limits that can be expected among healthy individuals, while the short horizontal bars indicate the average for the group. Thus a person 15 years old may have a normal resting rate between 59 and 98 beats per minute, and

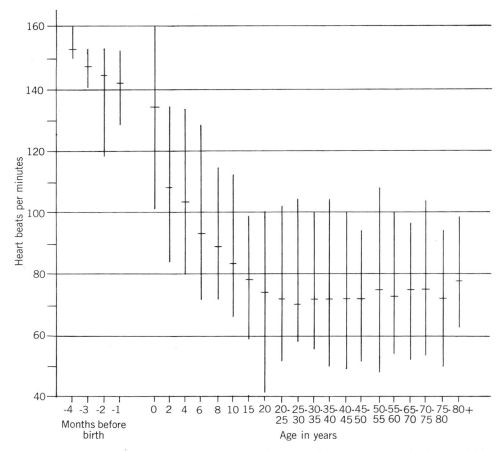

Fig. 24-8 The relation of age to heart rate. The vertical lines represent the limits within which the normal heart rates may be expected to fall when a person is resting. Mean (average) rates are shown by the small horizontal lines.

the average rate is 72 beats for the entire group. Since these data have not been separated on the basis of the sex of the individuals included, they do not show that women have a slightly higher average rate than men. A sleeping man may average about 59 beats per minute while a woman has a heart rate of approximately 78 while sleeping. On waking, both sexes show an increase,

the man's rate going up to an average of 75 beats, while the woman's increases to approximately 84 beats per minute.

Bioelectric Currents

In common with all muscles, the heart develops an electric current when it contracts. This arises within the muscle itself as a result of the movements of ions across cell membranes. Known as the *action current,* it flows between any active tissue such as a muscle, nerve, or gland and an inactive region. In the case of the heart, the production of this current is especially important from a medical standpoint because it means that the various stages in the heart beat can be studied with a great deal of precision.

The presence of an action current in the heart can be demonstrated in anesthetized animals. If an electrode is placed on the atrium of a beating heart, and another on the ventricle, and the two connected with a galvanometer, a definite flow of electrical current can be shown. Three stages in the passage of the current are evident. When the atrium contracts, the current flows from the atrium to the ventricle. This is followed by a short period of neutrality in which no current flows, and then the ventricle contracts and the current flows from the ventricle to the atrium.

The study of the electric changes in the human heart was made possible by the invention of the *string galvanometer* by Einthoven in 1903. This consists of a large electromagnet between the poles of which is suspended a very delicate string made of quartz. The surface of this thread, which is about three microns in diameter, is coated with silver so that any change in the electric field of the magnet will affect it. When the current through the coils of the magnet changes its direction of the flow, from the south pole toward the north or vice versa, the thread will be deflected rapidly. Since this thread is extremely delicate, its inertia is small and it responds quickly to any change

in the direction of the current. In addition to the string and the magnet, the instrument contains an optical system that throws a bright beam of light on the string. The shadow formed by the thread as it moves in the magnetic field is then projected on a piece of sensitive photographic paper that is moved behind the thread at a constant rate of speed. This photographic record of the electric changes can be analyzed and correlated with the activities of the heart. A record of this type is called an *electrocardiogram* (ECG), and the instrument that is used to obtain it is known as an *electrocardiograph.*

The currents that are generated by the heart pass through the tissues of the body and can be recorded at various points on the surface. To make an electrocardiogram, metal electrodes applied to the skin lead the currents present in the area into the electrocardiograph. Since the various tissues have differing abilities to conduct currents, the location of the electrodes is of extreme importance. Experience has shown that certain regions produce more uniform and easily interpreted results than others. The electrodes are usually applied in the places shown below; these combinations are known as the *standard leads.*

Lead I	right arm and left arm
Lead II	right arm and left leg
Lead III	left arm and left leg

Since each of these pairs of electrodes is in a different position, each shows certain characteristics of the heart beat that differ from each of the others. In Fig. 24-9 is seen a diagram of a typical wave of electricity as it is

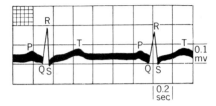

Fig. 24-9 The parts of a normal electro-cardiogram. Each vertical division represents 0.0001 volt (0.1 millivolt). The larger squares are divided into smaller units as shown in the upper left corner.

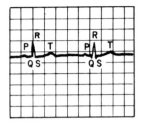

Lead I

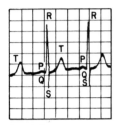

Lead II

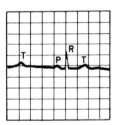

Lead III

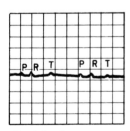

Chest lead

recorded by Lead I electrodes. Each of the parts of the wave has been assigned a letter, which is purely arbitrary, to identify specific events occurring during the passage of the electric current. The parts of the tracing that rise above the base line are formed by the passage of a positive wave, while those that drop below it indicate a negative wave. Thus, the P wave shows the passage of a positive current over the atria. The Q wave, which may be absent in a normal electrocardiogram, shows a period of electric neutrality or slight negativity. The R wave is tall, upright (positive) and sharply pointed. This represents the period of the contraction of the ventricles. Following this is another brief moment of neutrality which is followed by the T wave showing a positive electric condition. In Fig. 24-10 are shown the characteristic forms taken from the three standard leads.

In addition to these standard leads, other regions of the body may be used to produce their own characteristic electric waves. A record from a commonly used chest lead is shown in Fig.

Fig. 24-10 Drawings of photographic reproductions of normal electrocardiograms (Edward F. Bland, M.D.)

24-10. To obtain a tracing of this type, one electrode is placed on the left leg and the other on the chest in the region of the apex of the heart. The tracing is quite different from the standard leads Here we have a striking example of the result of the difference in the conductivity of body tissues.

Electrocardiograms are of great value to the physician in diagnosing defects of the heart. Some heart diseases can be identified by the abnormal sounds the heart makes, while others can only be detected by the slight changes that occur in the electrical conductivity of the heart muscle. For the diagnosis of such conditions, the electrocardiogram is of great importance.

In Fig. 24-11 is seen an electrocardiogram of a person afflicted with *auricular fibrillation*. This is the most common of all serious heart irregularities. In this condition, the atria are never completely emptied of blood and

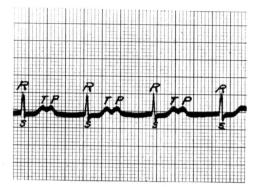

Fig. 24-12   Electrocardiogram of a case of delayed conduction. Note the interval between the P *and* R *waves and compare this interval with that shown in the normal Lead I tracing in Fig. 24-10. (Edward F. Bland, M.D.)*

their walls quiver instead of giving the pronounced contraction typical of a normal heart beat. The impulses which the atria may develop in this condition range from 400 to 600 flutters per minute. Only a few of these impulses pass through the atrioventricular bundle to activate the ventricle. This accounts for the fact that the *R* wave appears only after a succession of many minor waves that take the place of the normal *P* wave.

A second heart ailment that can be identified by the electrocardiogram is *delayed conduction* of the impulse from the atria to the ventricles. In this, the time interval between the contraction of the atria and the ventricles is greatly lengthened. Instead of the normal 0.2 second, the delay may be as much as 0.5 second. In this condition, the atrioventricular bundle fails to conduct the impulse in a normal manner (Fig. 24-12).

Fig. 24-11   An electrocardiogram of a person suffering from auricular fibrillation. Note the series of P *waves before the appearance of the* R *wave. (Edward F. Bland, M.D.)*

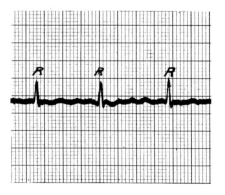

Heart Sounds

A physician listens to the heart beat for two reasons. First, by noting the beginning and end of the ventricular systole, he can time other events in the cardiac cycle. Second, the sounds produced by the heart indicate the condition of the atrioventricular and semilunar valves.

There are two heart sounds, each of which is related to a phase in the cardiac cycle. The first sound is the result of the closing of the atrioventricular valves and the contraction of the muscles in the walls of the ventricles. This is a low-pitched sound of some duration that coincides with the ventricular systole. Actually, it starts a fraction of a second before there is visible evidence of contraction in the ventricles and represents the instant at which the atrioventricular valves close. In relation to the electrocardiogram, the beginning of this sound corresponds to the upstroke of the *R* wave and continues to a point between the *S* and *T* waves. It sounds like the word *dub,* and can be heard best over the region of the apex of the heart, between the fifth and sixth ribs to the left of the sternum.

The second sound is much shorter in duration and of a higher pitch. This sounds like the word *dip,* and represents the sudden closing of the semilunar valves. It is heard most distinctly in the second intercostal space, between the second and third ribs, near the sternum. When it is correlated with an electrocardiogram, it is heard just before the appearance of the *T* wave.

Certain abnormal sounds called *murmurs* may be heard occasionally. They are generally low in intensity and quite difficult to identify without considerable practice. These unusual sounds may indicate some imperfection in the valves of the heart and may take the form of indistinct gurgling sounds, or hissings, as the valves fail to close properly. If the murmurs occur during systole, they may have little significance. They may be the result of a fever or undue physical exertion. If, however, they are heard during the diastolic phases of the heart beat, they may be due to failure of valves to close properly (incompetence) or a tightened condition of valves (stenosis).

VOCABULARY REVIEW

Match the phrase in the left column with the correct word in the right column. *Do not write in this book.*

1	The first chamber to receive deoxygenated blood.	**a** aorta
2	The major artery carrying deoxygenated blood.	**b** chorda tendina
3	Prevents the backflow of blood to a ventricle.	**c** diastole
4	Branches will eventually carry oxygenated blood to the legs.	**d** inferior vena cava
		e left atrium
5	Brings blood to the heart from the brain.	**f** left ventricle
6	Prevents an atrioventricular valve from bulging into the atrium.	**g** mitral valve
		h pulmonary artery

7 Vein carrying oxygenated blood.
8 Contains the most cardiac muscle tissue.
9 The period of filling of a chamber of the heart.
10 Carries blood from most body organs to the heart.

i pulmonary vein
j right atrium
k right ventricle
l semilunar valve
m septum
n superior vena cava
o systole
p tricuspid valve

TEST YOUR KNOWLEDGE

Group A

Select the correct statement. *Do not write in this book.*

1 A double-walled membrane surrounding the heart is called the (a) periosteum (b) perilymph (c) pericardium (d) peritoneum.
2 The coronary arteries supply blood to the (a) heart (b) stomach (c) pancreas (d) spleen.
3 The heart is divided into two halves by the (a) foramen ovalis (b) foramen magnum (c) septum (d) sacrum.
4 The valve between the right atrium and the right ventricle is the (a) bicuspid valve (b) incisor valve (c) tricuspid valve (d) mitral valve.
5 Blood enters the left atrium of the human heart through the (a) venae cavae (b) pulmonary veins (c) sinus venosus (d) atrial appendage.
6 The artery that carries blood to the head is the (a) innominate (b) carotid (c) subclavian (d) brachial.
7 Papillary muscles are found in the (a) skin (b) atria (c) ventricles (d) chordae tendineae.
8 The period of filling of the atria and ventricles is called (a) diastole (b) diastase (c) systole (d) systemic.
9 Oxygenated blood enters the heart in the (a) right atrium (b) left atrium (c) right ventricle (d) left ventricle.
10 The normal period of time for one complete cardiac cycle in an adult is about (a) 0.05 seconds (b) 0.5 seconds (c) 1.5 seconds (d) 5 seconds.

Group B

1 Trace the path of the blood as it passes through the heart naming the chambers and vessels involved. Indicate the regions of the body from which the blood enters the heart and the regions to which it passes after leaving the heart.
2 What tissue is responsible for the beat of the heart?
3 What effects does the nervous system have on the rate of the heart beat?
4 How is the rhythm of the heart beat maintained?
5 What effect does each of the following have on the heart: (a) adrenalin (b) alcohol (c) nicotine (d) muscarine (e) atropine (belladonna) (f) acetylcholine?

Chapter 25 Circulation of Body Fluids

THE CIRCULATION OF BLOOD

After the blood leaves the heart, it passes through two main channels. One goes to the lungs and forms the *pulmonary circulation*. The other carries blood to all other parts of the body and is called the *systemic circulation*. As we have stated previously, arteries always carry blood away from the heart while veins bring blood to the heart. These two principal types of vessels are connected by the capillaries.

Of the three types of blood vessels, the arteries are the strongest because of the thickness of their walls. Furthermore, since most of the larger arteries are closely surrounded by muscles and other tissues, they are able to withstand sudden large increases in internal pressure better than the other vessels. The walls of arteries are composed of three layers of tissues. On the outside is a layer of connective tissue cells throughout which are scattered bundles of smooth muscle cells. This outermost covering of the artery (Fig. 25-1) constitutes the *tunica externa*. In the larger arteries there is a thin layer of connective tissue which at-taches the tunica externa to the middle layer of smooth muscle fibers. The cells making up the muscular layer of the *tunica media* are arranged in a circular pattern so that the diameter of the artery is decreased or enlarged as they contract or relax. The innermost *tunica intima* is an extremely thin layer of endothelial cells. These give the lining of the arteries a smooth surface which does not appreciably obstruct the flow of blood.

The tunica media has the most important function. Because of the smooth muscle fibers and connective tissue in this layer, the arteries and arterioles are highly elastic. As a wave of blood flows through them, they dilate, and after the wave has passed, they constrict. This action is important because it allows the blood to pass freely through the vessels and keeps the flow even and steady.

The walls of some of the larger arteries are so thick that their cells are unable to obtain sufficient food and oxygen from the blood passing through them. It is necessary, therefore, for them to have their own system of capillaries. The arteries also have an extensive nerve network to control their dilation and thereby limit the amount of blood passing through them.

The pulmonary artery and the aorta branch to form the arterioles, some of which have internal diameters scarcely larger than the width of red blood corpuscles. These smallest arterial branches have the same wall structure as the great arteries, although their size makes it difficult to distinguish them from capillaries. The arterioles give rise to the smallest vessels, the capillaries.

Capillaries are branches of the finest divisions of the arterioles, the *metarterioles*. These small vessels have lost the majority of their muscle and connective tissue layers. When the last traces of these tissues have disappeared and there remains only a single

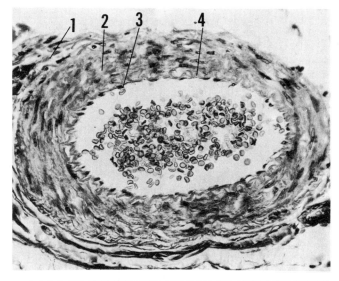

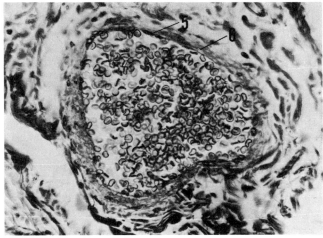

Fig. 25-1 Cross sections through an artery (above) and a vein (below). Tissues of the artery wall: 1, tunica externa; 2, circular muscle layer, the tunica media; 3, endothelial lining, the tunica intima; 4, elastic tissue. Tissues of the vein wall: 5, lining of endothelium; 6, thin layer of circular muscle.

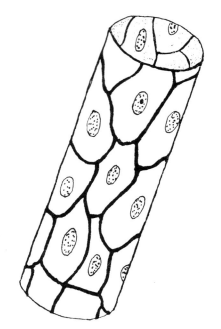

Fig. 25-2   The cellular structure of a capillary.

layer of endothelial cells (Fig. 25-2), the capillary bed is established. Endothelial cells which form the walls of the capillaries may be considered a continuation of the tunica intima of the larger vessels. These cells are held together by a cementing substance through which leukocytes and other small particles can enter and leave the blood stream. Although the cement is permeable to some large particles, this permeability is highly selective. There is no loss of proteins through the capillary walls.

The internal diameter of the capillaries averages approximately 8 microns, but some may be 15 or 20 microns in diameter while others are five microns, or less. In these narrow chan-

nels there is no rapidly flowing blood stream as there is in the larger vessels. This slow movement of blood is advantageous in that it permits an exchange of materials between the blood and the cells. As Fig. 25-3 indicates, the rate of the flow of blood decreases as the vessels become smaller and more numerous. The length of each capillary is between 0.4 and 0.7 millimeters.

It has been estimated that the total area of the capillary walls in human voluntary muscle approximates 6,000 square meters (1 square meter = 10.764 square feet). With this amount of surface exposed in this one type of active tissue, the exchange of materials between the blood and the cells is greatly accelerated.

Although the walls of the capillaries lack muscle cells, the flow of blood through them can be controlled to meet the requirements of the tissue they serve. This control is exercised by small bands of muscle, the *precapillary sphincters,* located at the point where the capillary branches from an arteriole or metarteriole (Fig. 25-4).

One of the striking features of the capillary bed is that not all capillaries are open at the same time. This adjusts the flow of blood to those tissues that are most active. In the brain, for example, the majority of the capillaries remain open, but in a resting muscle only 1/20 to 1/50 of the capillaries carry blood to the fibers. It has been found that in actively moving muscles there may be as many as 190 capillaries per square millimeter open to supply food and oxygen to them. When these same muscles are resting, there may be only 5 capillaries per square millimeter in active use.

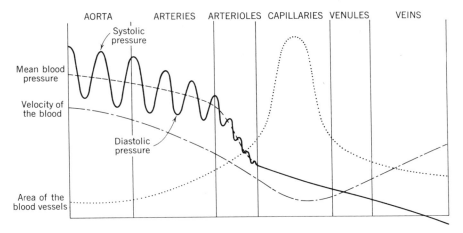

Fig. 25-3 The relation between blood pressure and velocity, and the area of blood vessels.

Some of these differences can be accounted for by the nature of the capillaries of which there are two types. One type connects the arterioles and venules directly. These capillaries are always open and offer little resistance to the flow of blood. The other type is formed by offshoots of the arteriole-venule system, and the flow of blood through these capillaries can be regulated by the requirements of the particular tissue.

The veins are formed by the joining together of capillaries. The distinction

Fig. 25-4 Capillary bed. The thickened walls of some of the vessels indicates the presence of muscle.

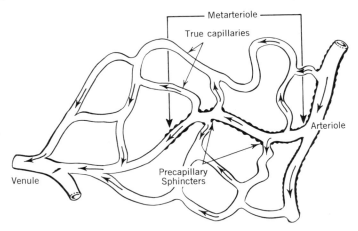

between a capillary and the smallest part of the venous system, the *venule*, is in the nature of the vessel walls. The smallest venules are scarcely larger than a capillary, but they have muscle fiber which is entirely lacking in capillaries. The venules rapidly converge to form larger vessels which eventually unite with others to form the great veins entering the heart. The walls of the veins are thinner than those of the arteries because they do not have such thick layers of muscles. This is not a handicap, since veins are not required to withstand the high internal pressures to which arteries are subjected.

The inner lining of veins is endothelial tissue. It differs from the lining of arteries in that it forms *valves* which help to direct the flow of blood. These are composed of semilunar folds of tissue with their free ends pointing toward the heart. When blood flows toward the heart, they are pressed back against the walls of the vessels so that the blood can flow freely. If a back-pressure develops, the valves close. This prevents the blood from flowing back toward the capillaries.

Valves in the veins are of special importance in those regions of the body where blood has to be raised against the force of gravity, as in the lower extremities. Valves in the superficial veins of the legs may occasionally become abnormally lengthened and dilated. This condition, known as *varicose veins*, usually develops spontaneously, presumably as a result of an hereditary weakness in vein structure. Man's upright posture, especially when accompanied by physical effort may cause the valves in the superficial veins to become weakened and enlarged. Aging contributes to a lessening of the support given these veins by the surrounding tissues.

The table below compares the structure and functions of the three types of blood vessels.

Blood Pressure

The human circulatory system, like that of all other vertebrates, is a closed system in which blood travels through the arteries and veins under pressure. The pressure level is fairly constant, although it varies in different parts of the system. The blood pressure is the result of several factors. The five main ones are discussed below. Variations from the normal in any of these factors cause changes in the blood pressure.

Elasticity of the vessel walls. Blood pumped from the heart keeps flowing in a steady stream throughout the system. If the walls of the arteries

	Arteries	Capillaries	Veins
Direction of Flow	Away from heart	Between arteries and veins	Toward the heart
Walls	Relatively thick muscle layer	Single layer of endothelium	Less muscle than arteries
Valves in Walls	Absent	Absent	Present

were rigid, the blood would flow in sudden spurts through them as the ventricles contracted. During the time when the ventricles were filling, there would be almost no flow of blood in the vessels. Because the walls of the arteries are elastic, they tend to expand as the blood pressure increases and contract as it decreases. This alternate expansion and contraction of the artery walls slows or hastens the flow of blood, keeping it steady. In a child the walls of the vessels are highly elastic. With advanced age, the vessels slowly become more rigid and greater internal pressure develops.

The pumping action of the heart. This is a major factor in blood pressure. It is difficult to determine just how much blood is forced from the heart with each beat, but it is estimated at approximately four ounces. If we assume that this figure is correct, a heart beating at the rate of 70 beats per minute would discharge about 8.2 kilograms of blood per minute, or about 11,430 kilograms in the course of 24 hours. This volume increases in direct relation to the activity of the heart. It has been estimated that the heart of an athlete rowing a strenuous race pumps about 57 liters (approximately 57 kilograms) of blood per minute. *The cardiac output,* or amount of blood expelled with each beat, depends on the rate and force of the beat. Measurements taken in the aorta show that the pressure of the wave of blood as it leaves the heart may vary between 90 and 130 millimeters of mercury. This is the *systolic pressure,* since it is produced by the systole of the ventricles. During the relaxation of the heart, the *diastolic pressure* in the aorta drops to

between 60 and 90 millimeters of mercury. Because of the action of the semilunar valves, there is always some blood in the aorta. In ventricular diastole, however, the pressure within the ventricles may drop to between 2 and 8 millimeters of mercury.

Blood volume. Blood pressure is partially determined by the amount of blood actually present in the system. Normally, this quantity varies only slightly, but conditions may arise in which the volume of the blood may be either increased or decreased. An extra amount of blood may be added to the general circulation when the spleen contracts slightly and gives up some of its stored blood. In this case the blood pressure rises. A fall in blood pressure may result from the loss of blood in hemorrhage. If the blood volume is lowered, the heart is unable to pump blood efficiently. In the preceding chapter, we studied methods to replace the volume of blood fluids.

Peripheral resistance. Blood pressure is determined by the resistance of the narrower arterioles to the flow of blood from the arteries. When a fluid travels from a large tube to several smaller ones, pressure is greater in the smaller tubes because of friction. The slowing up of the flow builds up back pressure in the larger tubes. The constriction of the arterioles and the precapillary sphincters under various stimuli increases the pressure. An example is the response to the secretion of the adrenal glands. In moments of excitement, these glands produce adrenaline, which enters the blood stream and causes the arterioles to contract. The result is a reduced flow of blood to the capillaries and an increased pressure

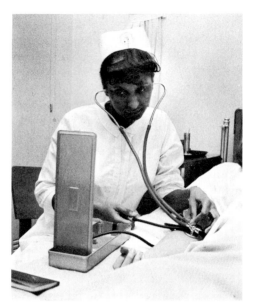

Fig. 25-5 The measurement of blood pressure using a sphygmomanometer. (National Institutes of Health)

within the arteries. The peripheral resistance is lowered when the arterioles and the sphincters expand, as under phychological stimuli. Thus the pain from an injury or an unpleasant experience may cause the capillaries to be suddenly filled with blood. Pressure in the brain may then fall to a point where the cells are not receiving a normal supply of blood and the person faints.

Viscosity of the blood. The viscosity of a liquid is the degree to which it resists flow. Viscosity results from internal friction produced by the rubbing together of the tiny particles in the liquid. Blood is about six times more viscid than water. In health, the viscosity of blood does not change. If there is a pronounced increase in the number of red cells, as in the condition known as *polycythemia vera,* the viscosity of the blood increases because of friction among the red corpuscles as well as between them and the walls of vessels. This causes the blood pressure to rise.

The Measurement of Blood Pressure

The first attempt to measure the pressure of blood as it flowed through an artery was made in 1733 by an English clergyman, Stephen Hales, who experimented on horses. He inserted a small brass tube into the aorta of a horse at a point near its junction with the ventricle. To this he attached a long glass tube and found that the blood in the tube rose to a height of about 2.5 meters. He also noted that the level of the blood in the tube rose and fell several millimeters with each heart beat. This indicated that there was considerable pressure in the aorta and that it fluctuated with the heart beat. In 1896, the modern method of measuring blood pressure was developed by Dr. Ricca Rocci.

This method of measuring blood pressure in the arteries is based on the principle of applying just enough pressure to the artery to stop the flow of blood through it. The apparatus used is called a *sphygmomanometer.* It consists of a cuff, containing an inflatable rubber bladder, which is wrapped around the upper arm. Connected to the rubber bladder are a bulb by which it can be inflated and a pressure gauge (Fig. 25-5). A valve in the bulb allows air to escape slowly from the bladder. A column of mercury is used in some sphygmomanometers, while in others there is a modified aneroid barometer scaled to measure pressure in millimeters of mercury. The use of these devices makes possible a small compact instrument.

When blood pressure is measured, the cuff is adjusted around the upper arm and a stethoscope is placed under the edge of the cuff, above the brachial artery. Air is then pumped into the bladder until the sound of the pulse in the artery can no longer be heard through the stethoscope. Air is then released slowly. The point at which the sound reappears is noted on the pressure gauge. This gives the *systolic pressure,* because it is the pressure exerted by the wave of blood that has been ejected during the ventricular systole.

As air is allowed to escape a change in the quality of the sound is heard just before it disappears. At this point the pressure recorded on the gauge is again noted. This gives the *diastolic pressure,* which is the pressure during the period when the ventricles are filling. Fig. 25-6 shows the usual limits of the average blood pressure at varying ages. About a five-percent variation from the averages is still considered normal.

The Pulse

We are all familiar with the fact that blood flows through the arteries in a series of waves and we have all felt the pulse beat at the wrist and elsewhere in the body. However, the motion of the artery that can be felt is not the

Fig. 25-6   The relation of blood pressure to age. The lines showing systolic and diastolic pressures are averages for each age group. Some variation from these averages is considered to be within the normal range.

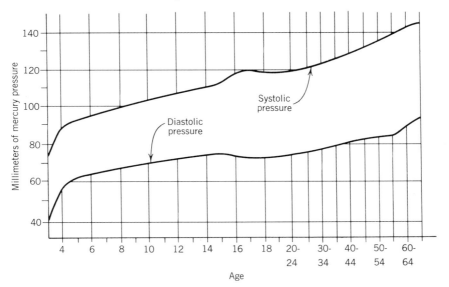

actual passage of a wave of blood that has started from the heart a short time before. There are two distinct processes involved in the pulsing of the arteries. First, there is a wave of muscular contraction and relaxation that constitutes the *pulse wave*. This is followed by the wave of blood.

The pulse wave originates in the aorta with each systole of the heart and spreads to all the arteries. This is a wavelike contraction that passes along the muscular coat of the arteries at a rate of between 6 and 9 meters per second (approximately 20 to 30 feet per second). The wave of blood, expelled from the heart at the same instant, travels at a rate of only 1 to 5 meters per second. The pulse we feel does not, therefore, correspond to the wave of blood that has just left the heart. A pulse wave will travel from the heart to the arterioles in the sole of a foot in about 0.25 second, while about 7 seconds are required for the wave of blood that left the heart at the same instant to reach this region.

The pulse wave forces the blood through the arteries. In a young person with highly elastic artery walls, the rate of the pulse wave is relatively slow. With age, the elasticity of the arterial walls decreases so that the wave travels more rapidly. This is comparable to the condition found in certain nonliving objects. For example, if a violin string is drawn slightly taut and then plucked, the wave motion is much slower than when the string is drawn more tightly. Regardless of how rapidly the violin string is plucked the note will be of the same pitch at any given tension of the string, due to its rate of vibration. The same general

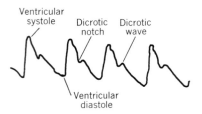

Fig. 25-7   Tracing of an arterial pulse. Note the irregular nature of the wave after the peak is passed.

principle applies to the pulse wave. If the walls of the arteries retain a great amount of elasticity, the pulse wave will travel at a definite rate of speed, regardless of the heart rate. If they become less elastic, comparable to the tightly stretched violin string, the pulse wave travels more rapidly.

A graphic record of an arterial pulse is shown in Fig. 25-7. The upstroke of the curve is smooth and steep. This portion of the curve is formed by the sudden increase in muscular pressure due to the systole of the ventricle. After reaching a peak, the pressure drops and the curve begins to fall. Partway down the descending line is an uneven portion known as the *dicrotic notch,* followed by the *dicrotic wave.* The notch is produced by the closing of the semilunar valves at the opening of the aorta; the dicrotic wave is the result of their bulging back into the ventricle because of the increased pressure of the blood above them.

A venous pulse similar to the arterial pulse cannot be detected in most veins. The reason for this is that the pulse wave is lost when the arteries lose their muscular coat as they merge into the capillaries. The veins arise from the convergence of the capillaries.

Although they contain muscle tissue, the pulse wave is not transmitted to them through the capillary bed due to the lack of muscle tissue in capillaries. There is one region where a pulse exists in veins. This is in the large veins leading to the heart, but the origin of this pulse is quite different from that of the arterial pulse. Its cause is purely mechanical and depends on the alternate flow of blood into the atria and its sudden stoppage during systolic periods of these chambers.

Blood Distribution

The flow of blood to different parts of the body is controlled by nervous impulses to the vessels. If there is an increase in activity in a certain region, more blood passes to that region and another is deprived of a maximum supply. Since there is only a certain amount of blood to supply the entire body, its distribution must be carefully controlled. A good example of this mechanism is seen when a person goes swimming in cold water too soon after a heavy meal. The digestive processes demand a large supply of blood, with the result that other regions of the body are deprived of an adequate supply to meet emergency conditions. In such a case, the muscles may not be supplied with enough blood to maintain the temperatures they require for oxidation. Then they will contract violently under the stimulus of the cold, and a cramp will result.

The normal distribution of blood to various regions of the body is controlled by the action of two nerve centers in the medulla of the brain. One of these, the *vasoconstrictor center,* is constantly sending out a series of impulses which keep the arterioles in a state of slight contraction, or tonus. These impulses reach the arterioles along paths that originate in the sympathetic ganglia. Impulses which result in the further constriction of the arterioles are generally stimulated by emotional states such as anger or fear, or by changes in normal requirements of some part of the body for an added supply of blood.

Vasodilation, or the relaxation of muscle fibers in the blood vessels, may be controlled by the *vasodilator center* which lies two or three millimeters distant from the vasoconstrictor center. We cannot say definitely that impulses arising in this center are the sole cause for the dilation of the arterioles. There appears to be a reciprocal action between the constrictor and dilator centers, so that when one is stimulated the other is inhibited. At present there is no easy method of distinguishing between the active role of the constrictor center and the more passive part played by the dilator center. Nerves from the vasodilator center pass out to the parasympathetic ganglia and thence to the vessels. Stimulation of these nerves results in the relaxation of the muscle fibers in the arterioles and the precapillary sphincters so that blood flows more readily into the capillaries. This results in a drop in pressure in the arteries which may reach an alarmingly low level.

Several factors control the flow of blood to various regions of the body.

The influence of the cerebral cortex. In the cortex of the cerebrum are centers that control a wide variety of reactions. For example, if we meet with an embarrassing situation, our face is apt to become flushed. The steps leading to this reaction are extremely complicated because they involve memory patterns, previous training, and our peculiarly individual reactions to certain situations. The flush is caused by inhibition of the vasoconstrictor center, with the result that the surface of the skin becomes filled with blood as the skin arterioles dilate. Other emotional stresses may have the opposite effect. If one becomes angry, the temperature of the fingers and hands drops because of stimulation of the vasoconstrictor center. (It is interesting to note that there is a decrease in the electrical resistance of the skin accompanying the constriction of the capillaries.) This change in fingertip temperatures has been used in studying the effect of drugs on the nervous system. Nicotine, for example, stimulates the vasoconstrictor center and its effect on the circulatory system can be observed in a lowering of the circulatory rate through the fingers.

The effect of carbon dioxide. We have already studied the effect of an increased amount of carbon dioxide on the rate of respiration. While there is some debate about the effect of this gas on the vasomotor system, there are certain well-recognized facts connected with its appearance in the blood in greater than normal quantities. It has been found, for example, that in early stages of asphyxiation, blood pressure rises. On the other hand, if the carbon dioxide content of blood is lowered as a result of forced breathing, a feeling of dizziness results which may be attributed to a decrease in blood pressure in the cerebrum. The mechanism involved in these cases appears to be the presence of chemoreceptors in the carotid sinus and the aortic bodies. Impulses arising in these regions result in the stimulation of the vasoconstrictor center. If the activities of these centers decrease as the result of too little carbon dioxide in the blood, the blood pressure is lowered and fainting follows. Conversely, an increase in the amount of carbon dioxide in the blood results in rising blood pressure which has a reflex action on the heart and increases its rate of beating.

Effect of drugs on vasomotor action. *Adrenaline,* a product of the adrenal glands, has a profound effect on the action of the precapillary sphincters. Under the influence of this hormone, the vasoconstrictor center is stimulated and the blood pressure rises sharply but remains elevated for only a short time. This effect is used by physicians, who may apply a dilute solution of adrenaline (about 1:10,000 parts of water) locally when performing minor operations on the eye, nasal membranes, or some other local area. This causes constriction of the arterioles in the region and helps to staunch the flow of blood.

Acetylcholine is produced at the junction of nerves and muscles. In most muscles, it has a stimulating action that results in contraction. In the heart and blood vessels it acts as a relaxing agent and causes dilation of vessels.

Histamine is a compound that is found in all tissues. If formed in larger than normal quantities, it dilates the arterioles and capillaries. The congestion of mucous membranes, a common sympton of asthma, the common cold, and certain allergies, is the result of this action. *Antihistamine* drugs have been developed to combat this congestion. These drugs do not affect the underlying causes of the symptoms, and therefore should not be considered a cure.

Alcohol is essentially a vasodilator which may produce a feeling of warmth if taken on a cold day. This effect is caused by dilation of the skin capillaries and may result in loss of heat over the surface of the body. The sensation of warmth may be misleading if the body is later subjected to extremely low external temperatures. Some extremely low internal body temperatures ($13–18.5°C = 55–65°F$) have been reported following excessive use of alcohol and later exposure to cold.

Nicotine has an effect opposite to that of alcohol: it is a vasoconstrictor. Smoking lowers the skin temperature of the fingers by one or two degrees due to the effect of the nicotine on the vasoconstrictor center. In the treatment of certain types of heart disease, smoking is prohibited because of the effect of nicotine on the coronary vessels that supply an already overtaxed heart.

Ephedrine, a drug obtained from plants, is frequently used in nose drops. It reduces the swelling of nasal membranes by the constriction of blood vessels locally. It is doubtful whether this affects the vasoconstrictor center. The effect seems to be mainly local, confined to stimulation of the constrictor nerves in the region.

External Pressure Changes

Since the heart is a muscular pump that forces blood through the body under normal pressure, any pronounced variation from the normal may have a profound effect on the flow of blood. The blood vessels, also, are designed to withstand pressures at sea level, or within reasonable distances above or below it. If the body is subjected to forces that exceed those normally encountered, the circulatory system has no means of adjusting to these changes. The result may be disastrous unless proper precautions are taken.

For instance, an aviator may be subjected to abnormal forces when he pulls his plane out of a dive or banks it sharply around a curve. In these maneuvers, the blood is forced into the lower parts of the body with a corresponding deficiency in circulation through the brain. This may cause loss of consciousness, commonly known as a *blackout.* If the maneuver forces blood away from the lower organs (as in an outside loop), the aviator may experience a condition called *redout,* in which he sees everything through a red haze. This effect is probably caused by excess blood in the vessels of the eye, and may be rapidly followed by unconsciousness.

Both of these conditions result from the inability of the circulatory system to cope with the abnormal forces to which the body is subjected. The reactions may be prevented to some degree by the use of pressure suits that can be inflated with air to produce counterpressures that tend to neutralize the effects of these unusual forces. Placing the pilot in a prone position is also helpful.

The Spleen

This is an important organ of the circulatory system. It is not essential, since a person can continue in good health after it has been surgically removed.

The location of this organ is shown in the Trans-Vision insert following page 212. Although its size is variable in different individuals, its average dimensions are about 12 centimeters long, 7 centimeters wide, and 4.6 centimeters thick. It is slightly concave on its front surface and correspondingly rounded in back. Its blood supply is very extensive, giving it a somewhat purple color. Before birth and for a short time afterwards, the spleen produces both red and white corpuscles, but in the adult this function is lost. The walls of the spleen are composed of several layers of connective tissue which make up its capsule. In the walls are scattered bundles of smooth muscle tissue which enable the spleen to contract slightly. The interior of the organ is composed of loosely connected cells that support many blood vessels, with numerous spaces (sinuses) in which excess blood is stored. This inner region is subdivided irregularly by strands of fibrous tissue, the *trabeculae*.

There are three major functions of the spleen. First, the spleen serves as a reservoir for blood. When it is stimulated by nervous impulses from the sympathetic nervous system, or when there is an increase in the amount of adrenaline in the blood, the muscle fibers contract and blood is forced into the general circulation. Thus in an emergency blood is supplied with an increased number of red corpuscles to carry more oxygen to the tissues.

A second function of the spleen is the destruction of aged or damaged red corpuscles. This activity is also carried out in the liver and bone marrow. The destruction is accomplished by the action of large specialized white corpuscles, the *macrophages,* which surround and engulf the erythrocytes in much the same manner that the neutrophils engulf bacteria. Just how the macrophages determine which of the red cells are to be destroyed and which will remain unharmed is unknown. It may be that as hemoglobin is used, it changes in its chemical composition so that it becomes "attractive" to the macrophages.

The third function of the spleen is the production of lymphocytes. This function it shares with lymphatic tissue.

EXTRAVASCULAR FLUIDS

As mentioned previously all of the fluids of the body that are not definitely parts of blood are known as extravascular fluids. This refers to the fact that they are outside the blood vessels. Those fluids which are contained within the cells are called intracellular fluids, while those lying between the cells are known as the intercellular or interstitial fluids. In a man of average weight (70 kilograms or 154 pounds) the distribution of these fluids would be approximately as follows:

Fluid	Percent of Body Weight
Plasma of the blood	5
Intracellular fluid	50
Intercellular fluid	15

The intracellular fluid contains approximately 25 percent protein materials by volume; the remaining 75 percent is an aqueous solution of both organic and inorganic salts. Plasma, on the other hand, contains about 2.6 percent protein materials, while there is less than 1 percent protein in the intercellular fluid. The reason for these differences is that protein molecules are generally too large to pass easily through cell membranes. This concentration of protein materials within cells is a highly important factor in maintaining osmotic pressure in the cells.

The intercellular fluid that bathes the cells serves as a middleman between blood and cells. Since capillaries do not come in direct contact with all cells, it is necessary for some fluid to act as a carrier between the blood and cells. This is the primary function of intercellular fluid. It originates from the portion of the blood plasma that diffuses through the walls of the capillaries. This may pass through the walls of the endothelial cells, but most of the fluid is thought to escape through the intercellular cement. Since the cement behaves as a semipermeable membrane, few of the blood proteins are present in the intercellular fluid. Thus fibrinogen is retained by the blood, with the result that intercellular fluid does not coagulate. We are all familiar with the appearance of this fluid in the form of the clear liquid that collects in a blister or that oozes from a slight abrasion.

The fluid that leaves the blood gradually filters through the spaces among the cells. It carries to the cells those materials that are essential to their maintenance and removes wastes formed during their metabolic activities. Oxygen, for example, is released into the plasma by hemoglobin as it reaches an area of low oxygen tension.

The gas diffuses through the plasma in solution. The resulting solution then passes through the capillary wall and is carried to cells that lack oxygen. Other materials pass to cells in the same way.

In addition to the required materials which pass into cells, various metabolic wastes diffuse out of the cells into the intercellular fluid. Some of these enter the blood stream directly because of the difference in osmotic pressure of the fluid and plasma. As we

Fig. 25-8   Valve in a lymphatic. The flaps of the valve are closed to prevent the backflow of lymph. (General Biological Supply House, Inc.)

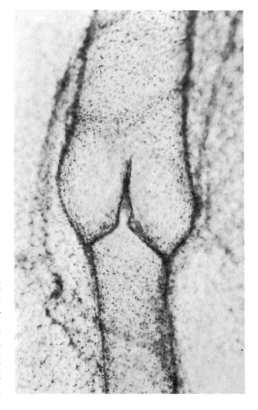

have mentioned, osmotic pressure of plasma is lower at the beginning of the capillary where it leaves the arteriole than it is at the end where it unites with others to form a venule. This gradient in the osmotic pressure along the course of the capillary hastens the exit of materials from the blood and then later permits the reabsorption of other substances. Some materials, however, do not re-enter the blood stream. These continue to flow between the cells until they enter a thin-walled lymph vessel that is a part of the *lymphatic system.*

The Lymphatic System

The lymph vessels, or *lymphatics,* form an extensive network throughout the body. The material in these vessels is known as *lymph,* although the same term is applied loosely to include the intercellular fluid as well. The vessels begin as extremely small tubes composed of endothelial tissue. These branch freely and intricately through all the tissues, gradually running together to form larger tubes. In the walls of the larger lymphatics are valves (Fig. 25-8) to control the flow of lymph. Fluids push through these tubes by the massaging or milking action of the tissues in which they lie. In inactive tissues, lymph flows very slowly or is completely stagnant. When activity increases, the fluid flows faster. The valves direct the flow of the lymph away from the tissue toward a central collecting point.

Fig. 25-9   A lymph node. Part of the diagram shows the node with the lymphocytes removed to make the internal structure clearer. Lymph flows through the coarse mesh, in which there are normally large numbers of lymphocytes.

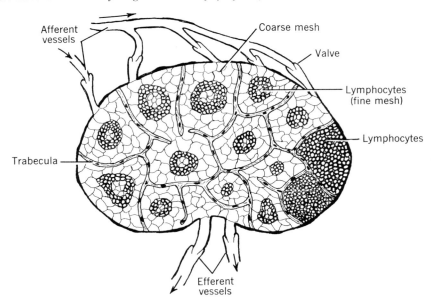

Afferent
vessels

Coarse mesh

Valve

Lymphocytes
(fine mesh)

Lymphocytes

Trabecula

Efferent
vessels

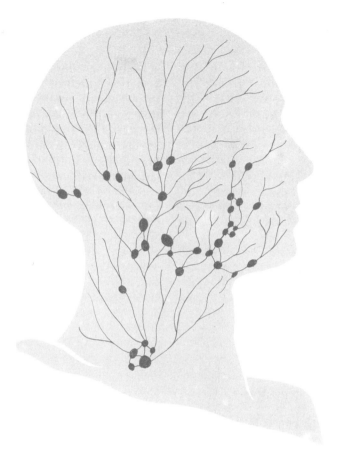

Fig. 25-10   Distribution of some of the lymphatics in the head region.

Smaller lymphatics empty into larger lymph cavities known as the *lymph nodes* (Fig. 25-9). Their cells are large and loosely arranged, containing numerous lymphocytes. These white corpuscles are responsible for the destruction of bacteria and other foreign bodies that may have invaded the tissues and been drained into the lymphatics. For example, during an attack of German measles, the lymph nodes in the sides of the neck become enlarged as the result of being invaded by numerous virus particles. See Fig. 25-10.

The larger vessels leaving the nodes flow together so that eventually the fluid is returned to the blood stream. These large vessels follow two paths. The lymphatics draining the right side of the head, right half of the thorax, and right arm drain into the *right lymphatic duct*. This duct enters the right subclavian vein through an opening guarded by a semilunar valve to prevent the entrance of venous blood into the duct.

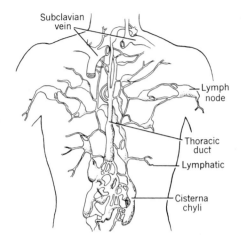

Subclavian vein

Lymph node

Thoracic duct

Lymphatic

Cisterna chyli

Lymph from the other body regions flows into a large series of vessels that collect to form the *cisterna chyli,* which is located in front of the second lumbar vertebra. Much of the lymph reaching this chamber contains globules of fat that give it a milky appearance. From the cisterna, the *thoracic duct* (Fig. 25-11) passes upward near the vertebral column until it empties into the left subclavian vein.

Fig. 25-11   Lymph flows toward the heart through the main lymph ducts, which collect it from the lymph nodes.

VOCABULARY REVIEW

Match the phrase in the left column with the correct word in the right column. *Do not write in this book.*

1   A muscle which controls the flow of blood through a capillary.

2   A colorless fluid collected in vessels from between cells.

3   The smallest division of the venous system that contains smooth muscle fibers.

4   Leukocytes can migrate to the tissues through its walls.

5   A large, thin-walled vessel in the abdomen that collects lymph.

6   A characteristic of blood that affects blood pressure.

7   Cells in the spleen that destroy old erythrocytes.

8   The cause for the reduction of blood flow to a region of the body.

a   arteriole
b   capillary
c   cisterna chyli
d   diastolic pressure
e   lymph
f   macrophages
g   metarteriole
h   precapillary sphincter
i   systolic pressure
j   tunica intima
k   vasoconstriction
l   vasodilation
m   venule
n   viscosity

9 The force exerted on the walls of an artery by a wave of blood flowing through it.

10 The part of the wall of an artery that is composed of endothelial cells.

TEST YOUR KNOWLEDGE

Group A

Select the correct statement. *Do not write in this book.*

1 The muscle fibers in the walls of arteries are composed of (a) striated muscle (b) cardiac muscle (c) smooth muscle (d) papillary muscle.

2 The main circulation system that supplies blood to most parts of the body is called the (a) pulmonary (b) systemic (c) portal (d) renal.

3 The blood vessels with the smallest diameter are called (a) capillaries (b) arterioles (c) venules (d) lymphatics.

4 The blood pressure during the time the ventricles are filling is called (a) systolic (b) systemic (c) diastolic (d) dialysis.

5 Enlargement of a blood vessel is called (a) vasodilation (b) vasoconstriction (c) vasopressin (d) edema.

6 Increased amounts of carbon dioxide in the blood cause the rate of the heartbeat to (a) decrease (b) increase (c) remain the same (d) vary widely.

7 Of the following, the one that is *not* a function of the spleen is (a) storage of blood (b) destruction of damaged erythrocytes (c) production of lymphocytes (d) storage of glycogen.

8 The muscular layer of an artery is called the (a) tunica intima (b) tunica media (c) tunica externa (d) tunica albuginea.

9 Valves are found in the walls of (a) arteries (b) arterioles (c) lymphatics (d) capillaries.

10 As a person becomes older, his blood pressure generally (a) increases (b) decreases (c) remains the same (d) varies widely.

Group B

1 How is the structure of the capillaries adapted to their function?

2 (a) What is meant by blood pressure? (b) How is the pressure of the blood maintained?

3 What is the pulse wave and how does it differ from the wave of blood that flows through an artery?

4 Identify each of the following (a) sphygmomanometer (b) diastolic pressure (c) spleen (d) tunica media (e) metarteriole.

Chapter 26 The Skin

Functions of the Skin

The covering of an animal's body is called its *integumentary system*. This may consist of a single layer of cells or may be a quite complex system that includes various protective devices. We find, for example, that the scales of fishes and reptiles assume quite strange and unusual forms that may be protective, either by disguising the animal or by otherwise helping to protect it from a predator's attack. The scales of fish and reptiles appear to be similar, but their origin is quite different, even though both are products of the skin. Birds have feathers, a highly specialized skin derivative, while mammals have a body covering of hair. Although it is beyond the scope of this text to explain the origin of all these body coverings, it should be pointed out that the scales of reptiles, the feathers of birds, and the hair of mammals are somewhat alike in that they develop from the same general types of tissues.

In man, the skin is the primary covering of the body surface. Hair is present over parts of the body, but its distribution is so variable among different peo-

ple that it cannot be considered as distinctive a feature of the human race as it is of other groups of mammals. The human skin has a variety of different functions.

The skin has several *protective* functions. Its unbroken surface presents a barrier against invasion by bacteria and other foreign bodies. This barrier is chiefly mechanical, but there also appears to be a chemical substance produced by the outer layers of the skin which destroys most of the bacteria that come in contact with it. The skin protects the body against the absorption of excessive amounts of water or the loss of fluid through the body surface. There is a small osmotic flow between the cells of the skin and the surrounding air or water, but this is of minor consequence in maintaining the water content of the body. Apparently, this osmotic action is of primary importance only in people who hereditarily lack sweat glands. The underlying tissues are protected by pigments formed in the skin against the injurious effects of ultraviolet radiations in sunlight.

The skin plays a very important role in the *regulation of body temperature*.

Heat is lost by the evaporation of sweat, and by the radiation of heat through the skin. Both of these activities will be discussed in detail in the next chapter.

Because it is well supplied with nerves, the skin is *sensitive* to changes in its surroundings. These may be temperature changes or mechanical changes, received in the form of pressure, touch, or pain sensations.

The skin *eliminates* excess water and salts. Although it disposes of these substances, it can hardly be considered a part of the excretory system because the quantity of waste materials liberated is negligible.

The skin has tissues for the temporary *storage* of water, fats, glucose, and some inorganic salts. Most of these are later reabsorbed by the blood and carried to other regions of the body where they may be used.

Structure of the Skin

The skin consists of two principal layers; the *epidermis,* or the *cuticle,* and the *dermis, corium,* or *true skin.* The word dermis is derived from the Greek word *derma* meaning skin. The prefix, *epi,* meaning on top of, is added to designate the outer layer. Below the dermis is a layer of varying thickness, the subdermal layer, which is not properly a part of the skin.

The epidermis is composed of two layers of cells, as shown in Fig. 26-1. The outer of these layers is made up of cells in which the living material has

Fig. 26-1   Photomicrograph of a section through hairless human skin. Note the thickness of the epidermal layer, especially that of the stratum corneum. (General Biological Supply House, Inc.)

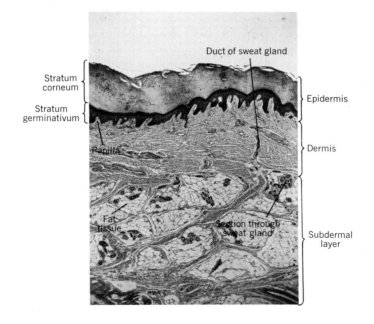

been replaced by a hard, nonliving protein substance, *keratin.* The outermost cells are flattened and scalelike and are constantly worn off by rubbing on clothing or on other objects in contact with the skin. This horny layer, the *stratum corneum,* is replaced by the division of cells that constitute the lower layer of the epidermis, the *stratum germinativum.*

The stratum corneum is one of the principal defenses of the body against invasion by bacteria and against mechanical injury to the delicate underlying tissues. So long as this outer layer remains unbroken, bacteria cannot easily invade the body. In fact, the protective action afforded by this keratinized layer is reportedly equal to the immunizing action of the blood in preventing infections. The thickness of the horny layer varies in different regions of the body. In some areas it is quite thin, while in others it becomes thick as the result of friction against some particular area. When the thickening develops outwardly in a local region, such as the palms of the hands, the enlargement is known as a *callus.* If the thickening results from inward growth of the stratum corneum, as it does occasionally on the toes, the growth is called a *corn.*

Between the bottom of the horny layer and the surface of the skin is an area of specialized material which helps to prevent the absorption of many substances by the skin. This region, called the *barrier layer,* shows some interesting characteristics when stained by special methods. It appears to be a sandwich-type of structure containing both lipids and water. Thus, substances which dissolve only in water or only in lipids will not enter the skin. On the other hand, those substances which are soluble in both lipids and water can enter. This layer may explain why water does not penetrate the skin easily, while chemical substances, such as acetone, ethyl alcohol, and methyl alcohol which are soluble in both lipids and water, can enter quite readily. Much more work remains to be done on this recently identified barrier layer before its full part in skin function can be determined.

The stratum germinativum is made up of cells of various forms. Those at the lowest level are columnar in shape, but this typical form is lost by repeated divisions, as these cells gradually become flattened to replace those lost from the horny layer. Within the columnar cells are the granules of pigment which give the skin its characteristic color. Among the substances that are responsible for the color are *melanin, carotene,* and *hemoglobin.* The various proportions of these substances are responsible for the different shades of color. White races are those in which the amount of melanin is normally reduced, while the colored races have a larger percentage of this material in the skin. An absence of pigments, except for hemoglobin, results in the condition known as *albinism.* The only cells in the human body capable of producing melanin arise from the region of the early embryo from which also develop parts of the nervous system. As a result of the later pattern of division of cells, those that form melanin, the melanocytes, are present in only the skin, the eyes, and the meninges of the brain—the pia mater and arachnoid. Within the melanocytes a series of quite complex chemical reactions occur which result in the formation of melanin. The final step in the series involves the conversion of a compound, dihydroxyphenylalanine, into the pigment melanin. This reaction requires the presence of a specific enzyme which albino individuals fail to synthesize. This is due to a defect in the DNA code for this protein and albinism is inherited as a recessive characteristic.

The production of excess pigment is stimulated by ultraviolet rays. It is these solar radiations that are responsible for sunburning and the subsequent tanning of the skin. Ultraviolet rays in the neighborhood of 3200 Ångstrom units (the lowest visible violet rays are about 4000 Ångstrom units in length) cause the familiar reddening of the exposed skin. Following the sunburn, a moderate tanning will appear. Continued exposure will darken the skin color as more pigment is formed in the cells. It should be remembered that this effect is a defense mechanism to protect the lower-lying tissues from injury by ultraviolet rays. The wisdom of prolonged exposure to sunlight is being questioned at present because of its possible connection with the development of skin cancers.

With the development of new electron microscope techniques, the unique structure of the epidermal cells is more clearly seen. It has been found that each of the cells contains innumerable minute strands of what is believed to be fibrous protein. These pass through the cytoplasm of the cell and become concentrated on the inner surface of the cell membrane. The strands do not pass from one cell to the next, but the fibers in one cell end exactly opposite the fibers in an adjacent cell. Between cells is a very small amount of noncellular material containing fibers which form intracellular bridges between the cells. By means of these bridges, the cells of the epidermis are held together. If these bridges disappear, the cells literally fall apart. This separation of the epidermal cells occurs in the rare but usually fatal condition known as *pemphigus vulgaris*. Microscopic examination of the skin in this disease shows the cells to be quite normal except for the loss of the intracellular bridges.

As is seen in the photomicrograph (Fig. 26-1), the lower edge of the stratum germinativum is very irregular in outline. The indentations are known as the *papillae* of the skin. In certain areas of the body the papillae are so pronounced that they raise the surface into permanent ridges. This occurs in the skin of the fingers, the palms of the hands, and the soles of the feet. The ridges are so arranged that they offer maximum resistance to slipping, either in walking or in grasping an object. They are therefore known as friction ridges. Those on the inner surfaces of the fingers are used for purposes of identification by means of fingerprints (Fig. 26-2).

Between the epidermis and the dermis there is a definite *boundary membrane*. It holds the epidermis firmly to the dermis so that the two layers cannot easily be separated. Strong shearing forces have little effect on this membrane, but if the skin is heated locally to 50°C (122°F), separation of the epidermis from the dermis does occur. However, when the skin is then cooled, the two layers become once more firmly

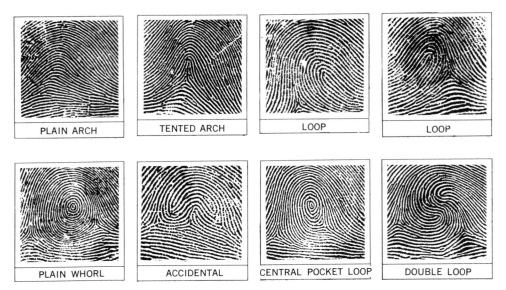

PLAIN ARCH	TENTED ARCH	LOOP	LOOP

PLAIN WHORL	ACCIDENTAL	CENTRAL POCKET LOOP	DOUBLE LOOP

Fig. 26-2   The friction ridges on the skin of the fingers and the feet form a characteristic pattern in each individual. (U.S. Dept. of Justice)

attached to each other. This membrane may possibly serve as an additional barrier to invasion by bacteria. Further investigation should give a clearer understanding of the functions of the boundary membrane.

The corium, or dermis, like the epidermis, varies in thickness in different regions of the body. It is especially thick over the palms of the hands and the soles of the feet. Also, it is thicker in the skin covering the back of the body than it is over the front of the thorax and abdomen. This layer is composed of matted masses of connective tissue and elastic fibers through which pass numerous blood vessels, lymphatics, and nerves.

Appendages of the Skin

Certain well-defined structures form what are called the appendages of the skin. These are *hairs,* the *nails* of fingers and toes, the *sweat glands* and the *sebaceous* (oil) *glands.* In Fig. 26-3 the relation of these to other parts of the skin is shown diagrammatically.

Although a part of each hair lies within the dermis, it is actually surrounded by an inward projection of the epidermis which forms a sloping tube, the *hair follicle.* At the lower end of this tube is an upward projection of the surrounding tissue, the *papilla* of the hair. The papilla contains capillaries to nourish the cells of the follicle, from which the hair is formed by the rapid division of cells at its lower end.

Microscopic examination of a hair shows that it is composed of three distinct parts. On the outside is the *cuticle,* a single layer composed of flat scalelike structures in which the protoplasm has been replaced by keratin. These overlap each other slightly and give the surface a scaly appearance. Within the cuticle is a keratinized layer of elongated, nonliving cells that make up the *cortex* of the hair. The pigment of the hair is in the cortex and the central cavity, or *medulla,* if one is present.

The distribution of hair over the body, its type, and its color are the result of hereditary factors that are beyond the scope of this text. However, two common conditions should be discussed briefly: these are the graying of hair and the loss of hair. White hair has pigment only in the medulla while blond hair has no medulla and just a small amount of color in the cortex. The period of life at which graying of the hair occurs appears to be genetically determined. There is no evidence that it is connected with any vitamin deficiency, as is the case in some animals. The loss of hair may be due to several causes. First, if the scalp is too tightly drawn over the bony framework of the skull, a temporary loss of hair may occur. This condition is the result of interference with the blood circulation to the hair follicles and can be relieved by massage. Illness, when accompanied by a very high fever, may result in the loss of hair. If the body

Fig. 26-3 A section through the skin showing the structures in the epidermis and dermis.

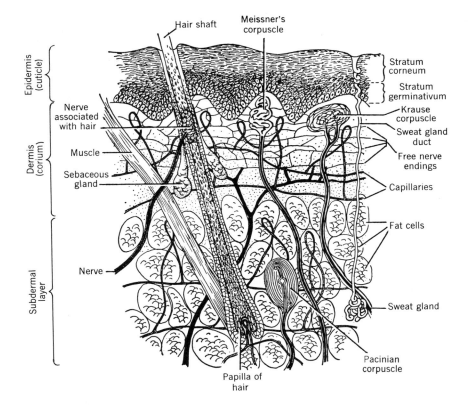

Hair shaft

Meissner's corpuscle

Epidermis (cuticle)

Dermis (corium)

Subdermal layer

Nerve associated with hair

Muscle

Sebaceous gland

Nerve

Stratum corneum

Stratum germinativum

Krause corpuscle

Sweat gland duct

Free nerve endings

Capillaries

Fat cells

Sweat gland

Pacinian corpuscle

Papilla of hair

temperature does not reach a point where there is destruction of the cells in the hair follicles, the hair will return. Local diseases of the scalp may also cause the loss of hair over small areas. A common infection of this type is caused by the growth of a fungus and is known as ringworm of the scalp. The most common type of baldness seems to be hereditary. Inheritance of certain factors determines not only the loss of hair but also the approximate time of life when this occurs and the pattern that it follows. Aging and the activities of the male sex hormones are also factors connected with baldness.

Attached to each hair follicle on the side toward which it slopes is a small bundle of smooth muscle fibers, the *arrector pili muscle*. This has its origin on the upper surface of the dermis and its insertion on the wall of the follicle. In some animals, the contraction of this muscle causes the hair to be erected, but in man the hair is too weak to allow this action to occur. When this muscle of the skin is stimulated, as by sudden chilling, it contracts and causes the skin to pucker around the hair. This produces the condition commonly called *goose flesh*. Contraction of the muscle also results in pressure on the sebaceous glands and the secretion of a small amount of oil.

The *sebaceous glands* lie along the walls of the hair follicle. They may also be present in hairless regions of the skin. They are especially numerous on the face and scalp, but absent from the palms of the hands and the soles of the feet. The product of these glands is composed of fats, proteins, salts, and water. Due to its oily nature, this secretion keeps the hair and the surface of the skin flexible. It also helps to prevent excessive absorption or loss of water over the surface of the body.

A third group of skin appendages is the *sweat glands*. These are small glands, the body of which consists of a coiled tube which lies in either the corium or the subdermal layer. From this a duct extends upward and opens on the surface as a pore. In regions of the body where the epidermis is thick, the outer end of the duct may be twisted or coiled. Elsewhere, in thinner skin, this tube is usually straight. The number of sweat glands varies in different regions of the body. They are very plentiful on the palms and the soles, where the openings of the ducts are evenly distributed along the friction ridges. Their approximate numbers are as follows:

Palms of the hand	370 per sq. cm.
Back of the hand	200 per sq. cm.
Forehead	175 per sq. cm.
Breast, abdomen, and forearm	155 per sq. cm.
Leg and back	70 per sq. cm.

It has been estimated that the sweat glands total approximately 2 million.

Sweat is composed principally of water (99 percent) and salts, of which sodium chloride is the most common. There are traces of other substances, including urea. The chemical composition can vary considerably in different regions of the body and under differing circumstances. Sweat may at times be very acid, with a pH of about 3, or it may shift toward the alkaline side. The quantity of sweat also varies, depending on the temperature of the air and the amount of physical activity. The glands are continually secreting sweat, but it is usually in such small quantities or is removed so quickly by evaporation or absorption by clothing that we are not conscious of it. This is called insensible perspiration. The total amount of sweat formed during twenty-four hours may be quite small (about a half liter) or it may reach two liters. Under extreme conditions of high temperature, a person may secrete about 10 liters per day. As a general rule, when the loss of water by sweating is great, the quantity of urine produced by the kidneys is reduced.

The activities of the sweat glands appear to be under the control of a sweat center located in the hypothalamus of the brain. Impulses from this center pass along sympathetic nerve fibers to the glands. The center is stimulated when the temperature of the blood passing over it rises by 0.17–0.5°C (0.3–0.9°F), due to increased muscular activity or a rise in external temperature. It may also be stimulated when the skin temperature rises above 34.4°C (94°F). Emotional conditions may also increase the production of sweat, as we know when we experience the cold sweat of fear. The role of sweat in helping to control the body temperature will be discussed in the next chapter.

The *nails* are also classified as appendages of the skin. These are hard structures, slightly convex on their upper surfaces and concave on the lower surfaces. They are formed by the epidermis, where they first appear as elongated cells which then fuse together into plates. During this process their protoplasm is replaced by keratin. The area producing the nail is called the *nail bed,* or matrix. As long as any part of this region remains intact, it will produce the nail substance. It is possible, therefore, for a fingernail to be lost and later on replaced by a new one. At the back of the nail bed, the skin forms a fold in which the base of the nail is embedded.

Sense Receptors in the Skin

Among the functions of the skin mentioned at the beginning of this chapter was its sensitivity to stimuli. Both afferent and efferent nerves pass through the skin, and some of these end in receptors which lie either in the dermis or the subdermal layer. Motor nerve fibers are distributed to the walls of the blood vessels and the arrector pili muscles. The sensory nerves end in receptors sensitive to heat (*end organs of Ruffini*) and cold (*end organs of Krause*). The sense of touch is the result of stimulation of other specialized nerve cells, *Meissner's corpuscles,* while pressure can be detected by the deeper lying *Pacinian corpuscles.* In addition to these specialized receptors, there are nerve endings for the sensation of pain just under the epidermis and around the hair follicles.

Skin sensitivity varies from one part of the body to another. Pain receptors are quite plentiful on the forehead, breast, and lower arm. On the forehead, for example, there are about 200 of these per square centimeter of skin surface. On the nose and thumb there are only about 50 per square centimeter, while a small area on the inside of each cheek just opposite the second molar tooth lacks them completely. Touch spots also differ in their distribution. On the lips these spots are less than one millimeter apart but in the middle of the back this distance is increased to over 10 millimeters. Sensitivity to heat or cold is not due to a temperature sense but to the presence of definite receptors that are sensitive to an increase in either heat or cold. The cold receptors are more numerous than those sensitive to warmth.

Subdermal Tissues

The subdermal tissues contain fat cells, connective tissue, blood vessels, lymphatics, and nerves. This region serves primarily as a connection between the skin and the deeper tissues. In some areas of the body, such as the neck, the connecting fibers are loose, so that the skin can be moved quite freely. In other regions, like the palms and the soles, the skin is attached much more firmly, so that very little motion is possible. The adipose tissue in the subdermal area stores fats. The subdermal layer of fat is generally thicker in women than in men, due to the action of sex hormones.

Care of the skin. Bathing removes the secretions of the sebaceous and sweat glands that adhere to the body surface. The oily nature of the sebaceous secretion is responsible for the clinging of dirt to the skin, and its removal is a physiological as well as esthetic necessity. Water with a temperature between about 26.5 and 32°C (80 and 90°F) is the most satisfactory for bathing. The widely held idea that cold baths (below 18°C) will toughen the body and thus make it more resistant to colds has been disproved.

The most common type of skin eruption is *acne*. The exact cause of this distressing condition is currently unknown, but it results in the formation of a keratinized plug at the openings of the sebaceous glands. This prevents the escape of the oily secretions, and the area becomes filled with leukocytes (pus). One of the baffling facts is the way in which the condition will become quiet for a time and then reappear. Diet may have something to do with these changes, but the primary cause for the condition is more fundamental than this. While there is no very definite proof to support the theory, it is thought that the appearance of acne may be associated with the secretion of sex hormones.

The skin is subject to a variety of injuries, cuts and burns being the most common. First aid treatment for a simple cut is to clean the wound, apply a mild antiseptic, and cover with a sterile dressing. The use of harsh antiseptics is not recommended because they may damage the surrounding tissues and delay the healing process.

If a burn is of the first degree, with an unbroken reddened area of skin, the application of cold water and then a mild ointment and covering with several layers of sterile gauze is usually all that is required. Second degree burns, in which the skin becomes blistered, should also be treated with cold water and then covered with moist sterile gauze. The blisters should not be opened because the enclosed tissue fluid helps the healing process. Third degree burns, involving serious destruction of tissues, should receive immediate medical care. The burned area should be covered lightly with a moistened sterile cloth until such time as a physician can treat the victim. Greases should not be applied to second or third degree burns unless prescribed by a physician. Dressings may be moistened with some mild antiseptic, such as that made by dissolving three table-spoonfuls of either sodium bicarbonate (baking soda) or Epsom salts in a quart of warm water.

VOCABULARY REVIEW

Match the phrase in the left column with the correct word in the right column. *Do not write in this book.*

1 A minute opening on the surface of the skin through which secretions can leave.	**a** barrier layer
	b boundary membrane
2 The result of the action of the arrector pilae muscles.	**c** cuticle
	d dermis
3 A nonliving material which replaces the protoplasm of the cells of the stratum corneum.	**e** epidermis
	f goose flesh
4 Produces a secretion that has a high lipid content.	**g** keratin
	h melanin
5 A layer of the skin which can become thickened.	**i** pore
6 Has papillae that project into the stratum germinativum.	**j** sebaceous gland
	k sweat gland
7 Connects the epidermis to the dermis.	
8 A layer of protective material composed of fat and water.	

TEST YOUR KNOWLEDGE

Group A

Select the correct statement. *Do not write in this book.*

1 Of the following, the one that has the *least* effect on skin color is (a) melanin (b) keratin (c) carotene (d) hemoglobin.
2 The production of excess pigment in the skin is stimulated mainly by (a) infrared rays (b) ultraviolet rays (c) vitamin D (d) vitamin C.
3 The outer layer of skin is called (a) epidermis (b) ectoderm (c) dermis (d) endoderm.
4 The appendages of the skin include all of the following *except* (a) hair (b) fingernails (c) sebaceous glands (d) sublingual glands.

5   The hair follicle is made up of cells of the (a) dermis (b) epidermis (c) boundary membrane (d) subdermal layer.

6   Another name for the epidermis is (a) papilla (b) corium (c) true skin (d) cuticle.

7   The living part of the epidermis is called (a) keratin (b) stratum corneum (c) stratum germinativum (d) conjunctiva.

8   Nerve endings in the skin are located chiefly in the (a) cuticle (b) epidermis (c) dermis (d) stratum corneum.

9   Sweat glands occur in greatest number in the skin of the (a) forehead (b) forearm (c) palm of the hand (d) back.

10   Adipose tissue in the skin is found mainly in the (a) epidermis (b) dermis (c) sebaceous glands (d) subdermal layer.

Group B

1   (a) What are the functions of the skin? (b) What structure, or structures, are mainly responsible for each of these?

2   (a) Briefly describe the structure of the epidermis and its method of growth. (b) What structures of epidermal origin are present in the dermis?

3   Describe the structure of a hair and the various associated parts.

4   Identify each of the following (a) friction ridge (b) callus (c) subdermal tissue (d) pore (e) barrier layer.

Chapter 27 Temperature Regulation

Temperature Variations

The members of the Animal Kingdom can be roughly classified in two groups, depending on their ability to regulate their internal body temperatures. The temperature of cold-blooded, *poikilothermic* animals changes with that of their environment, while warm-blooded, *homoiothermic* animals maintain a relatively constant temperature. This grouping does not take into consideration those animals that hibernate. The temperature of the big brown bat *(Eptescius fuscus)* varies from 38.55°C when active to approximately 10°C during hibernation.

At birth, the human infant has an imperfectly developed heat-regulatory mechanism. Infants are, therefore, quite susceptible to cold air, which is apt to lower their body temperature to a dangerous level. Even children of school age may show large variations in body temperature as a result of exercise or emotional disturbance. In old age, there is a similar inability to regulate the body temperature, with the result that the aged are easily affected by changes in temperature.

We are apt to think of the body temperature as remaining at a constant 37°C (98.6°F). This is the average normal temperature shown by a clinical thermometer placed under the tongue. Normal average rectal temperature is 37.5°C (99.6°F). Most probably the oral temperature is closer to the real average of the whole body. Studies show there is considerable variation in temperature among the members of any group. For example, 276 medical students, all members of one class, took their oral temperatures while seated in the classroom. The temperatures ranged from 35.8 to 37.4°C with an average of 36.7°C (96.5 to 99.3°F with an average of 98.1°F). Of these, 95 percent fell between 36.3 and 37.2°C (97.3 and 98.9°F).

During the course of 24 hours, the oral temperature of an individual fluctuates to a greater degree than is generally realized. As Fig. 27-1 shows, the body temperature drops below the generally accepted average during the night and then rises above this point during the day. The data for this graph were obtained from one person, and undoubtedly show some slight variations

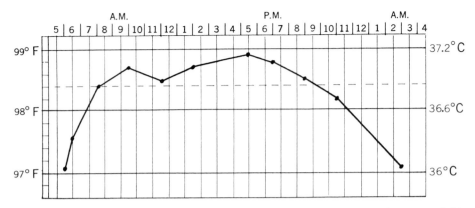

Fig. 27-1   A graph of a person's body temperatures during a twenty-four period. Note how the temperatures deviate from the usually accepted average oral temperature.

from similar data that might be obtained from another individual. Some interesting observations have been made on people who have changed their employment so that their working hours come during the night instead of the day. Such persons show a reversal of the normal temperature patterns after a few weeks so that their maxima are reached during the early hours of the morning.

The idea that the blood is of a uniform temperature throughout the body has been shown to be incorrect. Some relatively recent work by Dr. H. C. Bazett indicates that the temperature of the blood within arteries varies to a considerable degree. He placed thermocouples (electrical devices for measuring small differences in temperature) in the brachial and radial arteries (those of the upper and lower parts of the arm). Dr. Bazett found that the temperature of the blood in the brachial artery varied between 36 and 37°C (96.8 and 98.6°F). In the radial artery the temperature of the blood was between 35.5 and 37.2°C (95.9 and 98.9°F). These measurements were made in a room where the temperature was comfortable. If, however, he exposed himself to cold, but "without the subject's being unduly cold or the rectal temperature particularly low," the temperature in the brachial artery fell to 31°C and that in the radial artery to 29.5°C (87.9 and 70.7°F).

The maintenance of normal body temperature may be compared to the balancing of a scales. We can imagine that on one side of the scales we have the factors that are basic to the production of heat in the body and on the other those that are basic to its elimination. Under ideal conditions, the two sides would balance (as illustrated in Fig. 27-2). On the one side of the scales we have the heat that is produced by the oxidation of proteins, carbohydrates, and fats. On the other side we have the heat that is lost by convection, radiation, and vaporization, since these are the three most important ways in which heat is lost from the body. Physical processes maintain the balance. Any change in activity will alter the rate at which the nutrient materials are oxidized. Likewise, a change in the environment will affect the rate at which heat is lost.

In the next chapter the function of each of the organic nutrients in the production of heat will be discussed. Oxidation of nutrients in the body cells accounts for virtually all of the heat that is produced. Some tissues, however, are more active than others, and it is in these that the major production of heat occurs. Muscles make up over one half of the soft structures of the body, and oxidation in muscle tissue supplies the largest quantity of heat. If, for example, a warm-blooded animal receives an injection of curare, which renders its muscles incapable of movement, its body temperature varies in the same way as that of a cold-blooded animal. Active glands add a little heat, the liver being the greatest heat-pro-

ducing gland. The amount of heat supplied by warm food or drinks plays a negligible role in this process.

Temperature Measurement

The idea of measuring body temperature as a part of the treatment of disease was first used in the early part of the seventeenth century, only a few years after Galileo had invented the thermometer. It was not, however, used in the same way it is today. This later relationship between a specific disease and its temperature effects did not appear until the middle of the nineteenth century when standards for body temperatures were established. Since then, the use of the thermometer in medical procedures has become widespread. The oral thermometer was the first type used, and even today remains the most common method of measuring body temperature. Its reliability can be questioned because of the changing temperatures in the

Fig. 27-2 The balance between heat loss and heat gain. The weights on the left represent the relative amounts of heat produced by fats (extreme left), carbohydrates (middle), and proteins (right). The weights on the right represent the heat lost by radiation, convection and conduction; the three principal methods of heat loss. (Fundamental Photographs)

mouth. For more constant and reliable readings, rectal temperatures are preferred.

Not only is heat produced by healthy active tissues, but there are also certain types of diseases that are responsible for differences in heat production in local areas. Since all heat is a type of radiation in which the wave lengths are longer than the red of the visible spectrum, these radiations cannot be seen. They are called *infrared radiations*. It is possible, however, to photograph them if a special type of camera and film sensitive to infrared are used. This principle has been adapted recently to the diagnosing of certain diseases such as a cancer which may be near the surface of the body or an obstruction in a superficial blood vessel. If a sufficiently sensitive apparatus is used, the site of the disorder can be accurately located by the slight difference in temperature between it and the surrounding healthy tissue. Differences of as little as 2°C can be detected by this means. The term, *thermography*, is applied to this method.

Temperature Regulation

Body temperature is influenced by several different processes operating within the body or over its surface. Some of these are internal adjustments while others are responses to the state of the air surrounding the body. We can summarize these processes as follows:

Nervous regulation. In previous chapters we have mentioned several factors which influence body temperature. These include the dilation and constriction of the blood vessels under nervous control, the change in the tonus of the muscles and the production of sweat. The centers controlling these activities lie in the lower part of the medulla, but their function in temperature regulation appears to be under the control of a *heat-regulating center* located in the hypothalamus. It has been found, for example, that if the hypothalamus is separated from these lower centers in an experimental animal, all control over heat regulation is lost.

There are two distinct areas in this center. One of these controls the loss of heat and the other its production. If the anterior part of the center is destroyed or injured, a person adjusts his temperature normally in cool surroundings. However, in a warm place his body temperature rises because he cannot dispose of excess heat. When this anterior region is warmed in an experimental animal, all of the outward signs of heat loss are shown: the animal pants, sweats, and its body temperature drops. If the posterior part of the center is injured, heat production lags behind heat loss and the body temperature falls. Under normal conditions the hypothalamus is influenced either directly via the autonomic nervous system from sense receptors in the skin or via the blood.

Physical factors in regulation. There are three principal means by which the body loses heat: through the skin, from the lungs, and through excretions. Of these three, the skin plays the most important role because 87.5 percent of body heat is lost through its surface. The expired air is warmed as the result of contact with heated lung tissue. This accounts for 10.7 percent of the heat lost. The excretions, urine and feces, are responsible for only 1.8 percent of the heat loss. This quantity is so small that it is taken into consideration only when making the most exact measurements of loss of body heat.

Loss of heat over the surface of the body may occur in any of several ways, depending on environmental conditions. *Conduction* of heat may occur when the skin comes in contact with a surface of lower temperature, as when a lightly dressed person lies on cold ground. In general, however, we are protected from heat loss by conduction. We wear clothing that is made from materials that are poor conductors of heat because of numerous small pockets of air trapped within the fibers. The body does not normally lose much heat by conduction.

Body heat is lost by *radiation,* in the form of invisible waves in the infrared region of the spectrum (Fig. 27-3). Radiation occurs whenever an object is warmer than its surroundings. A radiator in a cold room gives off infrared (heat) waves which are absorbed by the walls and furniture. The air itself is not appreciably warmed by radiation, however. The body is constantly radiating heat.

Convection is partly responsible for loss of body heat. If a person is at rest in a cool room, he loses heat as a result of the convection currents that are set up over his skin. The layers of warmed air next to his skin are constantly replaced by cooler air that flows in as the lighter warm air rises. The same process occurs when a person swims in cold water. The thin layer of warmed water that has been in con-

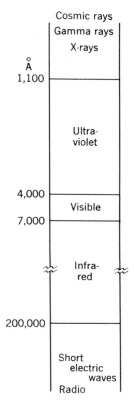

Fig. 27-3 *The parts of the spectrum that ordinarily affect humans.*

tact with the skin is replaced by cold water. In this condition the loss of body heat by convection is very great.

Vaporization, or evaporation, of sweat from the surface of the skin is the method we are most apt to think of as the principal means of heat loss. This is because we are more conscious of the cooling effects of the evaporation of sweat than we are of the other methods.

The capacity of a fluid to lower the temperature of the surface from which it is evaporating depends on its ability to absorb heat from the surface. This heat is the energy which converts the fluid into a vapor, and is known as the *heat of vaporization*. It can be defined as the amount of heat needed to vaporize one gram of a liquid without changing the temperature of the liquid. The amount of heat absorbed is measured in terms of small calories. It varies with each fluid and also with its temperature. Water has a very high heat of vaporization in comparison to other liquids, which means that its rate of evaporation is sufficiently slow to prevent too rapid chilling. At a skin temperature of 30°C each gram of water that evaporates requires 579.5 calories of heat. However, the rate of evaporation of sweat depends on factors such as the rate of the movement of air, the temperature of the air, and the relative humidity.

It must be kept in mind that relative humidity refers to the amount of moisture present in the air at a given temperature in relation to the absolute amount of moisture it could hold at that same temperature. As air becomes warmer it can hold more moisture. Therefore on a hot and humid day the body is still able to lose heat by evaporation of sweat, although there is a general feeling of discomfort. A relative humidity between 40 and 60 percent is generally considered to be the most desirable for health and comfort.

In a series of ingenious experiments conducted by Drs. Winslow and Herrington, the relative roles of radiation, convection, and vaporization under differing conditions of temperature were demonstrated. The experimenters constructed a booth in which air could be circulated freely, and in which the temperature of the air could be changed independently of that of the walls. Unclothed subjects were then placed in the booth and the conditions altered as shown in the table.

In *Series A* the temperature of the air and the walls was considerably below that of the skin, with the result that evaporation played a minor part in cooling the body but radiation and convection were important factors. In *Series B* the high temperature of the walls prevented the radiation of heat, as it also did in the last two series. In *Series B,* convection again was important because the air temperature was too low to permit much evaporation of sweat. In *Series C* we see that when the air temperature and the wall temperature were equal, but still somewhat below normal skin temperature, vaporization and radiation were about equally responsible for heat loss. *Series D* shows that as the temperature of the air was increased, the effect of sweating became more marked while convection was decreased. With a high

Series	Air Temp. °C	Wall Temp. °C	Percentage of Total Body Heat Lost Due to:		
			Vaporization	Radiation	Convection
A	17	19	10	40	50
B	16	49	21	0	79
C	23	23	17	13	70
D	29	52	78	0	22
E	35.5	33.5	100	0	0

wall temperature, such as the one found here, the body may even have absorbed some heat from the walls. In *Series E* the evaporation of sweat was the only method of losing heat because the air and wall were at about the same temperature as the body.

Chemical regulation. The range of external temperature in which man can regulate his body temperature effectively is 25–35°C (77–95°F), depending on the clothing worn and the presence of fat deposits. In this range of temperatures he loses the least heat and his rate of oxidation is lowest. In a resting state, the most economical temperature for the human body in terms of the balance between heat loss and oxidations is about 18°C (65°F).

If the body temperature falls because physical processes cannot keep heat loss at a normal level, chemical processes are initiated to increase production of heat. We have already mentioned some of these in connection with the neural control of heat regulation. When an excessive heat loss occurs, we have such activities as shivering, the chattering of the teeth, and the involuntary tensing of muscles which increase the rate of oxidation. These may be under the control of the nervous system, but it is the chemical reactions in the muscles that are chiefly responsible for the rise in temperature.

Some Disturbances of Heat Balance

One of the most common types of disturbance of the heat-regulating mechanism is *fever*. In the past, some people held the opinion that fever was the result of an increased rate of oxidation in the cells. There is little evidence to support this idea, since a fever produces an increase of only 20 percent in the rate of oxidation while muscular activity may raise the rate as much as 200 percent with no symptoms of a fever. Currently, it is thought that a fever is a result of the temporary failure of the "thermostat" in the hypothalamus to function properly, due to the action of bacterial or viral toxins, or to serious emotional upsets. The hypothalamus controls activities of the lower centers that are directly responsible for the operation of the heat-regulating processes. This theory is supported by the fact that a young child develops fevers very easily under conditions which cause a much slower response in adults. Nervous control of the heat-regulating mechanisms is not well established in babies and young children. A fever is accompanied by lack of sweat and constriction of the skin capillaries. This latter condition may cause the victim to feel cold, even though his temperature is higher than normal. Fever is the usual indication that infection of some kind is present in the body.

A second type of disturbance of heat regulation is *heatstroke,* or *sunstroke.* Both of these terms refer to the same condition, which is brought about by high external temperature and high humidity. The body is unable to lose heat normally and the temperature rises. Heatstroke is characterized by a dry, hot skin, a rapid pulse, and high blood pressure. The internal body temperature may rise to 43.3°C (110°F). At this temperature, destruction of brain cells results and death follows. Care for a victim of heatstroke may include external application of ice packs or wrapping the person in sheets soaked in cold water in order to reduce the body temperature rapidly. Overweight, the drinking of alcohol, and the wearing of unsuitably heavy clothing are all factors which may contribute to heatstroke.

Heat exhaustion, or *heat prostration,* occurs when the circulatory system fails to function properly under adverse temperature conditions. In this condition the pulse is weak, the blood pressure low, and the skin moist and clammy. A person suffering from heat exhaustion should be kept warm. Complete rest is the best method of caring for this condition.

Heat cramps develop when a person has been subjected to a high temperature and has been sweating profusely, with the loss of an excessive amount of sodium chloride from the body. The cramps are not accompanied by any rise in body temperature. The main characteristic of this condition is the painful cramps in the muscles that have been used most. People can tolerate surprisingly high temperatures if they take in enough salt. At zero humidity, experimenters have withstood temperatures of 98–121°C (200–250°F) for a few hours without any harmful effects. A 0.2 percent solution of salt in the drinking water will safeguard a person against heat cramps. Many industries encourage their employees to avoid heat cramps by providing salt tablets near the drinking fountains.

VOCABULARY REVIEW

Match the phrase in the left column with the correct word in the right column. *Do not write in this book.*

1 An animal whose body temperature varies with that of its surroundings.	a conduction
2 If properly stimulated, makes an experimental animal either sweat or shiver.	b convection
3 Results in heat loss by transfer of heat to a body (with which there is no contact) that is at a lower temperature.	c heat-regulating center
4 The Weather Bureau may report this as an index of bodily comfort.	d homoiothermic
5 An animal with a fairly constant body temperature.	e oxidation
6 May cause air disturbances over a radiating surface in a cool room.	f poikilothermic
7 The feeling of cooling experienced while holding an ice cube is due to this.	g radiation
	h relative humidity

TEST YOUR KNOWLEDGE

Group A

Select the correct statement. *Do not write in this book.*

1 The normal internal temperature of the human body (a) varies within narrow limits (b) varies over a range of 6°C (11°F) (c) is always the same (d) usually decreases during daylight hours.

2 The heat-regulating center of the brain is in the (a) cerebrum (b) cerebellum (c) hypothalamus (d) frontal lobes.

3 The human body loses heat mainly through (a) the skin (b) the lungs (c) excretions (d) secretions.

4 Of the following, the term that is *not* related to heat loss by the body is (a) convection (b) conduction (c) induction (d) radiation.

5 Heat is produced in the body by the process of (a) evaporation (b) condensation (c) oxidation (d) osmosis.

6 The largest quantity of heat is produced in (a) nerve tissue (b) epithelial tissue (c) muscle tissue (d) connective tissue.

7 Body temperature is usually lowest in the (a) mouth (b) rectum (c) hands (d) stomach.

8 If the air temperature is considerably lower than that of the body, heat is lost chiefly by (a) evaporation (b) sweating (c) radiation and convection (d) excretion.

9 Generally speaking, the most dangerous type of disturbance of heat balance in the body is (a) heat exhaustion (b) heat stroke (c) fever (d) heat cramps.

10 Warm-blooded animals are said to be (a) poikilothermic (b) homoiothermic (c) exothermic (d) endothermic.

Group B

1 Explain why running up a flight of stairs may make you sweat.

2 What role does each of the following play in regulating body temperature on a day when the air temperature is 33.9°C (93°F): (a) radiation (b) vaporization (c) conduction (e) convection.

3 Referring to Question 2, how does each of these conditions help to control the body temperature when the air temperature is 4.4°C (40°F)?

4 (a) What is the primary physiological difference between a cold-blooded animal and a warm-blooded animal? (b) Under what conditions may the physiology of animals that are normally warm-blooded behave like that of cold-blooded animals?

5 Relate each of the following to the regulation of body temperature: (a) wearing clothing (b) drinking salty water (c) drinking alcohol (d) staying in the shade when the air temperature is 43.3°C (110°F) (e) swimming in water that is 18.3°C (65°F).

Chapter 28 Metabolism

The term *metabolism* includes all those chemical processes within cells from which any living thing derives its energy. The processes are of two kinds. If a process is primarily constructive and results in the building of protoplasm, it is classed as *anabolism*. If it is a destructive process, it is called *catabolism*. As a result of anabolic processes, the protoplasmic materials of the cells either increase in quantity or are replaced. The catabolic processes, on the other hand, result in the destruction of substances and the conversion of the potential energy they contained into kinetic energy. Youth is the period in life when anabolism is the characteristic process; middle age is represented by a balance between the two, and old age is a period in which the catabolic processes are dominant.

We have already studied the nature of the various food materials and their preparation for distribution by the blood. We have also become acquainted with some of the processes involved in muscular contraction, the secretion of materials by glands, and the development of various parts of the body. All of these processes require transforma-

tions of energy which are basic to the study of metabolism.

The Energy Value of Food

In Chapter 13 we mentioned the uses of organic nutrients and the role each plays in producing energy or building protoplasm. The unit used in measuring the energy production of foods is the Calorie. The energy value of any food is measured by an apparatus such as the one diagrammed in Fig. 28-1. This is known as a *bomb calorimeter*. It consists of several vessels placed one inside another for complete insulation from changes in the outside temperature. The innermost chamber is surrounded by a water bath fitted with a stirring rod to keep the water in circulation. An accurate thermometer is inserted into the water bath. The material to be tested is put into a crucible in the innermost compartment, the bomb. By means of electrical connections, an arc is produced which ignites the food. As the food burns, the water outside the bomb is heated and any rise in temperature is read on the thermometer. After subtraction of the heat produced by the arc, the rise in temperature of the water

376

can then be translated into Calories produced by the complete combustion of the food.

The heat produced by an animal can also be measured directly in a similar device. This method of *direct calorimetry* employs a chamber large enough to accommodate the animal under investigation. Direct measurements of the heat produced by animals as large as an elephant have been made. Essentially, a calorimeter of this type consists of a fully insulated, airtight chamber through the walls of which pass tubes containing water in constant circulation. The temperature of the water is read as it enters the tubes and again as it leaves. The outgoing air is passed through sulfuric acid to remove moisture, and through soda lime to absorb the carbon dioxide. The air from which the water and the carbon dioxide have been removed is then mixed with a known amount of oxygen and recirculated through the chamber. At the conclusion of the test, the sulfuric acid and the soda lime are weighed to determine the amount of water and carbon dioxide that the animal has produced. Since the amount of oxygen entering the chamber during any given period of time is known, its conversion into water and carbon dioxide can be measured. These substances are the products of oxidations occurring within the cells, as shown in the following chemical equation for the oxidation of glucose:

$$\underset{\text{glucose}}{C_6H_{12}O_6} + \underset{\text{oxygen}}{6O_2} \rightarrow$$

$$\underset{\substack{\text{carbon} \\ \text{dioxide}}}{6CO_2} + \underset{\text{water}}{6H_2O} \quad \underset{\text{gram of glucose}}{(+4.1 \text{ Calories per}}$$

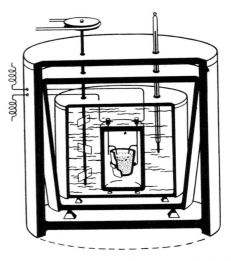

Fig. 28-1   A bomb calorimeter. The food is placed in the crucible and ignited by an electric arc. The heat generated by burning the food can be calculated from the rise in temperature of the water in the surrounding water bath. (Bawden, Matter and Energy, *Holt, 1957)*

Direct calorimetry requires the use of cumbersome and complicated apparatus. To simplify the procedure, a method of *indirect calorimetry* has been developed to measure human metabolic rates. In Fig. 28-2 is shown a person undergoing a metabolism test in a hospital. The apparatus pictured here is a modification of the spirometer mentioned in Chapter 19. The patient breathes into a closed system. The floating bell of the spirometer in the photograph is filled with oxygen which the patient inhales. Provision is made for the absorption of carbon dioxide by soda lime. Attached to the respiratory calorimeter is a revolving drum on which is recorded the rate at which

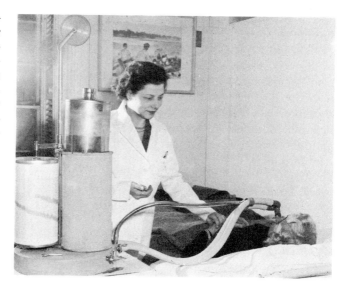

Fig. 28-2 A person undergoing a BMR test. The kymograph at the left records the amount of oxygen consumed during a measured period of time. Next to the kymograph is the spirometer. (Massachusetts General Hospital)

oxygen is used. Fig. 28-3 is a reproduction of a graph obtained from this apparatus.

A person who is scheduled to undergo a metabolism test is required to rest quietly in a comfortable room for at least an hour before starting the test. Also, he should not have eaten for a period of at least 12 to 14 hours so that the presence of food in the digestive system will not affect the metabolic rate.

Since the test determines the amount of oxygen consumed in a given period of time (usually six minutes), certain facts regarding the oxidation of the organic nutrients must be taken into consideration. These are:

To oxidize 1 gram of carbohydrates and liberate 4.1 Calories, 0.812 liter of oxygen is required. Hence when 1 liter of oxygen is used, 5 Calories will be generated.

To oxidize 1 gram of fat and liberate 9.3 Calories, 0.71 liter of oxygen is needed. One liter of oxygen will therefore produce 13.098 Calories.

The oxidation value for the proteins must be reached by indirect methods because of the extreme complexity of their chemical structure. It has been determined, however, that 1 gram of protein requires 0.95 liter of oxygen for its combustion and will supply 4.1 Calories. One liter of oxygen therefore produces 4.5 Calories of heat when proteins are oxidized.

If we assume that all three nutrients are present in the patient's diet in the usual proportions (carbohydrates, three parts; fats, two parts; proteins, one part), and are completely oxidized, the consumption of 1 liter of oxygen will produce a total of 7.616 Calories. If, therefore, in the six minutes required by this test, he uses 1.25 liters of oxygen, the patient will generate 9.59 Calories ($7.616 \times 1.25 = 9.59$). If we then assume that the patient remains quiet for a total of twenty-four hours, his

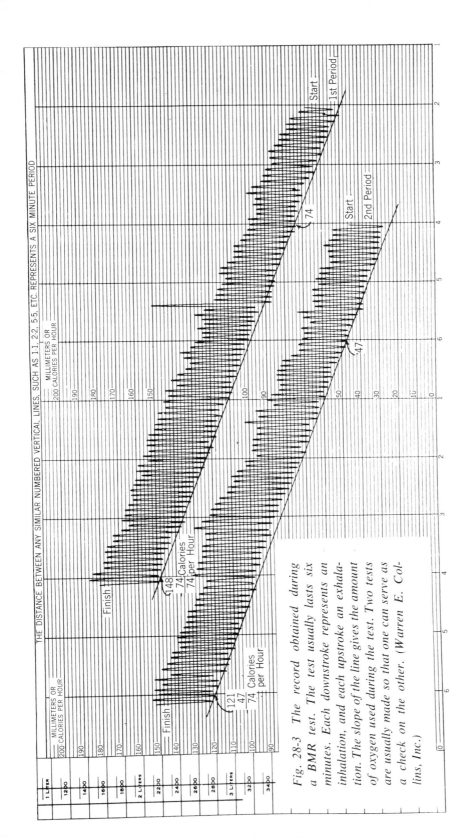

Fig. 28-3 The record obtained during a BMR test. The test usually lasts six minutes. Each downstroke represents an inhalation, and each upstroke an exhalation. The slope of the line gives the amount of oxygen used during the test. Two tests are usually made so that one can serve as a check on the other. (Warren E. Collins, Inc.)

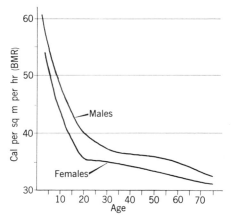

Fig. 28-4 The relation between BMR and age. These are smoothed curves, based on averages obtained by three separate research groups who studied results from 1762 men and 2254 women.

total heat production during this period will be 2301.6 Calories.

Basal Metabolic Rate

The results obtained by the method just described constitute the *basal metabolic rate (BMR)*. Since the activity of the person undergoing such a test is controlled, the rate of oxidation is considered to be the fundamental rate for that individual.

Two factors which influence the BMR are the age and sex of the person. Fig. 28-4 shows the effect of age on the basal rate. It is seen that the rate is high among young children and decreases as they grow older, until the period of rapid growth has been completed. From that point on, there is relatively little decline. The reason for this pattern is not clearly understood at present.

The rate of metabolism is given in terms of Calories per square meter of body surface per hour. Most energy is lost in the form of heat, and the surface of the body is the principal radiator. Because body surface varies with weight and height, the metabolic rate is described in terms of the heat lost per unit of body surface. An adult male who weighs 68 kilograms and is 1.6 meters tall (150 pounds and 5 feet 7 inches) tall has a body surface area of about 1.7 square meters. He will have a BMR of approximately 38 Calories per square meter of body surface per hour, or 912 Calories per square meter per day. The chart in Fig. 28-5 is used to find the surface area of a person of known height and weight.

If we consider two animals of different sizes, we will get a clearer idea of the significance of the area of body surface in relation to heat loss. A mouse, for example, will lose more heat in proportion to the amount of heat it produces than a larger animal, such as a dog. The reason for this is that the mouse has a greater surface area in relation to its weight than the dog. If, however, we calculate the heat loss in terms of a unit of body surface, we find that all mammals are approximately alike in this respect. The table of averages below illustrates this.

	Body Surface Area (In Square Meters)		Basal Metabolism (Calories per Square Meter per Day)	
	Male	Female	Male	Female
Man	1.83	1.65	910	790
Dog	0.65	0.58	800	770
Rat	0.031		905	760

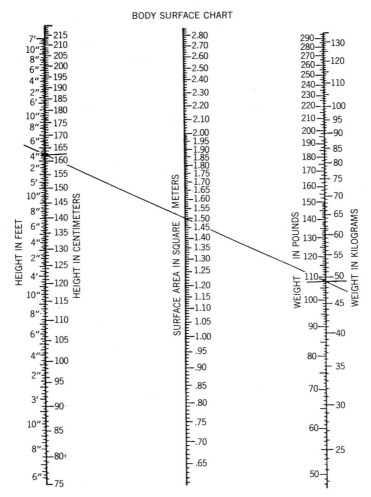

Fig. 28-5 A chart showing the relationship of weight, height, and body surface area. To use the chart, lay a ruler across it so that one end intersects the weight line at your own weight and the other end intersects the height line at your own height. You then read your body surface area (in square meters) where the ruler intersects the middle line. The line drawn on this chart gives the body surface area of a person who was 5 feet 4 inches tall and who weighed 108 pounds. (Warren E. Collins, Inc.)

Changes of Rate

As we have mentioned, one of the results of muscular activity is the production of heat within the muscles.

Since muscle tissue forms a large percentage of the body's mass, it is reasonable to expect that muscular activity will have a proportionately large relation to heat production. The list of

Activity	Percent Increase in Basal Rate
MEN	
Sleeping	− 0.1
Working as draftsman	+ 54
Working as radio mechanic	131
Driving car	139
Dressing	242
Using pick to dig earth	498
Walking at 4.2 mph	678
Farming: hoeing, deep ridging	712
Lumbering: horizontal chopping	1028
Walking in loose snow carrying 40-lb weight at 2.4 mph	1627
WOMEN	
Sleeping	− 0.1
Darning	+ 29
Typing (40 words per minute)	51
Washing dishes	54
Washing floor on knees	66
Sweeping	89
Carrying tray in hands	130
Washing clothes by hand	175
Changing bed linen	451
Skiing on level hard snow, moderate speed	1002

activities on this page shows the effect of muscular activity on the BMR.

Changes in *external temperature* also affect metabolic processes. If the BMR of a dog is measured at an external temperature of 7.6°C (45.7°F) and then the temperature is raised to 30°C (86°F), we find that the BMR drops by 65 percent. The higher rate of metabolism at the low temperature can be explained by the fact that muscle tonus (and, hence, rate of oxidations) has increased. At low temperatures, there is an increased rate of heat loss over the surface of the body and this must be balanced by an increase in the rate of heat production. This applies to humans also, and accounts for our feeling of general well-being on a cool, clear day. Although our reaction may be pleasurable, from a physiological standpoint we are paying for our pleasure by increasing our rate of cellular oxidations to offset the loss of heat from the surface of the body. It was found that American soldiers in tropical countries with daily air temperatures near to 32.2°C (90°F) needed only 3100 Calories per day, while soldiers serving under Arctic conditions required a much higher average of 4900 Calories per day.

A consistently high external temperature influences basal metabolism. If we examine the basal metabolism of people who are acclimatized to tropical conditions, we find that there is considerable variation from the expected response. Studies made on a group of male Brazilians showed that their BMR was less than that of a comparable group of men in northern parts of the United States. On the other hand, the BMR of natives of Yucatan averaged about 8 percent above that of a comparable group of Americans. In India, a group of women showed an average BMR of 17.4 percent lower than that of a comparable group of American women. Since Yucatan, Brazil, and India are tropical countries with high temperatures, the differences in results obtained in these regions indicate that heat production is not dependent on body surface area alone.

Altitude evidently plays no part in determining the basal metabolic rate of an individual. Many BMR tests have been made on people who live at high altitudes and are acclimatized to them. When their test results are compared with those of people living at lower elevations, no significant differences appear. Likewise, no change in the rate has been observed in aviators flying at different altitudes nor in people who have been placed in low-pressure chambers where the effects of increased altitudes could be imitated.

Food itself has an effect on the BMR. Experimental evidence of this was obtained by putting a person on a starvation diet. At the end of the first day it was found that he liberated only 718 Calories. On the second day of the experiment he received 400 grams of proteins in addition to his starvation diet and his heat liberation jumped to 1046 Calories. The third day he again went on starvation rations and his Calorie output dropped to 746, but rose to 1105 on the fourth day when he again received an additional 400 grams of protein. Other experiments have shown conclusively that this is not due to the presence of food in the digestive system because such inert materials as agar (a complex carbohydrate that cannot be digested) do not increase the BMR. This action of foods, especially proteins, is known as the *specific dynamic action* (SDA) of foods.

The reason for the SDA of foods is not clearly understood at present. Currently, experimental evidence points to the liver as the controlling factor in this action. Studies indicate that if this organ is removed from an experimental animal, the addition of large amounts of protein to its diet does not result in an increase in the BMR. These experiments lead us to the conclusion that the specific dynamic action of foods is connected with deamination (removal of NH_2 from the protein molecule). It is possible that the presence in the body of proteins that cannot be immediately utilized will help to keep the body warm.

Thyroxin also influences the BMR. This hormone is a secretion of the thyroid gland, which will be discussed later. The malfunctioning of this gland is one of the abnormal conditions revealed by a basal metabolism test. If the gland produces too little thyroxin, the BMR will be low, but if it is overactive, the BMR will be above normal. Other conditions which may lower the metabolic rate of the body are lack of proper amount of one of the pituitary secretions, starvation, and kidney disorders. An elevation of the rate may be the result of an overproduction of the pituitary secretion, severe anemia, certain types of leukemia, or too rapid growth.

Protein Bound Iodine

Another method of determining the efficiency of the body's chemical reactions has recently been developed. This method involves the thyroid gland, which plays an important part in the regulation of metabolism, and is also the principal storehouse of iodine in the body.

Iodine is present in two forms in the blood. In one form, *protein bound iodine (PBI)*, it is part of a protein molecule produced by the thyroid gland. Iodine is also found as inorganic salts dissolved in minute quantities in the blood plasma. These two forms of iodine can be readily separated by chemical methods. Because PBI is produced by the thyroid gland, the amount of this compound in the blood may be considered an index of the rate of thyroid activity, and, hence, of the rate of metabolism. Since the two types of iodine can be easily separated, the amount of inorganic iodine does not influence the analysis of PBI in the blood.

The PBI method of determining the basic efficiency of the body has some important advantages. First, the PBI determination is made on a small sample of blood. It does not require the physician to maintain and operate a highly specialized piece of apparatus which also might cause alarm or discomfort to the person being examined.

The PBI determination indicates the rates at which bodily processes are proceeding. It should not be confused with the BMR, which shows how varying factors affect the rate of these chemical processes.

The method of determining the BMR mentioned earlier in this chapter should not be eliminated, since it gives factual information about changes in metabolic rate under varying environmental conditions. However, determination of PBI does supplement the older method by supplying more information about the body's metabolism.

Relative Energy Values

Carbohydrates, fats, and proteins, as we have seen, vary in their ability to produce heat on oxidation. Other factors in the metabolism of these nutrients must be considered, however, in judging their value as energy-giving foods.

The idea that a good day's work requires a diet that is primarily composed of proteins has little basis in observed fact. Physiologists working on this problem report that the protein content of the diet has little effect on the physical vigor or efficiency of the individual. This is due to the fact that proteins are rarely oxidized to produce energy.

A person with a diet low in protein (50 grams per kilogram of body weight per day) did not show any difference in physical vigor or efficiency when he was changed to one rich in protein (160 grams per kilogram of body weight per day).

As we have previously said, the function of proteins is to build tissue. They can be replaced by carbohydrates, the protein-savers, as sources of energy. Fats are the best sources of energy, but when they serve as the exclusive energy-producing materials, there is an increase in the amount of ketones formed. Since ketones are acid in their reaction, they threaten the acid-base equilibrium of the blood, and the condition of ketosis develops. We will have more to say about this later in connection with diabetes and the action of insulin in Chapter 31.

It is a well recognized fact that eating sugar before an athletic contest increases one's ability to combat exhaustion. This is because carbohydrates are quickly digested and made available for oxidation. We can conclude, therefore, that while all three types of organic nutrients are capable of supplying energy, carbohydrates are preferred. In this connection, it is interesting to note that the majority of the most muscular animals are her-bivorous in their diets. It appears, therefore, that plant nutrients can supply the major part of the carbohydrates needed for energy.

VOCABULARY REVIEW

Match the phrase in the left column with the correct word in the right column. *Do not write in this book.*

1 A general term including all of the chemical reactions occurring in the body.
2 The rate at which chemical reactions proceed when the body is completely rested.
3 An apparatus for measuring the caloric content of food.
4 A chemical test which will determine the activity of the thyroid gland.
5 The building-up processes which result in the growth of the body.
6 A method of determining the BMR by measuring the amount of oxygen used during a given period of time.
7 An organ in the neck region which can determine the rate of metabolism.
8 The breaking-down processes resulting in the release of energy.

a anabolism
b BMR
c bomb calorimeter
d calorimetry
e catabolism
f direct calorimetry
g indirect calorimetry
h metabolism
i PBI
j thyroid gland

TEST YOUR KNOWLEDGE

Group A

Select the correct statement. *Do not write in this book.*

1 The nutrient that provides the greatest amount of energy per unit of weight is (a) starch (b) protein (c) sugar (d) fat.
2 The caloric requirements of the body can be determined indirectly by measuring its intake of (a) nitrogen (b) carbon dioxide (c) oxygen (d) carbohydrates.

3 Of the following, the term that is *not* closely related to the chemical processes within living cells is (a) catabolism (b) anabolism (c) embolism (d) metabolism.

4 One product of oxidation of nutrients in the cells is (a) water (b) carbon monoxide (c) glycerol (d) amino acid.

5 The rate of metabolism is greatest in (a) young children (b) teen agers (c) adult males (d) adult females.

6 The rate at which a person performs his fundamental chemical reactions is abbreviated by the letters (a) DNA (b) BTI (c) BMR (d) ATP.

7 The activity that increases the metabolic rate the most is (a) sleeping (b) sitting (c) standing (d) walking.

8 Of the following factors, the one that has *least* effect on the rate of metabolism is (a) age (b) sex (c) altitude (d) external temperature.

9 The basal metabolic rate is usually calculated in terms of Calories per (a) centimeter of body height (b) kilogram of body weight (c) degree of body temperature (d) square meter of body surface.

10 A hormone which greatly affects the basal metabolic rate is (a) thyroxin (b) pepsin (c) actin (d) myosin.

Group B

1 What relation does a diet of seal blubber have on the BMR of the Eskimos?

2 (a) List the various factors that can affect a person's BMR. (b) Which of these may be affecting your BMR under the present conditions?

3 What effect does eating a diet that is largely composed of proteins have on the physical efficiency of an individual?

4 Briefly describe the procedure used in making a BMR test and the physiological basis for each part of the test.

Chapter 29 The Kidneys

The elimination of waste materials from the cells and the body as a whole is the function of the excretory system. The catabolic activities of cells produce waste materials of various types, most important of which are carbon dioxide and nitrogenous end products of protein metabolism. If these were allowed to accumulate, serious injury would result to the tissues.

These wastes are dissolved in water and carried to the appropriate excretory organ; carbon dioxide to the lungs, nitrogenous and other wastes to the kidneys. It follows that the volume of water in the body must be maintained at a constant level. The intake must obviously equal the output otherwise the balance will be upset and the cells seriously affected. As the following table shows, there are several means of

eliminating the excess. The values given here are only approximations and depend on average conditions of temperature, humidity, and diet each day.

Kidney Functions

The *kidneys*, with their associated structures, are the most important organs of excretion. Their primary excretory function is the elimination from the body of the products of protein catabolism. Only small amounts of nitrogenous wastes leave by any other pathway. So long as the kidneys continue to excrete nitrogenous compounds, the processes of anabolism and catabolism are kept in balance. A dynamic equilibrium is thus established in the body.

Water Intake		Water Output	
Solid and semi-solid food	1200 ml	Skin (insensible	
Oxidation of food	300 ml	perspiration)	500 ml
Drinking liquids	1000 ml	Urine	1500 ml
Total	2500 ml	Expired air	350 ml
		Feces	150 ml
		Total	2500 ml

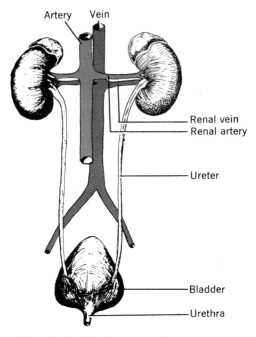

Artery Vein

Renal vein
Renal artery

Ureter

Bladder

Urethra

Fig. 29-1   The human urinary system as seen from the back. The kidney capsules are shown at the top, each side of the midline of the body with the suprarenal glands resting on their upper edges.

The kidneys also regulate the exact volume and composition of the blood by selective removal of any substance present in excess and conservation of others. In this way they maintain the acid-base equilibrium of the body, by not only excreting fixed acids, but also manufacturing materials which help to neutralize excess acids and bases. They regulate the osmotic action of the blood, maintaining a balance between its salt content and that of the cells by the excretion of excess amounts of various inorganic salts. They regulate the blood volume by eliminating excess water.

Kidney Structure

The kidneys (as shown in the Trans-Vision which follows page 212) lie against the dorsal wall of the abdominal cavity between the peritoneum and the muscles of the back. They are enclosed in capsules of fatty tissue. Each kidney is about 11 centimeters long, 5–7.5 centimeters broad, and about 2.5 centimeters thick. Their combined weight is approximately 280 grams. The *renal artery* and the *renal vein* enter the concave side of the kidney at the *hilum*. The *ureters* also leave at the hilum and carry the urine to the *urinary bladder* which lies on the floor of the abdominal cavity. From the bladder, a tube called the *urethra* carries the urine to the outside (Fig. 29-1).

If a kidney is cut lengthwise (Fig. 29-2), a definite arrangement of tissues can be seen. On the outer surface is a layer with a fine granular texture called the *cortex* of the kidney. Within it is a layer composed of tissues arranged in lines that radiate outward from the central part toward the cortex. This layer is the *medulla* of the kidney. On its inner margin are numerous rounded projections, the *pyramids*. The cavity into which the pyramids protrude forms the *pelvis* of the kidney, which is the uppermost end of the ureter.

A microscopic examination of the various layers shows that they have a very extensive blood supply. The renal artery branches extensively immediately after it enters the kidney. The branches give rise to numerous capillaries. The blood supply to the kidneys is so great that when the body is at rest, about one fourth of the total output of the heart passes through them. This means that a volume equal to the total blood supply of the entire body enters and leaves the kidneys every four or five minutes.

The Nephron

There are approximately 1 million highly specialized tubules, nephrons, within each kidney. Each tubule is about 14 millimeters long and 0.055 millimeter in diameter. The upper end of the tubule is expanded into a sac-like structure known as Bowman's capsule. It is composed of two layers of cells and its shape is like that obtained by pushing a blunt object against one side of a thin-walled rubber ball to form an indentation. The afferent blood vessel, a minute branch of the renal artery, enters at this indentation. The afferent vessel then forms a knot of about fifty separate capillaries, the *glomerulus,* which fills the cavity of

Fig. 29-2   Model of a longitudinal section through the kidney. (Model designed by Dr. Justus F. Mueller. Photograph, Ward's Natural Science Establishment Inc.)

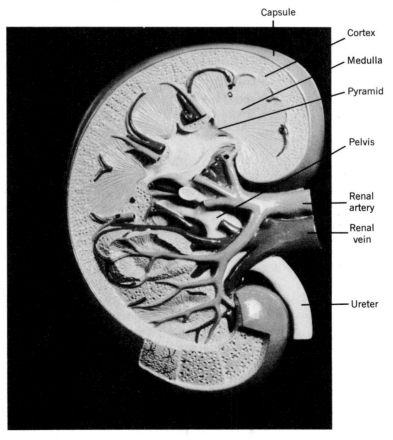

Capsule

Cortex

Medulla

Pyramid

Pelvis

Renal artery

Renal vein

Ureter

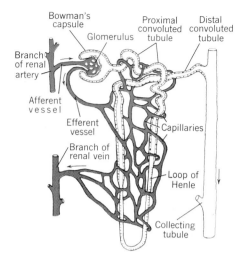

Fig. 29-3 A diagram showing the close relationship of the nephron and the blood vessels by which materials are reabsorbed into the blood.

the capsule. Although the diagram of the nephron in Fig. 29-3 does not show it, the inner wall of Bowman's capsule dips between the capillaries to surround them. This brings each blood vessel into direct and intimate contact with the capsule wall.

Beyond the capsule, the nephron becomes tubular and quite convoluted (twisted) for a short distance. Both the capsule and the convoluted tubule lie in the cortex, but the tube suddenly dips into the medulla to form the *loop of Henle* and then returns to the cortex. It once more becomes convoluted before opening into a *collecting tubule*. This larger vessel is joined by other tubules which open into the pelvis from the surface of a pyramid.

As Fig. 29-3 shows, the walls of the tubule are surrounded by capillaries. After the afferent vessel has formed the

glomerulus, it continues out of the capsule as the efferent vessel and this again branches to form the capillaries which surround the tubules. These eventually flow together to form a small branch of the renal vein which takes the blood from the kidney.

Filtration

We have already mentioned that the renal artery branches into smaller vessels immediately upon entering the kidney. The result of the sudden lessening in size of the vessels is a sudden increase in pressure, like that which would occur if we attached many small rubber tubes to the end of a garden hose. Thus, when the blood flows into the glomerulus, the pressure rises as the artery branches into capillaries. In most of the capillaries throughout the body, pressure is about 25 millimeters of mercury, but within the glomerulus it rises to between 60 and 70 millimeters. As a result of this high pressure, blood fluids are filtered through the capillary wall and pass into the capsule.

The total amount of fluid passing from the capillaries is astounding. It is estimated that approximately 125 milliliters of fluid leave the blood each minute. In the course of 24 hours this amounts to about 180 liters, or 190 quarts. In spite of this tremendous withdrawal of fluid from the blood, the body loses only 1.4–1.9 liters of fluid per day in the form of urine. The reason may be found in the relationship of the blood capillaries to the tubular parts of the nephron.

Due to the loss of large amounts of fluid in the capsule, the blood in the efferent vessel is highly concentrated and viscous. The various blood proteins do not filter through the walls of the capillaries. Since these proteins have a high colloidal osmotic action, much of the fluid is reabsorbed by the blood as it passes through the capillaries surrounding the convoluted tubule. With the water that reenters the blood go other materials valuable to the body, some of which are selectively reabsorbed by active transport which requires expenditure of energy. Conversely some wastes such as urea are selectively removed from the blood by active transport and are concentrated in the urine. Thus all the glucose and all the amino acids that left the blood in the capsule are normally returned to it so that there are none present in the urine.

Under normal conditions, very few proteins leave the blood in the kidneys. As numerous experiments have shown, this is due to the large size of the protein molecules. Their retention by the blood is of primary importance in maintaining the protein balance of the body. It has been found that if gelatin (molecular weight: 35,000) or egg albumin (molecular weight: 34,500) are injected into the blood, they pass out of the blood into the urine quite easily. The same is true of other proteins with molecular weights lower than that of hemoglobin (molecular weight: 68,000). Hemoglobin, however, may appear in the urine when it is present in plasma in excessively large quantities. Such a condition arises when there is an abnormally great destruction of red blood corpuscles, as in a disease such as malignant malaria.

Excess albumin in the urine produces a condition known as *albuminuria,* found in four to five percent of healthy individuals. The amount present is very small (about 0.2 percent). If a person with no previous history of albuminuria takes a cold bath or performs severe muscular exercise, the excretion of albumin in the urine increases. In *nephritis,* a condition of inflammation and degeneration of the kidneys, the glomeruli fail to prevent the loss of albumin from the blood. Therefore albumin appears in the urine in large amounts. Other conditions such as scarlet fever, pneumonia, decaying teeth, streptococcic sore throat, and sinus infections may result in temporary albuminuria. The loss of blood protein lowers the osmotic action of the blood with the result that fluids tend to collect in the tissues and cause edema. The presence of glucose in the urine, called *glycosuria,* indicates either a loss of ability of the tubules to reabsorb sugar or an excess of sugar in the blood.

When glycosuria is *not* accompanied by an increase of sugar in the blood, it is called *renal glycosuria.* If there is an increase in both blood sugar and sugar in the urine, the condition is called *diabetes.* This disease will be discussed in a later chapter.

The kidney should not be considered as a simple filter that removes the end products of protein metabolism in an automatically uniform manner. It is a highly discriminating organ that is able to remove wastes and foreign substances. At the same time it makes adjustments which will conserve some materials that are in low supply or remove these same substances when they are in excess. There may be a great variability in the quantity and composition of urine samples from the same individual over even a short period of time. This indicates that the materials are handled individually in order to serve the body effectively.

The average concentration of various materials present in the urine as compared with their presence in the plasma is shown in the table below adapted from the work of Dr. A. R. Cushny.

Organic Excretory Products

The principal organic excretions are urea, uric acid, and creatinine. In addition to these, other organic substances may be present in the urine at various times and in differing amounts depending on the composition of the food ingested. Ammonia, in the form of ammonium salts, is also present. Though not one of the organic compounds, it is an intermediate product in their formation.

Urea $(CO(NH_2)_2)$. This substance is synthesized from the ammonia produced by the decomposition of amino acids. The disposition of the products of amino acid decomposition is shown in Fig. 29-4. The liver is the only place in the body where urea is formed. This can be demonstrated by removing the liver from an anesthetized experimental animal. When this is done, the volume of urea excreted by the kidneys drops to zero. The processes involved in the formation of urea from amino acids are quite complicated and are generally beyond the scope of this book. In brief the amino group (NH_2) is split off from the amino acid as free ammonia NH_3. Two molecules of ammonia are combined with one of carbon dioxide in a cyclic series of reactions. The overall reaction is as follows:

$$2NH_3 + CO_2 \rightarrow CO(NH_2)_2 + H_2O$$
$$\text{ammonia} \qquad \text{urea}$$

PERCENTAGE COMPOSITION OF PLASMA AND URINE

	Plasma %	Urine %	Change in Concentration in the Kidney
Water	90-93	95	____
Proteins, other colloids, and fats	8.0	____	____
Glucose	0.1	____	____
Sodium	0.32	0.35	1
Potassium	0.02	0.15	7
Chlorine	0.37	0.6	2
Phosphates	0.009	0.15	90
Ammonia	0.001	0.04	40
Urea	0.03	2.0	60
Uric acid	0.004	0.05	12
Creatinine	0.001	0.075	75

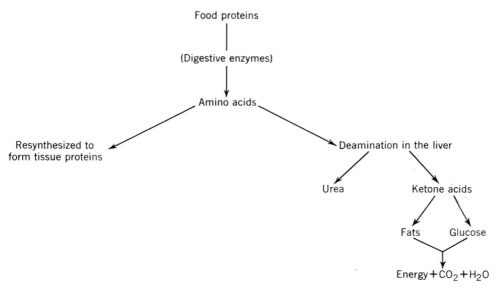

Fig. 29-4   Chart of the products resulting from the decomposition of amino acids.

Urea enters the systemic circulation through the hepatic vein. The deaminated amino acid is broken down stepwise to carbon dioxide and water releasing energy used in the synthesis of ATP.

Creatinine ($C_4H_7N_3O$). Creatinine is evidently formed in the muscles from creatine (N_3H_8COOH), a normal constituent of muscle tissue. It may also be ingested. By the use of radioactive isotopes of nitrogen (N^{15}) and hydrogen (H^3), the formation of creatinine from creatine has been traced. The amount excreted by any one individual remains quite constant but is dependent on his size. Its production is therefore reported in terms of body surface area in the same manner as the BMR, and for the same reason.

Uric acid ($C_5H_4O_3N_4$). Small quantities of uric acid are normally present in the blood. The source of this acid is the destruction of nucleoproteins. It varies in quantity in the urine depending on the amount of protein material ingested. Following a heavy protein meal, the kidneys will excrete an additional amount of uric acid. Leukemia, a disease characterized by the production of leukocytes, results in an increase of uric acid excretion. Gout also is believed to be the result of an upset in the uric acid balance in the body, but some question regarding this has been raised recently.

Ammonia (NH_3). This is a highly toxic material that the cells cannot tolerate, but its compounds help in the maintenance of the acid-base equilibrium of the blood. As we have learned, acids which might cause a shift in the pH of the blood are neutralized by basic ions. In the kidney, certain of these acids are combined with ammonia and excreted as ammonium salts. The basic ions thus freed help to maintain the body's alkaline reserve. When the diet is rich in acid-forming food, such as proteins, the excretion of ammonium salts increases. Excretion decreases when alkali-producing foods are eaten in quantity. Formation of ammonia in the kidneys is a result of the deamination of amino acids.

Factors Affecting Function

The blood supply to the kidneys is largely responsible for determining the amount of urine produced under normal conditions. When the nerves to the kidneys are cut, the kidney blood vessels dilate. If the sympathetic nerves are then stimulated, raising the blood pressure, more blood passes through the kidneys, resulting in greater production of urine by the kidney.

The amount and concentration of the urine is affected by ingestion of water or salt. As we all know, if we drink large quantities of fluid, the amount of urine increases. This urine is quite dilute due to the presence of an increased amount of water. On the other hand, if very salty substances are eaten, the quantity of urine is reduced, but the concentration of dissolved materials increases. The events which bring about these reactions involve both the nervous and endocrine systems. Intake of fluid or salt changes the osmotic pressure of the blood. This in turn apparently affects certain receptors (osmoreceptors) of the hypothalamus of the brain. The hypothalamus then sends impulses to the posterior lobe of the pituitary gland, which produces an *antidiuretic hormone (ADH)*. Diuretic refers to urine production. If too much fluid has been taken into the body, the production of this hormone is reduced and the liquid is excreted in order to maintain the correct osmotic state of the blood. On the other hand, if an excess of salt has been eaten, the osmoreceptors are stimulated and the amount of antidiuretic hormone is increased.

The effect of the osmotic pressure of the blood on the production of ADH and the reciprocal effect of ADH on the blood's osmotic pressure constitutes an excellent example of negative feedback maintaining a system in equilibrium. A rise in osmotic pressure stimulates production of ADH which prevents further rise in osmotic pressure by inhibiting excretion of water in the urine. As the osmotic pressure falls further the production of ADH is no longer stimulated. This is just one example of the many homeostatic (self-regulatory) mechanisms present in living organisms.

Urination

The elimination of urine, urination or micturition, is under the control of the nervous system and is brought about by stimulation of bundles of smooth muscles in the walls of the ureters, bladder, and urethra.

The ureters enter the bladder at a sharp angle which prevents the backflow of urine into them. Liquid flows along them in an almost continuous stream, although the quantity may vary greatly, depending on the factors mentioned earlier. The normal capacity of the bladder is about 470 milliliters. Its opening into the urethra is controlled by two groups of sphincter muscles which are under voluntary control.

The process of voiding the urine is the result of a combination of involuntary and voluntary processes. In the first place, the presence of urine in the pelvis of the kidney stimulates peristaltic waves in the walls of the ureters, which pass urine to the bladder. The bladder becomes distended as the quantity of fluid in it increases, and receptors in its walls are stimulated. A feeling of fullness results when the bladder contains between 250 and 300 milliliters of liquid. In the infant, the flow of urine is purely reflex in nature and will continue to be until proper training (conditioned reflex) has been established. In an adult, the nerve impulses from the bladder pass to a center in the spinal cord just as they do in an infant. However, in the adult they continue upward to the cerebral cortex and result in the control of the expulsion of urine through the urethra.

VOCABULARY REVIEW

Match the phrase in the left column with the correct word in the right column. *Do not write in this book.*

1	A vessel that branches over a short distance into about 50 smaller ones.	**a** albuminuria
2	The structural unit of filtration in the kidney.	**b** Bowman's capsule
3	A knot of capillaries through the walls of which filtration occurs.	**c** diuretic
4	A small saclike structure with walls composed of two layers of cells.	**d** glomerulus
5	The condition of having protein present in the urine.	**e** glycosuria
6	A product that is formed in the liver and excreted by the kidneys.	**f** nephron
7	A tube connecting a kidney and the urinary bladder.	**g** renal artery
8	The appearance of sugar in the urine.	**h** renal vein
9	A chemical material that increases the production of urine.	**i** urea
		j ureters
		k urethra

TEST YOUR KNOWLEDGE

Group A

Select the correct statement. *Do not write in this book.*

1 The pelvis of the kidney opens directly into the (a) urinary bladder (b) urea (c) ureter (d) urethra.

2 One product of the chemical decomposition of amino acids in the body cells is (a) urea (b) proteins (c) peptones (d) proteoses.

3 The plasma component which is normally *not* found in urine is (a) uric acid (b) ammonia compounds (c) creatinine (d) albumin.

4 The chief function of the kidneys is excretion of the decomposition products resulting from the catabolism of (a) starches (b) sugars (c) proteins (d) fats.

5 The region of the kidneys that is composed chiefly of blood capillaries and tubules is the (a) tunic (b) cortex (c) pyramids (d) pelvis.

6 The kidneys are a part of the (a) portal circulation (b) pulmonary circulation (c) systemic circulation (d) renal circulation.

7 The saclike end of a kidney tubule is called the (a) glomerulus (b) collecting tubule (c) Bowman's capsule (d) alveolus.

8 A condition of inflammation of the kidneys is (a) albuminuria (b) glycosuria (c) nephritis (d) diabetes.

9 When the body is at rest, the percentage of the total blood volume passing through the kidneys is about (a) 5% (b) 10% (c) 25% (d) 33%.

10 The part of a nephron that is most intimately related to Bowman's capsule is the (a) proximal tubule (b) distal tubule (c) glomerulus (d) loop of Henle.

Group B

1 What part does each of the following structures play in the process of excretion: (a) the liver (b) the hilum of the kidney (c) the pelvis of the kidney (d) Bowman's capsule (e) a glomerulus?

2 (a) How does the kidney remove waste materials from the blood? (b) Why are not all of the materials that are removed from the blood excreted in the urine?

3 (a) Why does the urine normally lack many proteins and carbohydrates? (b) Under what conditions may these substances infrequently appear in the urine?

4 (a) In what respects does the composition of the plasma differ from that of the urine? (b) How is the balance between these two fluids maintained?

Chapter 30 The Pituitary and Thyroid

Chemical Integration

Two different systems working separately or together integrate the activities of the body and maintain the constancy of its internal environment. One of these, the nervous system, brings about responses in muscles and glands often in such extremely short periods of time that they can be measured in thousandths of a second. The other integrating system is one in which the response is frequently slower and more generalized; in just a few cases it may be very rapid. This second system is operated by chemical products that are secreted by groups of specialized cells, the *endocrine glands*. These glands differ from other types such as the salivary and gastric glands in that their secretions enter the blood directly or diffuse to tissues and are not carried to a definite region through ducts. Glands with ducts are sometimes called *exocrine glands* to distinguish them from those lacking ducts. Some glands have the ability to behave as both exocrine and endocrine glands by producing some materials which pass to a local area through ducts while other substances enter the blood stream directly. Such *mixed glands* are represented by the pancreas and the reproductive organs.

The products of the endocrine glands are known by the general name of *hormones* (Greek, *hormodzein*, to arouse). The hormones have three principal functions. In the first place they play important roles in bone growth and in the control and development of the reproductive organs and development of secondary sex characteristics. A second function of the hormones is the regulation of metabolism; of the use and distribution of foodstuffs, salts, and water in the body. A third function lies in their ability to cooperate in various ways with the autonomic nervous system including the control of sex and maternal behavior.

Because the hormones enter the blood stream directly, their effects may be widespread and apparent in many different regions. Many of these effects may occur at considerable distances from the place where the hormone has been formed. This is a characteristic which distinguishes them from the secretions of exocrine glands. The latter usually bring about an effect in a localized region near the site of their formation. The action of hormones is always a regulatory one. That is, they do not initiate a process but must always act on normal tissues in which they regulate the rate of cell metabolic activities.

Exactly how hormones exert their regulatory effects on cells is not certain but there are currently three possible explanations. First, hormones might influence the permeability of the cell membrane and therefore control substances entering or leaving the cell. Second, they might influence cell metabolism by acting on the enzyme systems; either activating key enzymes or acting as coenzymes (p. 202). Third, they might influence cell metabolism at source by acting directly on the DNA and regulating its synthesis of messenger RNA and therefore regulate protein (including enzymes) synthesis. (Protein synthesis is discussed on p. 35). Some recent research supports this last explanation—certain hormones do stimulate protein synthesis and they are unable to have any effect on cells which have been treated with a substance which inhibits the DNA from synthesizing messenger RNA. The activity of some other hormones can be explained by the other theories and it may be that there are several forms of hormone action.

The amounts of hormones needed to bring about a response are unbelievably small. In the case of adrenaline (epinephrine), the amount required to cause appreciable activity is only one part in 400 million parts of the solvent. This is equivalent to 1 gram of adrenaline dissolved in 400,000 kilograms of solvent. In this respect, hormones resemble some of the enzymes. The effect of a hormone is not however dependent only on its concentration, the receptivity of the larger organ is also important.

It is fortunate that hormones are of such constant chemical composition that those produced by one animal can be safely transferred to another without causing injury. For example, an extract of the pancreas of a sheep or a steer can be injected into a human being in order to compensate for the lack of insulin in diabetes. The body does not store hormones to any known extent. If the normal secretions of endocrine glands are lacking, the required hormones must be administered regularly. The form -trophic is used to construct words which refer to a hormone's action, as in thyrotrophic hormone, one affecting the thyroid.

Fig. 30-1 shows the location of some of the more important endocrine glands. Some of these are known to have definite hormonal secretions, while the secretions of others are not well understood.

The entire field of endocrinology (the study of the secretions of endocrine glands) is very complex. These glands form an "interlocking directorate" over the activities of the body. Although each hormone has a specific function, this can be modified by the effects of hormones from some of the other glands. This interdependence of one gland on another results in the endocrine balance of the individual. When all glands are functioning properly, a person is considered to be normal. Hormonal balance is generally achieved by negative feedback whereby hormone A stimulates production of hormone *B*, while *B* inhibits secretion of *A* as its own level in the blood rises. Exceptions to this are certain neurosecretions; hormones which are secreted upon nervous stimulation. An excellent example of the workings of this homeostatic system of glands is shown by the action of the pituitary gland. As you will see, the pituitary influences other endocrine glands so that their activities are either accelerated or depressed and their activities in turn influence the pituitary.

Two prefixes are commonly used to refer to the relative amount of hormones produced by an endocrine gland. *Hypo-* refers to a subnormal secretion, the term being derived from the similar Greek word meaning "under." The prefix *hyper-* means "over," and is

Fig. 30-1   The location and general structure of the principal glands of the endocrine system. (From Johnson, Laubengayer, DeLanney, General Biology, *Holt, 1956)*

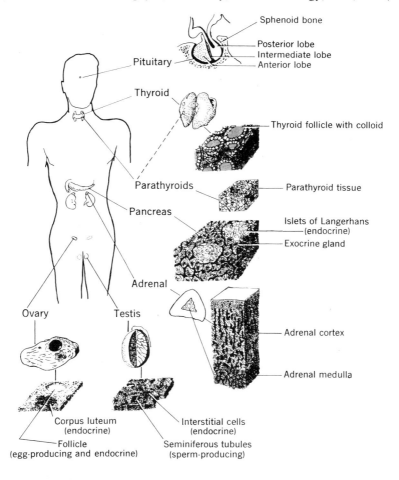

used to describe overproduction of a secretion. For example, undersecretion of the thyroid gland is hypothyroidism; oversecretion is hyperthyroidism.

The functions of endocrine glands may be determined by observing the effects on body functions of hypo- and hypersecretion, resulting respectively from atrophy (shrinkage) or hypertrophy (enlargement) of the glands. Another way to learn about the functions of endocrine glands is to feed portions of the glands to experimental animals, or to inject extracts of the glands into them. The effect of injury or surgical removal of all or part of a gland also gives clues to its function.

The Pituitary Gland

The *pituitary* or *hypophysis* is a small gland, weighing about 0.5 gram, attached to the lower surface of the brain just behind the optic chiasma (Fig. 11-1) and connected with the brain by a short stalk, the *infundibulum*. The pituitary body is partly enclosed by bone. The pituitary gland was named by the great anatomist, Vesalius (1514–1564) because he thought that it produced a lubricating substance (Italian, *pituita)* which found its way into the nose. Modern medical literature now prefers the term hypophysis in recognition of its location below the brain.

The gland, about the size of a small cherry, is divided into three distinct regions. Of these, the *anterior lobe* is the largest. This region arises in the early embryo from epithelial cells, some others of which later develop into the roof of the mouth. The *posterior lobe* is only slightly smaller and arises as an outgrowth of the floor of the third ventricle of the cerebrum and remains attached to it by the short infundibulum. Between these two lobes lies the smallest part, the *pars intermedia.* Each of these regions produces its own group of hormones. The anterior lobe is the most active, with the posterior lobe second in importance. The function of the pars intermedia in human beings is poorly understood at present, but it is known to control the production of skin pigment in some of the lower vertebrates.

In Fig. 30-2 are shown some of the far-reaching effects of this gland. Some of these will be discussed in this chapter, but others will be left for later chapters where pituitary influence on other glands can be more fully appreciated. In view of its many activities, the pituitary has sometimes been called the "master" gland of the body.

The anterior lobe. The anterior lobe secretes several different hormones which have two general functions. One group affects various parts of the body by influencing their metabolic activities. The other group of hormones exerts control over some of the other endocrine glands, thus regulating various bodily processes by remote control. When endocrine glands are affected by specific hormones produced by the anterior lobe, they are frequently called "target" areas. These targets, in turn, have some control over the production of the specific hormone secreted by the lobe. This will be recognized as another example of negative feedback maintaining the body's equilibrium.

In common with the majority of the cells that secrete materials, the products of the anterior lobe are first manu-

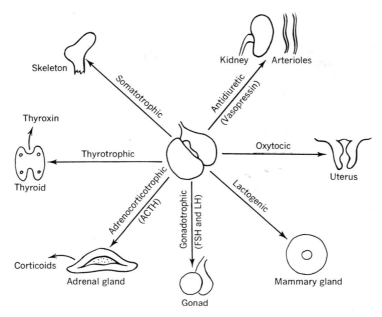

Fig. 30-2 A chart showing the relationship of the pituitary hormones to other body organs and functions.

factured in the form of granules within the Golgi apparatus of the cytoplasm. These solid particles then move toward the plasma membrane of the cell and are evidently converted into a fluid which diffuses through the membrane. In the case of the individual products formed by the lobe, each has its own particular density and shape when viewed under an electron microscope. Thus, each specific type of secretion has been identified with the possible exception of ACTH (page 403).

Abnormal variations in production of some hormones by the anterior lobe have startling effects. One of the most striking evidences of a disturbance in this portion of the gland is the effect of the *somatotrophic hormone* (growth

hormone) on physical development. If this hormone is produced in greater than normal quantities, the bones grow to unusual length. The hormone acts to prevent the epiphyseal junction from forming in a normal manner, thus making continued growth possible.

A child who is the victim of this condition may have a normal birth weight but quickly shows signs of excessive growth. This accelerated growth results in the condition which is known as *gigantism*. Giants have been reported from earliest times, in some instances fantastic claims being made about their height. These early records must be accepted with certain reservations. One of the tallest giants reported in medical literature was 9.2 feet tall.

Fig. 30-3 A giant and a dwarf dramatically show the effects of abnormal functioning of the pituitary gland.

Gigantism is accompanied by an increase in protein metabolism, so that there is corresponding growth of the internal organs. (See Fig. 30-3.)

If the growth hormone is injected into an experimental animal, its rate of growth exceeds that of litter mates who were not given the treatment (Fig.

30-4). Thus far, however, little success has been obtained when this hormone is administered to humans in an attempt to stimulate growth.

If increased secretion of the somatotrophic hormone occurs after adult size has been reached, the long bones are no longer affected, but those of the face, hands, and feet may show marked changes. The lower jaw enlarges and spaces appear between the teeth. Other bones of the face enlarge and the person's expression is greatly changed. The bones of the hands also become larger, especially in the region of the knuckles. This condition is called *acromegaly* (Greek *akron*, extremity + *megas*, large.)

Deficiency of the somatotrophic hormone gives rise to *dwarfism*. In this condition the skeletal structures reach their maximum development early in life and are unable to grow any larger. The result of this stoppage of growth is a miniature person (sometimes called a midget) whose bodily proportions and mental ability are generally normal.

In human beings there does not appear to be any inheritance factor that controls either gigantism or dwarfism. Among some breeds of animals, on the other hand, there is evidently a hereditary agent involved. We are all familiar with small breeds of horses called ponies, and various miniature types of dogs and chickens. In these animals there appears to be an inherited deficiency of the anterior lobe of the gland which is perpetuated by selective breeding.

A second hormone formed by the anterior lobe of the pituitary is *prolactin*, also called the *lactogenic* hormone. Some of the earliest experimental work on this hormone was done in studying the secretions of the esophageal glands of pigeons, the only vertebrates known to have these specialized glands. Normally, the gland is active only during the breeding season, since its secretion is used to feed the young squabs. It was found that if prolactin was injected into female pigeons during the non-breeding season, they showed all of the characteristics typical of the nesting period. Also, the secretion of the esophageal glands appeared and continued as long as injections were continued. In mammals, the injection of the lactogenic hormone stimulates the production of milk and arouses other reactions associated with the care of the young. If a female rat is given injections of this hormone, she will attempt to mother not only young rats belonging to another parent, but also mice, and even young birds. She will build an elaborate nest for them and hover around showing all the signs of a true mother. More recently it has been ascertained that the lactogenic hormone also promotes the functional activity of the corpora lutea of the ovaries which secrete progesterone.

A third hormone produced by the anterior lobe affects the thyroid gland. The absence of this *thyrotrophic hormone,* following the removal of the pituitary gland from an anesthetized experimental animal, results in the degeneration of the thyroid gland. Also, if a purified form of this secretion is injected into a normal animal, the thyroid gland enlarges and the animal shows all the effects of an increased amount of thyroid secretion.

The anterior lobe also produces a secretion known as the *adrenocorti-*

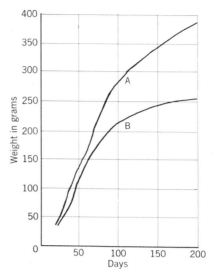

Fig. 30-4 Curve A shows the effect of giving growth hormone to growing rats. Curve B shows the normal increase in weight of untreated rats.

cotrophic hormone (ACTH). This influences the amount of various corticoids produced by the cortex of the adrenal glands. These in turn affect mineral and carbohydrate metabolism.

The Menstrual Cycle. The two gonadotrophic hormones of the anterior pituitary are indirectly responsible for a regular series of changes in the female sex organs, the *menstrual cycle.* The *follicle-stimulating hormone* (FSH) stimulates the follicle cells surrounding each maturing ovum and assures its development. In the male, FSH has a comparable effect in stimulating the seminiferous tubules to produce sperm. The *luteinizing hormone* (LH), affects only the ovary where it stimulates the development of the corpus luteum. These changes are shown

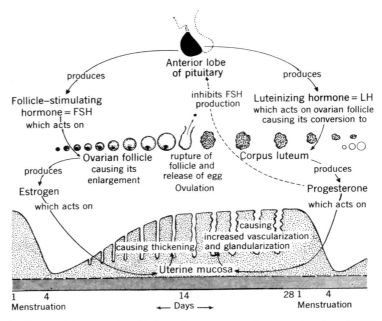

Fig. 30-5 Effects of the anterior pituitary hormones on the menstrual cycle. (From John-son, Laubengayer, and DeLanney, General Biology, Holt, 1956)

in Fig. 30-5. In the course of the menstrual cycle, FSH stimulates the maturation of the ovum, and with it, the production of the hormone *estradiol* by the follicle. Estradiol brings about repair and thickening of the uterine mucosa. Following ovulation, LH stimulates growth of a hormone producing mass within the follicle, the *corpus luteum*. The name of this body is derived from its yellow color (Latin, *luteus*, orange-yellow). The corpus luteum secretes progesterone. Under the influence of progesterone, the walls of the uterus become filled with blood and glandular tissues develop, ready to receive and nourish the fertilized ovum. These processes are known respectively as *vascularization* and *glandu-

larization,* and result in the thickening of the lining of the uterus. At the end of this cycle, unless the ovum has been fertilized and has become attached to the walls of the uterus, the *corpus luteum* degenerates. Without the stimulation of progesterone, the mucous lining of the uterus breaks down and external bleeding occurs. This is the period of *menstruation.* Considerable variation in the length of the menstrual cycle is common, especially during early puberty. Generally, however, it has an average length of 29.5 days which corresponds to a synodial month (Latin, *mensa,* month).

The posterior lobe. The posterior lobe of the pituitary secretes two hormones. One of these, the *antidiuretic* *hormone* (ADH) appears to control the reabsorption of water from the kidney tubules. When the blood stream fails to reabsorb water from the nephron, urine is formed in copious quantities but in a very dilute condition. This disease is known as *diabetes insipidus* and may result in the excretion of as much as 10 gallons of urine per day.

ADH also affects blood pressure by stimulating the contraction of smooth muscle sphincters of the arterioles. Some investigators also report that it stimulates the contraction of muscle fibers in the walls of the intestine. This second activity of ADH was previously thought to be caused by a separate hormone which was called vasopressin after the effect it produced on blood vessels. Both the antidiuretic and vasopressin effects have now been found to be caused by the same substance which can be called by either name.

A second hormone produced by the posterior lobe affects the uterus (womb). This is called the *oxytocic hormone,* and is normally produced in increased quantities at the time of childbirth. Its presence in the blood stream causes a series of contractions of the smooth-muscle bundles of the uterus which bring about the birth of the child. It also depresses the formation of urine and increases blood pressure and respiratory rates. Highly purified salts of this hormone have been found to produce an effect on uterine muscle when they were in a concentration of 1:15 billion parts.

The pars intermedia. The remaining part of the pituitary gland, the pars intermedia or intermediate lobe, is re-ported to produce a hormone called *intermedin.* Little is known about the effects of this material on human beings. In fish and amphibia, injection of this material increases the production of *melanophores,* the parts of the scales or skin responsible for dark colors. In some work on higher animals, intermedin has been reported to reduce the loss of water from the blood through the kidneys without changing the salt content of the urine.

The Thyroid Gland

The thyroid gland is a dark red structure in the neck, lying along the sides of the larynx (Fig. 30-6). Its general shape is that of the letter H with the upright members about 5 centimeters long and about 3 centimeters broad. These are joined by a bridge of tissue, the

Fig. 30-6 The thyroid gland is located in the neck region, partly surrounding the larynx.

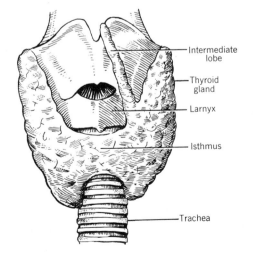

Intermediate lobe

Thyroid gland

Larnyx

Isthmus

Trachea

isthmus, which lies in front of the second and third cartilage rings of the trachea and is about 1.3 centimeters wide. From the isthmus, or from the region on either side near it, an *intermediate lobe* of varying length projects upward toward the floor of the mouth. This third part may extend to the hyoid bone, or it may be attached to the bone by means of a continuing cord of connective tissue. The thyroid gland produces several iodinated derivatives of the amino acid tyrosine: the iodothyronines. Only two show marked hormonal activity: thyroxin and triiodothyronine. The latter is the more active of the two, but its action is less prolonged. The effect of both is similar.

Microscopically, the thyroid gland is composed of groups of cells that surround open spaces, the *follicles.* These cells secrete a viscous material of uni-

Fig. 30-7 Photomicrograph of a section through the thyroid gland. (Walter Dawn)

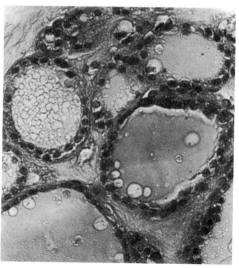

form structure known as *colloid,* which is deposited in the follicles. Each follicle is richly supplied with blood vessels through which cells of the gland extract materials they need. The colloid is composed of a protein-iodine combination known as *thyroglobulin.* This is changed into a more fluid form by an enzyme formed in the surrounding cells. In this less viscous condition it is absorbed by the blood. The formation of the enzyme appears to be controlled by the thyrotrophic hormone produced by the anterior lobe of the pituitary gland.

The microscopic structure of the thyroid shows such wide variations in detail that it is impossible to describe or picture a normal section through the gland. The supply of iodine in the diet or drinking water determines the microscopic appearance of the gland as well as its activites. Since a large percentage of the world's population lives in regions where there is a deficiency of iodine, the structure of a gland that might be considered normal in one region would be abnormal in another. These variations in structure may not be sufficiently great to cause external symptoms of thyroid disorder. A section through the thyroid gland is shown in Fig. 30-7.

The thyroglobulin is formed by a combination of the amino acid, tyrosine, with iodine. Although iodine in its pure form is a violent cellular poison, it is present in all body fluids in minute amounts in the form of iodides. Approximately one half of the body's total supply of iodine (about 50 milligrams) is concentrated in the thyroid in the form of thyroglobulin. Before it enters the blood, thyroglobulin undergoes several chemical changes to convert it into the hormone thyroxin (molecular formula, $C_{15}H_{11}I_4NO_4$). Thyroxin is not stored in the thyroid gland.

The formation of the hormones and the functions of the gland have been clarified in recent years by the use of the radioactive isotope, I^{131}. If this material is fed to an experimental animal, its method of incorporation into colloid can be followed by tracing the paths of the beta radiations which are emitted.

The functions of the thyroid gland are primarily concerned with cellular metabolism. It regulates the metabolic rate, heat production, and oxidation rate in all cells with the possible exception of those of the brain and the spleen. In the liver, the hormones influence the conversion of glycogen from sources other than sugar, and also the conversion of the glycogen to glucose with a resulting rise in blood sugar. Throughout the body, the secretions control the growth and differentiation of tissues thereby influencing both physical and mental growth in a child.

Hypothyroidism. The deficiency of thyroid secretion is characterized by a reduction in the metabolic rate: the fires of life burn low. If the thyroid gland is removed from an eight-day-old lamb, its gain in weight over a seven-month period will be only about one-third that of a normal lamb. In humans a somewhat similar condition develops in the absence of a normal amount of thyroxin. If a child is born with a defective thyroid gland or develops one early in life, a type of dwarfism appears. Whereas a pituitary dwarf is usually of normal intelligence and bodily proportions, a thyroid dwarf is of low mental ability and subnormal

physical development. Such an individual is called a *cretin,* or is said to show signs of *infantile myxedema.*

Hypothyroidism in childhood prevents the normal growth of the bones. Cretins are frequently bow-legged, and have dry scaly skin. The abdomen usually protrudes, due to weakness of the muscles in its walls, and the tongue projects slightly between the characteristically open lips. The face is puffy due to the formation of an excess amount of connective tissue. Mental ability is retarded, so that the victim is hardly able to take care of himself. All of these signs point to a marked retardation of the metabolic rate. If a child in this condition is treated regularly with thyroid extract, his improvement is usually spectacular. He will become normal in appearance and reactions, and will remain normal as long as treatment is continued.

Hypothyroidism in an adult results in the reduction of the BMR by between 30 and 45 percent. This causes a drop in body temperature so that the person feels cold, especially in his hands and feet, even when in comfortably warm surroundings. There is also a decrease in heart rate and blood pressure, and the muscle tone is lowered so that slight exertion quickly brings on fatigue. This combination of symptoms indicates the condition known as *adult myxedema.* There may also be changes in skin texture, puffiness of the face, and a decrease in mental capacity. As in the case of the cretin, administering thyroxin will relieve the symptoms.

Hyperthyroidism. The condition of hyperthyroidism results from overproduction of thyroxin. This condition may be caused by gland enlargement or

Fig. 30-8   Most of these villagers from Paraguay are suffering from simple goiter due to a dietary deficiency of iodine. (World Health Organization)

by oversecretion. Just as the undersecretion of thyroxin results in a lowering of the rate of metabolism, so an excess will lead to an increase in metabolism. We may therefore expect that a person suffering from hyperthyroidism will show symptoms that are almost the opposite from those displayed by a victim of hypothyroidism.

In hyperthyroidism the BMR may exceed the normal by 50 to 75 percent. Although there is a tendency to consume large amounts of food, there is usually a loss in body fat and weight. There is also an increase in heart rate and blood pressure. Since the BMR has risen, the person feels uncomfortably warm and sweats profusely. More glucose is freed by the liver, so that blood sugar content is slightly higher than normal. This produces a mild condition of glycosuria.

People who suffer from hyperthyroidism show nervous excitability and have a tendency to react to an emergency or unpleasant situation in an exaggerated manner. In some cases, the eyeballs become prominent and protrude from their sockets. This is accompanied by dilated pupils and widely opened eyelids. The immediate cause of this condition *(exophthalmos)* is not clearly understood, as it is apparently not the direct result of the hyperthyroid state. Thus, removal of the thyroid does not always cause the eyeballs to return to their normal condition. Treatment of hyperthyroidism may involve removal of all or part of the gland, or the use of certain drugs which reduce the secretion of thyroxin.

Goiter. Enlargement of the thyroid results in goiter, a swelling in the neck region. It may be associated with either hypothyroidism or hyperthyroidism *(toxic goiter).* If it is not accompanied by symptoms of either of these, it is *simple goiter* (Fig. 30-8). Experiments on animals show that if they are deprived of iodine, goiter develops soon after. However, if iodine is added to their drinking water, the goiter does not form. The amounts required to maintain the thyroid in proper condition are almost unbelievably small. Slightly more than 1 milligram of iodine in 1 million parts of water is all that is needed. In areas where iodine is lacking, thyroid defects are common.

Cretinism, myxedema, and goiter are especially prevalent in the Himalayas, the Alps, and the Andes. In the United States, such conditions are most common in the Pacific Northwest, the Great Lakes basin, and the St. Lawrence region. Most of the iodine has been leached out of the soil and rocks in these regions, probably by the action of the great continental glaciers that covered them in recent geologic times. Areas along the oceans are seldom affected in the same way because of the availability of saltwater fish and the presence of iodine in the soil. Goiter has become less common, due to modern methods of preventing this disease. The easiest method is the use of table salt to which about 0.02 percent of potassium iodide has been added. A good example of the effect of this iodized salt is seen in an experiment started in Denver in 1924. Here the number of goiter cases among the total school population fell from 36 to 2.1 percent following the introduction of iodized salt for cooking and table use. However, the use of excessive amounts of iodized salt in regions where it is not necessary may have a serious effect on the activities of the thyroid.

In the absence of sufficient iodine, the thyroid cannot supply the body's needs. Formation of a simple goiter is the result of an attempt by the thyroid gland to increase its production. The lowered concentration of thyroxin in the blood evidently stimulates the anterior lobe of the pituitary gland and an excess of thyrotrophic hormone is formed. This in turn brings about the enlargement of the follicles of the gland, and even the development of new follicles. Administration of proper amounts of iodine results in the shrinking of the gland to normal size.

The thyroid gland plays another role among the Amphibia. If bits of thyroid gland are fed to some of the tailless amphibians (frogs and toads) when they are in the tadpole stage, they rapidly pass through their metamorphosis and become adults (Fig. 30-9). Not all frogs and toads respond in this manner, but if the bullfrog *(Rana catesbiana)* tadpole is fed thyroid, it metamorphoses within about two weeks. Normally, this frog requires two to three years to accomplish these changes. It has also been found that if the thyroid is removed from this species in the tadpole stage, feeding the tadpole iodine or even adding a small amount of iodine to the water in which it swims is all that is required to bring about metamorphosis. The use of frog tadpoles is one of the most delicate tests for thyroid secretions and therefore a useful tool for research.

Fig. 30-9   The effect of feeding thyroid to a bullfrog tadpole (A). The change to the adult form (B) occurs rapidly without appreciable growth in size. Both figures are drawn to the same scale.

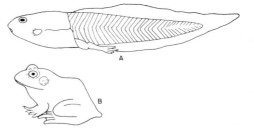

VOCABULARY REVIEW

Match the phrase in the left column with the correct word in the right column. *Do not write in this book.*

1 A condition resulting from the failure of the kidneys to reabsorb water in the nephrons.
2 The result of hypothyroidism at birth.
3 A chemical substance carried by the blood which can affect reactions at a distance from the place of its formation.
4 A group of specialized cells forming materials that are carried away in ducts.
5 Formed by the same embryonic tissues that form part of the roof of the mouth.
6 A glandular product that has a high iodine content.
7 A hormone that increases maternal instincts.
8 The result of the overgrowth of a gland as the result of an absence of iodine.
9 One characteristic of hyperthyroidism.
10 A hormone that affects an endocrine gland located at a distance from the site of its own formation.
11 A hormone which promotes the formation of ova and sperm.

a   ACTH
b   ADH
c   anterior lobe of the hypophysis
d   cretinism
e   diabetes insipidus
f   endocrine gland
g   exocrine gland
h   exophthalmos
i   FSH
j   hormone
k   infundibulum
l   myxedema
m   posterior lobe of the hypophysis
n   prolactin
o   simple goiter
p   thyroxin
q   vasopressin

TEST YOUR KNOWLEDGE

Group A

Select the correct statement. *Do not write in this book.*

1 Glands that lack ducts are said to be (a) exocrine glands (b) endocrine glands (c) parietal glands (d) heterocrine glands.
2 The thyroid gland is in the (a) head (b) neck (c) thorax (d) abdomen.
3 The enlargement of a gland is called (a) atrophy (b) hypertrophy (c) hypotrophy (d) autonomy.
4 A ductless gland attached to the lower surface of the brain is the (a) pituitary (b) infundibulum (c) adrenal (d) thyroid.
5 Of the following, the hormone *not* secreted by the anterior pituitary lobe is (a) ACTH (b) prolactin (c) thyrotrophic (d) ADH.

6 The somatotrophic hormone chiefly affects growth of (a) bones (b) muscles (c) hair (d) connective tissue.

7 Prolactin stimulates the production of (a) lactic acid (b) urine (c) milk (d) lactase.

8 The gonadotrophic hormone mainly affects the function of (a) digestive glands (b) ovaries (c) kidneys (d) pancreas.

9 Of the following, the one that is *not* affected by lack of iodine in the diet is (a) goiter (b) cretinism (c) myxedema (d) acromegaly.

10 In hyperthyroidism, the BMR is (a) lower than normal (b) higher than normal (c) about normal (d) difficult to measure.

Group B

1 For each of the following abnormalities, name the gland and the hormone in-involved and tell whether the condition is due to an over- or undersecretion of the hormone (a) acromegaly (b) cretinism (c) dwarfism (d) diabetes insipidus (e) exophthalmos (f) adult myxedema.

2 (a) What is the difference between an exocrine gland and an endocrine gland? (b) How do the secretions of these two glands differ?

3 The pituitary gland has been called the "master gland" of the body. How can this statement be substantiated?

4 (a) Why are cretins found in some regions of the world more frequently than in others? (b) What measures can be taken to prevent this condition?

5 (a) What successive changes take place in the uterine wall during the menstrual cycle? (b) Name the hormones that effect these changes and the site of production of each.

Chapter 31   The Other Endocrine Glands

The Parathyroid Glands

The parathyroid glands are usually embedded in the posterior part of the thyroid gland, but occasionally they may be located a short distance from it on either side. They are four small bodies, each about 8 millimeters in length.

The parathyroids were not well understood until recent times. Following the introduction of aseptic methods of surgery by Lister, removal of the thyroid gland became easier and safer. It was found that when the entire gland was removed, the patient frequently developed tetanic contractions of the muscles (convulsions), often followed by death. There was considerable confusion regarding these unfortunate incidents until 1881, when it was demonstrated that the parathyroids were usually removed with the thyroid gland. When there were no adverse symptoms, the parathyroids were evidently not embedded in the thyroid and hence escaped removal.

The parathyroids control the amount of calcium and phosphate ions that are present in the plasma of the blood. These two ions are of extreme impor-

tance in promoting proper nerve and muscle function as well as maintaining bone structure. The hormone which the parathyroids produce, *parathormone,* has the ability to release calcium and phosphate from the type of bone that is known as "exchangeable bone." This is newly formed bone resulting from the activities of the osteoblasts, but later this will become compact and less easily used as a source of these ions.

The places where the parathyroid hormone is most effective in maintaining the levels of calcium and phosphate are areas of exchangeable bone, the digestive tract, and the kidneys. As previously mentioned, parathormone is able to release calcium and phosphate ions from bone, but it is also able to control the absorption of calcium and phosphates through the walls of the small intestine. If the hormone is in low supply, less of these two ions are absorbed and a lack of them is apparent in the plasma and intercellular fluid. Bone formation is then stopped and the nervous system is adversely affected. The kidneys are also a site where the effects of the hormone can be noted. If there is an excess of parathormone, the excretion of phosphate is greatly increased and an imbalance between this ion and calcium then develops. It is, however, the exchangeable bone that is the principal factor in maintaining the proper balance of these two ions in the body fluids.

Vitamin D also plays a part in the retention of calcium and phosphate ions, by keeping them at a constant level. If vitamin D is absent from the diet, an increased excretion of calcium occurs. This is lost through the feces. Also, a large amount of phosphate is excreted in the urine. Without these, bones cannot develop; they become soft and bend, and the disease of *rickets* appears.

The Adrenal Glands

The adrenal glands resemble small caps, resting above each kidney, as shown in the Trans-Vision insert following page 212, and in Fig. 29-1. They are occasionally termed suprarenal glands (Latin, *supra,* above), because of their position over the kidneys in man, but this term is applicable only in those cases where the organism has assumed an upright position. In other animals, the adrenal glands bear various relations to the kidneys. In some of the lower fish, the tissue corresponding to the adrenals is embedded in the kidneys. In man, the right adrenal has the general outline of a cocked hat, while the left is somewhat more curved.

The human adrenal glands appear early in embryonic life and at one stage of development are the most conspicuous organs in the abdominal cavity. At birth they are approximately one-third the size of the kidneys, but in early infancy they become smaller. Each gland is composed of two distinct regions: an

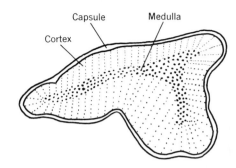

Fig. 31-1   A section through the adrenal gland showing the relationship of its parts.

outer area that makes up the *cortex* of the gland; and an inner part, the *medulla* (Fig. 31-1). These two parts have very different and distinct secretions, and their origins in the embryo are equally different. The cortex is derived from the same group of tissues which produces the reproductive organs, while the medulla develops from the embryonic tissues which also produce the sympathetic nervous system. The entire gland is surrounded by a firm layer of connective tissue, the *capsule.*

The adrenal cortex. This part of the adrenal gland is absolutely essential to life. Its removal, or extensive injury, in experimental animals invariably leads to death unless its secretions are replaced artificially. In man, the diseased cortex produces symptons known as Addison's disease, which will be discussed later.

The hormones formed by the cortex belong to a group of chemical compounds known as *steroids,* of which cholesterol is an example. As a result of research carried on during the past twenty-five years, 28 different steroids have been isolated from the cortex and as many more from urine. It is assumed

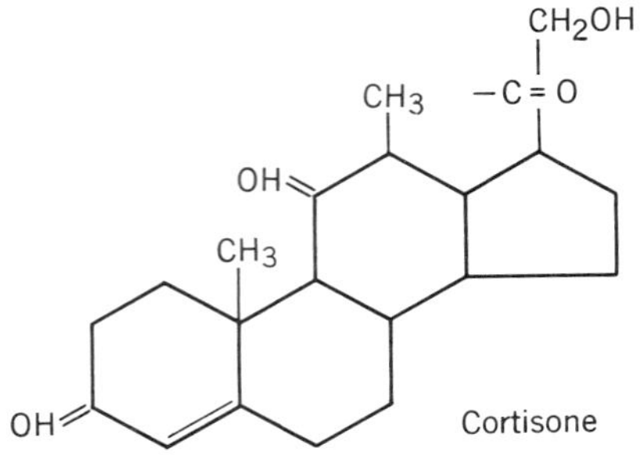

Fig. 31-2   Structural formula of the corti-sone molecule. This is typical of the structure of the other steroids.

that those in urine also originate in the adrenal glands. These steroids are frequently called *corticosteroids,* or *corticoids.* Of these, the best known is *cortisone.* The structural formula for cortisone is given in Fig. 31-2.

The removal of the adrenal glands from an experimental animal, or their hypofunction in man, results in a drop in the BMR. This is a symptom of Addison's disease. It is brought about by a decrease in the amount of glycogen stored in the liver and an accompanying loss of available glucose in the blood. Body temperature falls, and the weakened heart is unable to maintain normal blood pressure. Appetite declines and satisfactory absorption of food is impaired, with the result that the body weight drops alarmingly.

One of the most important functions of the adrenal cortex is to maintain a proper balance between the water content of the blood and its salt content. In man, lack of this balance is characteristic of Addison's disease. If the cortical hormones are lacking, the kidneys excrete an increased amount of sodium salts. Since these affect the osmotic pressure of the blood, there is gradual dehydration of the tissues because fluid passes from them into the blood. If an animal from which the adrenal glands have been removed is given the choice of drinking fresh water or a 3 percent solution of water and table salt, it will choose the latter. This compensates for the loss of sodium through the excretory system.

Another characteristic of Addison's disease is the failure of the kidneys to excrete a proper amount of potassium salts. Potassium ions cause the heart muscle to relax. If they are in too great concentration, they bring about the stoppage of the heart during diastole. Thus, insufficiency of cortical hormones directly affects the heart as well as influencing the composition of blood. Injections of cortisone will alleviate the acute effects of this disease.

One of the most obvious symptoms of Addison's disease is the bronze coloration of the skin of the face. As the disease progresses, the skin becomes darker in color due to formation of a greenish-yellow pigment in it. This condition has not been satisfactorily explained.

The cortical hormones have a definite effect on leukocytes. If cortisone or ACTH is injected into an experimental animal, there is a rapid destruction of eosinophils and lymphocytes. Lymphoid tissue is also affected. At one time, cortisone and/or ACTH were widely used as a treatment for various conditions, such as arthritis. Spectacular cures were reported following their administration, but their continued use had some undesirable side effects. One was that the feeling of well-being following their use sometimes hid the

Fig. 31-3 The structural formulas of adrenaline and noradrenaline molecules. Their functions are quite different even though their chemical structures are very similar.

underlying disease symptoms. Another was their effect on the blood and lymphoid tissue. Since this tissue is the principal source of gamma globulin, the carrier of antibodies, any lessening of its activities will decrease the person's general resistance to disease. Stomach ulcers, high blood pressure, edema, and even atrophy of the adrenal glands have followed excessive use of cortisone. However, with moderate dosage and under careful medical supervision, cortisone is a valuable weapon against a variety of human diseases.

The adrenal cortex appears to be concerned with the production of sex hormones (page 421), although its action is not clearly understood at present. There is some indication that the corticoids affect the maturation of the gametes in some animals. When the adrenals are removed from experimental animals, the cortical hormones must be given to keep the animals in a normal state. It has been noted that a smaller amount of the cortical hormone is necessary if the adrenals are removed from male animals just before or after the breeding season.

In humans, tumors on the adrenal cortex may cause changes in the secondary sex characteristics. For example, the appearance of heavy facial hair on a woman may be due to a tumor of the adrenal cortex, resulting in too much androgen. It may also be the

result of failure of the adrenal cortex to secrete hormones to maintain a balance between the estrogenic and androgenic substances formed by the ovaries. Both the ovaries and the testes are capable of producing substances that favor the appearance of some of the secondary sex characteristics of the opposite sex. If therefore, the estrogens are suppressed, the androgenic hormones may become more dominant, and masculine characteristics may appear. In a similar manner, if there is any disturbance in the adrenal secretions of a male, the androgenic hormones are depressed and the man becomes effeminate.

The adrenal medulla. This part of the adrenal glands is not as vitally important as the cortex. The loss of the hormones of the medulla does not result in death, nor does replacement of these hormones save the life of an animal from which the adrenal glands have been removed.

In earlier research, it was believed that the secretion of the medulla consisted of only one hormone, *adrenaline* (sometimes *epinephrine*). Thus the pronounced vasoconstrictor effect of extracts made from the medulla was understood to be due to adrenaline. More recently a second hormone, *noradrenaline (norepinephrine)* has been found to be responsible for the constriction of blood vessels (Fig. 31-3).

In some other respects, noradrenaline has an action similar to that of adrenaline, though weaker. Adrenaline is still listed in many texts as the most powerful vasoconstrictor in the body. Actually it is in general a vasodilator with only local constricting effects. The human medulla forms noradrenaline and adrenaline in the ratio of approximately 1 : 5. Commercial preparations of adrenaline contain from 10 to 20 percent noradrenaline.

As we have said, the over-all effect of adrenalin itself is vasodilation. However, it does cause constriction of the small vessels passing to the skin, with resulting pallor. It also causes constriction of the vessels in the kidneys and a corresponding reduction in the amount of urine. Its major effect is the dilation of the vessels in the liver, heart, brain, and skeletal muscles. Adrenaline also causes a rise in the systolic blood pressure, and an increase in the respiratory rate. Another important effect of adrenaline is that it converts glycogen, stored in the liver, into glucose, which then enters the blood stream. Thus there is an extra supply of energy available to the tissues in cases of emergency. Some of this excess glucose is excreted by the kidneys, resulting in temporary glycosuria.

Smoking influences the normal secretion of adrenaline and the release of sugar from the liver. Experiments have shown that the nicotine in tobacco stimulates the sympathetic nervous system. This stimulation then brings about a slight increase in the secretion of adrenaline, with a corresponding increase of blood sugar. The added blood sugar dulls the sensation of hunger slightly, since hunger is partly induced by a decreased supply of sugar in the blood. Thus we can explain, at least in part, why smoking sometimes causes a loss of weight. Conversely, we have a possible explanation of the observation that those who stop smoking often gain weight unless they consciously control their food intake.

Increased production of adrenaline results in increased muscle tone in skeletal muscles but decrease in tone of the smooth muscles of the stomach and intestines. Thus the injection of adrenaline into an experimental animal, such as a rabbit, causes the inhibition of intestinal movements and twitching of the skeletal muscles. In human physiology this effect plays an important role. If we are eating a meal when an unpleasant situation arises, the adrenaline content of the blood increases. We tense our muscles, become pale, and perhaps later feel some abdominal distress. Similar effects occur when we are frightened. We suddenly feel weak in the pit of the stomach (stoppage of normal peristalsis) and break out in a cold sweat due to the constriction of the skin capillaries. Noradrenaline has the same effect when injected into an animal, but to a lesser degree.

The amount of adrenaline present in the blood stream under normal resting conditions is minute. It has been determined that adrenaline is normally present in the blood in the ratio of 1 part of adrenaline in from 2 billion to 1 billion parts of blood. This concentration has no effect on the body. On stimulation of the adrenals, the concentration may rise to 1 : 4 million.

The presence of an appreciable amount of adrenaline in the blood increases the BMR by as much as 20 percent. Oxygen use may rise 20 to 40 percent above the resting rate, while CO_2 production may mount by 40 percent. If 0.5 cubic centimeters of a 1 : 1000 solution of adrenaline is injected into a human subject, these increases occur within a few minutes and last for about two hours. If adrenaline is injected into an experimental animal from which the liver has been removed, comparable results are not obtained. It is therefore thought that the increase in the BMR, use of oxygen, and carbon dioxide production are the result of the release of glycogen from the liver (glycogenesis). Other factors that contribute to these effects are the constriction of skin capillaries, which prevents the loss of heat, and an increase in muscle tone.

As we have already pointed out, the relation of the adrenal glands to the sympathetic nervous system is a close one. Adrenaline and noradrenaline are in fact secreted by the motor nerve endings of the sympathetic system. Also, it has been found that in moments of stress, such as those occasioned by fear, anger, pain, and exposure to cold, the adrenal medulla is stimulated by sympathetic nerves to produce more adrenaline. The effects of this increased supply of adrenaline seem to be independent of normal nerve control. For example, the rate of a cat's heart action increases by 15 to 20 beats per minute on hearing a dog's bark, even after the nerves to the heart have been cut. When the same cat struggles against a restraining device, its rate of heart beat may increase by 40 to 50 contractions per minute when the dog barks. This increase is usually accompanied by dilation of the pupils and erection of hair. On the other hand, if the nerves to the adrenal glands are cut or veins leading from the glands are tied off, these reactions are only very weak and are due to sympathetic nerve ending secretions.

Islets of Langerhans

Insulin. The failure of the islets of Langerhans in the pancreas is generally considered to be the cause of *diabetes mellitus*. These small but numerous groups of cells lie scattered among the other cells of the pancreas but have no connection with them. Furthermore, there are no ducts leading from them, so their secretion can be considered a hormone. There are two distinct types of cells in the islets. The cells of one type are larger but fewer than the other cells. These are called the *alpha cells,* and are known to produce a hormone that has been named *glucagon.* The smaller and more numerous *beta cells* have a different staining reaction and produce *insulin.*

The definite structure of the insulin molecule was discovered due to the brilliant work of the English biochemist, Dr. Frederick Sanger, and his associates. This was the first time that the exact sequence of the amino acids in a protein molecule had been determined. Insulin is a relatively small protein molecule with only a total of 51 amino acids, and yet it took 10 years to unravel its structure. The molecule was found to be composed of two chains of amino acids. In one chain (A chain) there are 21 amino acids, and in the other (B chain) there are 30. The two chains are bonded by two double sulfide linkages. Experimental work has shown that the hormonal activity of

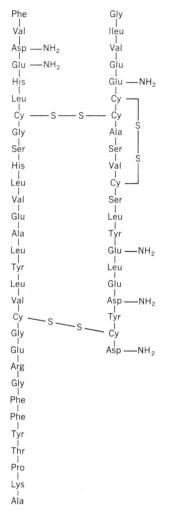

Fig. 31-4   Structure of a molecule of insulin showing the sequence of amino acids (each here represented by an abbreviation of its name).

insulin is due to the spatial configuration of the molecule in the region of these sulphide links. (See Fig. 31-4.)

Diabetes affects people of all ages and economic conditions. Children may be born with defective islets, and the type of diet, unless excessively heavy in carbohydrates, appears to have little effect on the disease. Inheritance may be a factor in its appearance. Early symptoms include pallor, loss of weight, excessive thirst, increase in the amount of urine, and a feeling of fatigue following moderate exercise. There is also an extensive itching of the skin. It is estimated that there are about a million known diabetics in this country, and a like number of people who have diabetes but are not aware of it. The prevalence of the disease and the lack of recognition in its early stages have prompted some local boards of health to establish mobile units to identify the condition and help the victims to receive early treatment. There are also simple test kits for sugar in the urine available at any drug store.

The manner in which insulin acts in controlling diabetes is not clear but it is known that insulin controls certain aspects of carbohydrate metabolism. Insulin promotes the formation and storage of glycogen in the liver, and then the later breakdown of it to form glucose for transport to tissues that require it. It is thought to exert this effect either by activation of a key enzyme or, more probably, by facilitating the penetration of glucose into cells. If insulin is not present in sufficient quantities, these conversions of glucose do not occur and the amount of sugar in the blood increases greatly because the liver cannot remove it. The excess of glucose becomes too great for the kidneys to reabsorb, and much is excreted in the urine. Excretion of this glucose requires accompanying excretion of an increased amount of water in order to keep the sugar concentration from rising too high. This water must be replaced with the result that a diabetic suffers from thirst.

Since the proper form of glucose is not available for cellular oxidations, proteins are oxidized. This generally results in a loss of weight because the body lacks the necessary materials from which to form protoplasm. Another serious result of the deficiency of insulin is the effect it has on the metabolism of fats. These compounds are rapidly and incompletely oxidized. One group of compounds formed by this abnormal rate of oxidation comprises the so-called ketone bodies. These have a highly toxic effect. Acetoacetic acid is the ketone body most commonly formed. It is thought that this acid is responsible for the *diabetic coma* (unconsciousness) which results from a lack of insulin. Many of the ketones have a sweet odor which can be detected in the breath of an untreated diabetic.

If a child is born with diabetes, it means that there are only a few islets present in the pancreas. On the other hand, if a person who is over forty develops the disease, his pancreas may still have a sufficient number of intact islets to supply an adequate amount of insulin and one is forced to ask what happened to the insulin. Thus far there has been no really satisfactory answer. It is now known, however, that the glucagon that is formed by the alpha cells of the islets is antagonistic to insulin and has the ability to destroy it in the liver. This action may involve the separation of the two chains (A and B) of the insulin molecule. Several other substances are known to act as insulin antagonists. One of these appears to be the somatotrophic hormone. It has been found, for example, that some individuals who suffer from diabetes have a larger amount of this hormone than is normal. They need not show any of the symptoms of gigantism. Also some of the secretions of the cortex of the adrenal gland appear to limit the action of insulin.

Among the group of older diabetics, relief from the disease can frequently be obtained by taking one of the new drugs in the form of tablets. For a child born with diabetes, these are not sufficient; injections of insulin must be used. This is due to the protein nature of the insulin molecule and the fact that it would be digested if taken by mouth. Once these injections have been started, they must be continued throughout life. When there is an upset in the hormonal balance of the body, the injection of the missing hormone does not cure the condition; it simply alleviates the symptoms. If the injections are discontinued, the sufferer may develop diabetic coma and die. An overdose of insulin may have serious effects. This is known as *insulin shock* and is the result of the amount of blood sugar being lowered to such levels that the central nervous system is seriously affected. Convulsions and death may result unless the sugar level is raised either by injections of glucose or eating some type of sugar that can be absorbed very easily. Many diabetics therefore carry tablets of sugar or pieces of candy with them to offset such an eventuality.

The Pineal Body

The pineal body is a small organ attached to the lower surface of the brain between the anterior corpora quadrigemina. It is about 8 millimeters long and is reddish gray in color. During its development, the cells are glandular in nature, but no specific secretion was isolated from them until 1958 when a compound with hormonal activity was finally isolated and named *melatonin* because of the blanching action it had on frog skin. In most lizard embryos, the pineal body develops briefly into a structure resembling a third eye, but this disappears before the animal hatches from the egg.

In higher vertebrates, including man, it does not show this curious development and its function has remained a mystery until recently. New work indicates that it is an endocrine organ of the neurosecretory type, like the adrenal medulla and the posterior pituitary, both of which produce secretions upon nervous stimulation. The pineal apparently converts cyclic nervous activity generated by light in the environment into secretion of melatonin. It is not yet certain what physiological processes are influenced by pineal activity but the present evidence suggests that it participates in regulation of some vertebrate gonad cycles.

The Thymus Gland

The role of the thymus gland and the hormone secreted by it in the development of immunity has already been discussed at length in Chapter 23, p. 312.

The Gonads

The primary sex characteristics of any individual are established at the time of the fertilization of the egg (ovum). As will be explained later (Chapter 33), the presence of certain chromosome combinations will determine whether the individual will be a male or female. The chromosomal type that an individual inherits controls the development of certain primitive and essentially similar embryonic tissues into the reproductive organs; testes in the males or ovaries in the female. A general term used to identify the reproductive organs is *gonads,* and thus, the testes can be referred to as the male gonads, while the ovaries are the female gonads. The former produce the sperm and the characteristic sex hormones, and the latter the ova and the female sex hormones. It is these hormones which control the development of the *secondary sex characteristics* which are indicative of the sex of the individual.

The gonads are generally classed as glands. The ova and sperm cells, although solids, are considered to be exocrine secretions because they must leave their point of origin by means of ducts before they can become active. The hormones produced by the gonads are endocrine secretions, passing directly into the blood. In view of the two different types of secretions produced, the gonads, like the pancreas, are called mixed glands.

The effects of human sex hormones usually become noticeable between the ages of 12 and 15 years, when a person is becoming physically mature. At that time a young person begins to show the secondary sex characteristics which so definitely distinguish a man from a woman. This period is frequently called the *age of puberty.* Prior to this, there is little structural difference between the two sexes (other than the type of gonad present). Environment and training account in large measure for differences in the reactions of very young boys and girls. With the onset of puberty and the production of increased amounts of sex hormones, the similarities in bodily structure and psychology disappear. The pitch of a boy's *voice* becomes lower as a result of an enlargement of the larynx, while that of a girl retains its higher register. *Hair* grows on a boy's face, but a girl's face remains relatively hairless. A girl begins to develop a layer of *subdermal fat,* while a boy does not and therefore keeps a more angular frame. The *skeleton* of a boy increases in size and weight to a greater degree than a girl's. The hips of a girl grow wider between the crests of the ilia (Fig. 5-17), while those of a boy continue their growth pattern without marked change. A girl's *mammary glands* develop while their growth in a boy is inhibited. *Psychologically,* there is a difference between the two sexes; boys generally show more drive and aggressiveness than girls.

All these secondary sex characteristics fail to develop if the gonadotrophic hormone of the anterior lobe of the pituitary gland (Chapter 30) is missing or greatly deficient in quantity. This hormone is responsible for the development of the gonads, which in turn produce male and female hormones.

Estrogenic hormones. Three different groups of sex hormones are produced by various organs of the female reproductive system. First, there is a group of steroids, the estrogenic hormones, secreted by the follicles of the ovary and by cells lying within the ovary but not associated with the formation of the ova. They cause repair and thickening of the uterine mucosa. Of the many substances produced by the follicle cells, *alpha-estradiol* is the most potent and maybe the true female sex hormone. Since over forty substances are known to have estrogenic effects (some of these substances are not formed by the human ovary), it is very difficult to isolate any one that is of primary importance. The chemistry of these compounds is imperfectly understood, and it is possible that some may undergo slight changes in the ovary that effect their potency.

A second group of steroid hormones is formed by the *corpus luteum* following ovulation. The hormone is known as *progesterone* and is concerned with changes in the lining of the uterus which favor the reception of the developing embryo.

The third group of female sex hormones is produced by cells in the membranes within which the embryo grows. One of these hormones is *chorionic gonadotrophin,* which resembles the anterior pituitary gonadotrophic hormone in some respects, but is not identical with it. Its presence in the body during pregnancy prevents the maturing of additional ova.

Androgenic hormones. The male sex hormones are considerably less complex than those of the female, since they are not concerned with the developing child or its birth. The hormones produced by the testes are known by the general name of androgenic hormones (Greek, *andros,* man). Several of them are steroids. The most important is *testosterone,* which is produced by the interstitial cells of the testes, and which is responsible for the development of the male secondary sex characteristics.

Experimental evidence regarding the effects of this hormone have been obtained by removing the testes from anesthetized animals at various periods in their development. If the gonads are removed early in the animal's life, it fails to undergo normal changes associated with maturation. In fact, a male animal from which the testes have been removed early in life shows many female characteristics. This reversion is seen following *castration* (removal of the testes) of animals. Castrated bull calves grow into steers, castrated stallions become geldings. Roosters grow into capons following castration. In these animals male characteristics are lacking so that they become more docile and easily tamed (geldings) or they add weight and are better sources of food because of their less active muscle tissue (steers and capons). In the final analysis, castration has the effect of returning the animal to a status intermediate between the true male and the true female.

In this chapter and in Chapter 30 we have tried to point out how some of the activities of the body are interrelated. Throughout the text this same theme has been emphasized, but there are few better examples of this correlation than those shown by the nervous and endocrine systems.

We must constantly remember that each living thing succeeds because of the coordination of its chemical and physical activities. If any of these fail, the organism suffers.

The greater the complexity of its body structure, the more dependent the organism becomes on the proper functioning of its body parts. Basically, the activities of the human body depend on the action of its individual cells. Although some of the structures mentioned in Chapter 3 may be lacking in some types of cells, their absence is balanced by the activities of other cells that are specialized for definite functions. Thus the body may be said to represent a community of interests in which each cell, tissue, organ, and system plays an important part.

VOCABULARY REVIEW

Match the phrase in the left column with the correct word in the right column. *Do not write in this book.*

1	A source of calcium and phosphate ions.	a	adrenaline
2	The product of over rapid and incomplete oxidation of fats.	b	alpha cells
		c	beta cells
3	Formed by the alpha cells of the islets of Langerhans.	d	cortisone
		e	estrogen
4	A group of chemical compounds represented by cortisone.	f	exchangeable bone
		g	glucagon
5	The site of the manufacture of insulin.	h	insulin
6	The most powerful vasoconstrictor in the body.	i	ketones

7 Is composed of A and B chains of amino acids.
8 A hormone that plays a large role in the development of rickets.
9 Changes in the lining of the uterus which prepare it to receive the embryo, are due to this hormone.
10 Responsible for the development of male secondary sex characteristics.

j noradrenaline
k parathormone
l progesterone
m steroids
n testosterone

TEST YOUR KNOWLEDGE

Group A

Select the correct statement. *Do not write in this book.*

1 Parathormone is chiefly concerned with the presence in the blood of (a) sodium (b) nitrogen (c) calcium (d) iron.
2 The adrenal glands are physically associated with the (a) liver (b) pancreas (c) spleen (d) kidneys.
3 Of the following, the term that is *not* related to the function of the adrenal cortex is (a) steroids (b) cortisone (c) sex hormones (d) adrenaline.
4 Another name for adrenaline is (a) ACTH (b) epinephrine (c) vasopressin (d) noradrenaline.
5 Insulin is associated with all of the following terms *except* (a) pancreas (b) diabetes (c) goiter (d) islets.
6 Insulin controls the metabolism of (a) fats (b) proteins (c) carbohydrates (d) hormones.
7 Reproductive organs are called (a) sperms (b) gonads (c) eggs (d) ova.
8 Of the following, the term *least* closely associated with androgenic hormones is (a) testis (b) testosterone (c) gonad (d) ovary.
9 An organ that is glandular in structure whose function is currently questionable is the (a) thyroid (b) parathyroid (c) pineal (d) testis.
10 An estrogenic hormone is (a) testosterone (b) progesterone (c) ergosterol (d) cortisone.

Group B

1 What hormone is related to each of the following conditions: (a) an increase in blood phosphorus (b) the release of glucose after smoking (c) reduction of blood calcium (d) local constriction of capillaries (e) Addison's disease?
2 Briefly describe the structure of (a) the adrenal glands (b) the pancreas.
3 In what ways are the adrenal glands related to the sympathetic (autonomic) nervous system?
4 Why is it a wise procedure for diabetics to carry sugar with them either in the form of tablets or candy?
5 (a) What are the main physical and psychological changes which take place in boys and girls at puberty? (b) What are the causes for these changes?

Chapter 32 **Human Reproduction**

One of the primary characteristics of living things is their ability to reproduce themselves. In multicellular animals a new individual develops from the union of two special cells, the ovum and the sperm. The ovum is produced in the ovary of the female and the sperm in the testes of the male. The union of the two is called fertilization. This is sexual reproduction; it requires the cooperation of two individuals of different sex. It is the usual method of reproduction in multicellular animals.

In prenatal development both male and female sex organs arise from tissues that are so alike that the sex of the individual cannot be visually determined in the early stages. They begin their development as ridges on the surfaces of the embryonic kidneys and then become separated from them as growth continues. In the female, the tissue becomes a pair of ovaries, capable of producing *ova* or egg cells; in the male, they become a pair of testes, capable of producing *sperm* cells.

The accessory structures in each sex aid in the process of reproduction. Depending on the animal species, they insure fertilization of the egg outside the body of the female, or internally, and provide for the nourishment and protection of the developing young.

Vertebrates produce their young by one of three general methods. In *oviparous* (Latin, *ovum,* egg + *parere,* to give birth to) reproduction, the ova may be fertilized internally or externally, depending on the species of the animal, and are deposited as eggs outside the body. Most fish, amphibians, and reptiles, all birds, and one group of mammals use this method. The eggs when laid may or may not be provided with a shell or other protective membranes.

In a few vertebrates, internal fertilization is followed by the formation of eggs protected by a membrane. The eggs are retained within the mother's body until they are hatched. This is known as *ovoviviparous* reproduction and is found in some sharks, a few bony fish, and certain snakes.

In the third method of sexual reproduction, there is no need for the formation of a protected egg because there is a very close structural relationship between the female and the developing young. After internal fertilization of the ovum, development proceeds within the body of the mother, in a specialized internal organ to which the young become attached. Among the pouched *(marsupial)* mammals, such as the opossum, kangaroo, and koala, the period of attachment is brief, so that the young are born in a very immature form. They complete their development within a pouch on the ventral surface of the mother's abdomen. In the higher *(placental)* mammals, such as horses, cats, dogs, monkeys, and man, this prenatal attachment continues for a relatively long period of time. In man, for example it normally lasts about nine months. This method of reproduction is known as *viviparous*.

Human Female Sex Organs

As we have said, the primary sex organs of the female are the ovaries. In the human being, these are two small, almond-shaped bodies about 32 millimeters long and about 16 millimeters wide. They lie on either side of the midline of the body in the lower part of the abdominal cavity. Each ovary contains many small groups of cells, the *follicles;* one ovum matures within each follicle. When completely developed and freed from the ovary, the human ovum measures about 0.09 millimeter in diameter and contains a large and distinct nucleus. The cytoplasm is plentiful and within it is a small amount of nonliving yolk material. This yolk, together with other materials that are present in the cytoplasm, serves as a source of food material for the early embryo until further provision is made for its development (Fig. 32-1). There is much more of this stored food material in the ova of oviparous animals than there is in the human ovum.

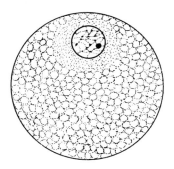

Fig. 32-1   A human ovum.

The ova develop continuously, but normally only one ovum is released at a time. Thus, at any given moment, ova are present in various stages of development (Fig. 32-2).

As shown in Fig. 32-3, the *oviducts,* or *Fallopian tubes,* partly cover the ovaries but are not directly connected to them. These tubes are lined with ciliated epithelial cells. The motion of the cilia carries the matured ova into the mouth of the oviduct and passes them along toward the *uterus* (womb).

The uterus is a hollow, thick-walled highly muscular organ. Its general location is shown in Figs. 32-3 and 4. In a woman who has borne no children, it measures about three inches in length by about two inches in breadth, and is approximately one inch thick. It is lined by mucous membranes covered with ciliated epithelial cells, and contains numerous glandular structures, the *uterine glands,* and many capillaries. In the uterus the ovum disintegrates, unless it has been fertilized. The lower opening of the uterus is slightly narrowed and forms the *cervix* (Latin, *cervix,* neck) which opens into the *vagina* (Latin, *vagina,* sheath). The

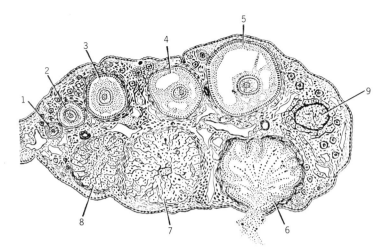

Fig. 32-2   Section through the human ovary. The maturation of the ovum is shown in stages 1–5, ovulation in stage 6, and the formation of the corpus luteum in stages 7–9. (Redrawn from Principles of Biology, *Whaley, Breland, Heimsch, Phelps, and Rabideau, 1954, permission of Harper and Brothers)*

relation of these parts to other abdominal structures is shown in Fig. 32-4. The external opening of the vagina lies between folds of skin called the *labia*.

The female accessory structures which provide for the nourishment of

Fig. 32-3   The human female reproductive organs as seen from the ventral side. Note the relationship of the oviducts to the ovaries. (From Miller and Haub, General Zoology, *Holt, 1956)*

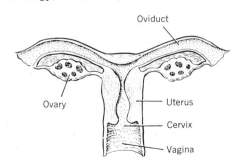

the child after birth are the *mammary glands*. Each gland (breast) is composed of numerous lobes of glandular tissue with their branching ducts, surrounded and embedded in adipose tissue. The mammary glands become active under the influence of the hormone prolactin and secrete milk after the birth of the child.

Human Male Sex Organs

Like the ovaries, the male gonads (testes) develop in the region of the embryonic kidneys. However, during the development of the embryo, they move through the body cavity to occupy a specialized fold of the skin, the *scrotum*, which hangs from the body in the ventral abdominal region between the thighs. During this movement, each testis passes through an opening in the muscular wall of the lower abdomen, called the *inguinal canal*. This canal

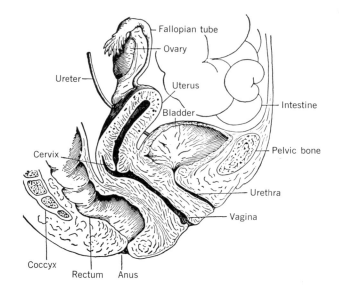

Fig. 32-4   Longitudinal section through the lower abdomen, showing the human female reproductive organs in relation to the other abdominal organs. (From   Haggard, The Science of Health and Disease, *copyright 1938, permission of Harper and Brothers)*

then normally closes up, although it may leave a weakened area in the muscular wall. Later in life, severe straining of this weakened area may result in a *hernia* (rupture) in which a fold of the intestine is pushed through the weakened spot. Hernia of this type is generally corrected by a surgical operation which draws the muscular walls together.

Apparently, the testes must descend into the scrotum in order to function properly. If they do not descend before puberty, they are unable to produce mature sperm. One possible explanation for this failure is that the temperature within the body cavity is too high for normal production of sperm.

The testes are glandular organs which are divided internally into many small subdivisions called *lobules,* each containing one or more convoluted tubules. A cross-section of one of these lobules and its tubule (Fig. 32-5) shows that they are composed of several dif-

ferent types of cells. The *interstitial cells* produce the male sex hormones. The *spermatogonia* are the cells from which the mature sperm develop. The function of the *Sertoli cells* is apparently to supply nourishment for the

Fig. 32-5   Section through a lobule of the human testis. (From Marsland, Principles of Modern Biology, *Holt, 1957)*

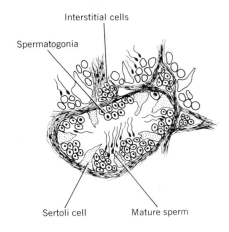

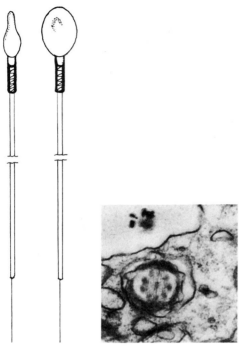

Fig. 32-6 Human sperm from the side (left) and top (right). The head at the top, is joined by a short neck to the middle piece, and the tail. The electron micrograph shows a cross section of a sperm tail. Compare this structure with the structure of a centriole, Chapter 3, (Photomicrograph, Drs. K. A. Siegesmund and C. A. Fox, Marquette University School of Medicine)

sperm cells that occupy the cavity of the lobule at some distance from the nearest blood vessels. Compared with most of the other cells of the body, the sperm are very minute. The total average length of a sperm cell is about 0.05 millimeter and the *head,* which contains the nuclear material, is only 0.003 millimeter in length. Behind the head region is a part usually referred to as the *middle piece,* which contains the mitochondria and the centrosome. Be-

hind the middle piece is the *tail,* which serves as a locomotor organ (Fig. 32-6). There is, in fact, very little cytoplasmic material in a sperm cell. The functions of the sperm are to add its chromatin material to the ovum at the time of fertilization and to initiate the development of the ovum into a new individual. The importance of the chromatin of these cells in the formation of chromosomes, and the functions of the chromosomes in determining hereditary characteristics will be discussed much more fully in Chapter 33. The mature sperm cells are stored in the vicinity of the testes and upon proper stimulation are discharged through a system of ducts to the outside.

The cavity shown in the lobule in Fig. 32-5 is a cross section of one of the long convoluted *seminiferous tubules* which occupy a large volume of the testis. These tubules, whose combined length may extend twenty feet, empty through other tiny tubules, the *vasa efferentia,* into the coiled *epididymis,* which in turn empties into the *vas deferens.* All of these tubules (like the oviducts) are lined with ciliated epithelium whose cilia constantly sweep the mature sperm toward the *seminal vesicle,* which secretes a thick fluid, thought to be vital to sperm viability. The relationship of these various structures is shown in Figs. 32-7 and 8. Beyond the seminal vesicle, other fluids are added to the sperm by the *prostate gland* and the small *Cowper's gland.* The action of these latter fluids is not completely understood but they do furnish a medium in which the sperm can swim. The combination of sperm and the fluid medium is known as the *semen.*

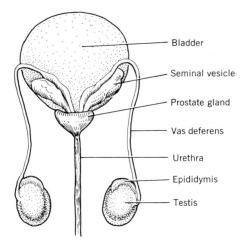

Fig. 32-7 Posterior view of the human male reproductive organs. (From Miller and Haub, General Zoology, *Holt, 1956)*

semen into the urethra of the penis. The penis is the organ by which sperm are introduced into the vagina of the female. The outer surface of the penis is covered by a layer of skin, the lower end of which is more or less free from the body of the penis. This is the *prepuce,* which is sometimes removed from the enlarged end of the penis, the *glans,* by a surgical operation called circumcision. It has been found that circumcision reduces the chances of an individual developing cancer of the penis. The body of the penis consists of very spongy tissues containing many blood sinuses. These erectile tissues are arranged in three groups: one is the *corpus spongiosum* which lies below the urethra, while the others form the *corpora cavernosa* and lie above and

It is rather interesting to note that the pH of the fluid in the testes and the prostate gland is slightly acid. The semen, however, shows a slightly alkaline reaction. It has been suggested that an acid medium retards the swimming activity of the sperm, but that the slightly alkaline fluid stimulates them to activity. When active, the life of a single sperm cell averages 12 to 36 hours, but there is some doubt whether it retains its ability to fertilize an egg during this entire period. Due to the extremely small amount of cytoplasm in the sperm cell, there is virtually no food supply from which it can derive energy.

Beyond the point of entrance of the prostate and Cowper's glands, the vas deferens opens into the urethra, which has its external opening at the end of the *penis* (Fig. 32-8). Contractions of the epididymis and vas deferens expel

Fig. 32-8 Longitudinal section of the human male reproductive organs and associated organs in the lower abdomen. (From Haggard, The Science of Health and Disease, *copyright 1938, permission of Harper and Brothers)*

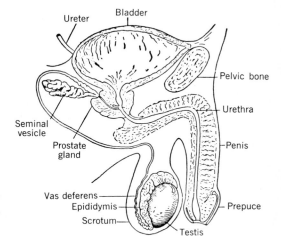

to the sides of the urethra. Erection of the penis occurs when the sinuses in these tissues become distended with blood.

Fertilization

As we have seen, the reproductive organs function normally under the influence of the sex hormones to produce the sperm and ova. The union of these cells results in development of a new individual. Human sperms and eggs differ greatly in size, and at the time of fertilization the tiny sperm must penetrate the cytoplasm of the ovum as far as the nucleus. The chromatin contained in the sperm head then combines with the chromatin in the ovum nucleus and fertilization is complete. In Fig. 32-9 it will be noted that the egg is surrounded by a clear area, the *zona pellucida.* A layer of epithelial cells, the *corona radiata,* is present outside this zone around the ovum when it leaves the ovary and must be broken

down before fertilization is possible. Some research workers in the field of embryology believe that an enzyme in the spermatic fluid brings about the disintegration of the corona.

The two types of gametes differ also in numbers. Normally, a single human ovum matures each month and leaves the ovary approximately 10 days before the onset of menstruation. Although both ovaries produce ova, the rate of maturation is such that ovulation occurs alternately, a single ovum usually being produced by either the right or left ovary. In the testis of the male, the maturation of the sperm cells is continuous. The sperm are collected (Fig. 32-8) and stored temporarily in the epididymis and vas deferens. This continuous activity on the part of the testes produces a larger number of sperm. It is estimated that several million motile sperm are necessary to insure fertilization of one egg. It must be remembered, however, that only one sperm actually fertilizes the ovum.

Fig. 32-9 A live human ovum. The light gray peripheral area is the zona pellucida, the large white central area contains the cytoplasm and nucleus of the ovum (the vitellus), and in between lies the perivitelline space. (Fig. 1, Z. Dickmann, T. H. Clewe, W. A. Bonney and R. W. Noyes, Anatomical Record, 152:297)

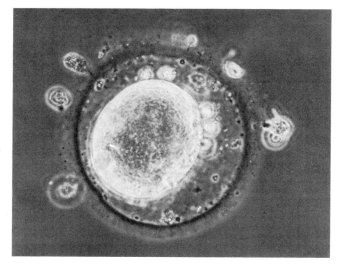

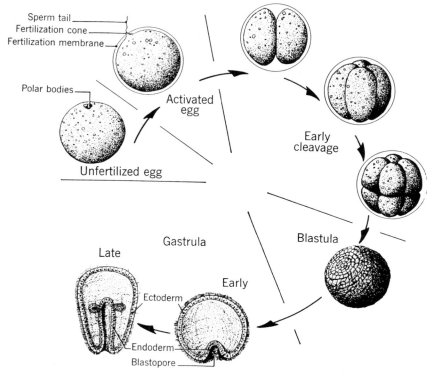

Sperm tail
Fertilization cone
Fertilization membrane

Polar bodies

Activated egg

Unfertilized egg

Early cleavage

Gastrula

Late

Blastula

Early

Ectoderm

Endoderm
Blastopore

Fig. 32-10 Fertilization and early cleavage (division) of the egg of a starfish. Note that there is little change in the volume of protoplasm during this process, even though there is rapid division of the cells. (From Marsland, Principles of Modern Biology, *Holt, 1957)*

In fertilization, the seminal fluid containing the sperm is forced into the urethra and from there is introduced into the vagina and passes upward through the uterus to the Fallopian tubes. A sperm cell meets and penetrates the ovum which is passing downward through the tube. Although the Fallopian tubes are lined with cilia that sweep the ovum toward the uterus, the upward progress of the swimming sperm is aided by a series of wavelike contractions of the tubes. It is estimated that it takes a sperm about three hours to pass from the vagina to the end of the Fallopian tubes.

Following the fertilization of the egg, the *zygote,* or fertilized egg cell, begins to divide. In all organisms which reproduce sexually, the repeated division of the egg (cleavage) results in the formation of a hollow sphere of cells, the *blastula* (Fig. 32-10). One side of the blastula then folds inward to form the *gastrula* (lower left, Fig. 32-10). In a later stage, the edges of the gastrula become fused together, so that a cross-section of it resembles two concentric

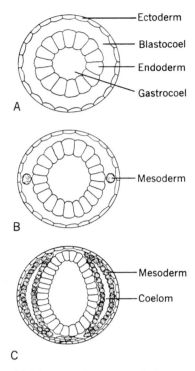

A

B

C

Fig. 32-11   Development of the primary germ layers in the gastrula. (From Marsland, Principles of Modern Biology, *Holt, 1957)*

tubes of cells with two cavities (Fig. 32-11A). In this stage, the outer layer of cells is called *ectoderm* (Greek, *ektos,* outside; *derma,* skin) and the inner layer is called the *endoderm* (Greek, *endon,* internal). The central cavity is the *gastrocoel,* which later becomes the cavity of the digestive tract. The space between the ectoderm and the endoderm is called the *blastocoel.* As the embryo continues to develop, cells from several points along the endoderm grow into the blastocoel to form still another layer of cells, the *mesoderm.* This in turn forms two distinct layers: the outer layer becomes the body wall, the inner develops into the wall of the digestive system (Fig. 32-11B and C). The space between these mesodermal layers becomes the *coelom,* or body cavity.

The ectoderm, endoderm, and mesoderm are called the *primary germ layers* and are responsible for the development of the various organs of the body as summarized in the table.

Ectoderm
  Outer skin, skin glands, hair
  Nervous system
  Parts of sense organs
  Lining of nose, mouth, and anus
  Salivary glands
  Tooth enamel
  Pituitary, and adrenal medulla

Mesoderm
  Heart, blood vessels, blood
  Lymphatic system
  Skeleton
  Muscles
  Kidneys and their ducts
  Gonads and their ducts
  All other connective tissue

Endoderm
  Lining of alimentary and respiratory
    tracts and of the bladder
  Secretory cells of the liver, pancreas,
    thyroid and parathyroids

During the early cleavage of the zygote, the individual cells of the two-cell stage sometimes become separated from each other. Each cell may then develop separately to form a new individual. As a result of this, *identical twins* are born. Since both of these individuals develop from a single zygote, they have received similar genes and hence the similarity between them is frequently very striking. Another type of twinning may occur when two ova mature and are fertilized at approximately the same time. Since these ova and the sperm that fertilize them are quite different in their gene content, the resulting twins are no more alike than any two other members of the family. Such twins are called *fraternal twins.*

If the parts of the dividing zygote happen to separate when they reach the four-cell stage, each can again start the cycle of division and give rise to identical *quadruplets.* This number of births is typical of one American mammal, the nine-banded armadillo of Texas. The females of this species normally produce four identical offspring at a time.

Beyond this four-cell stage, separation of the individual cells may occur, but if it does, the possibility that the young will develop to maturity is greatly decreased. In the later stages of cleavage, the cytoplasm of the cells begins to differentiate, so that the different areas of the developing blastula will determine the formation of the various body organs. It has been demonstrated that beyond the four-cell stage this differentiation of the substances within the cytoplasm reaches the point at which a separate single cell is no longer able to develop into a new organism.

In human beings, multiple births follow the so-called Law of 86. This means that twins (identical and fraternal) appear in the general population once in approximately every 86 births. Triplets are born about once in 86^2 births; quadruplets once in 86^3 births; and quintuplets once in 86^4 births. The birth of triplets and quintuplets is evidently the result of various possible combinations of fertilizations or separations of cells during cleavage. The possibilities involved in these are beyond the scope of this text, and appear to be conditioned by hereditary factors.

Embryo Implantation

In mammals, the zygote goes through the early stages of its development as it moves down the Fallopian tube toward the uterus. A period of between three and five days is required for the zygote to reach the stage which corresponds to the blastula. Between five and eight days following fertilization, it becomes implanted (attached) on the inner wall of the uterus.

Following the fertilization of the ovum, hormones are produced to prepare the uterus for the arrival of the developing embryo (Fig. 30-5). In the meantime, a membrane forms around the mass of dividing cells as it passes down the tube. This is the first of the *extraembryonic membranes* to appear and is known as the *chorion.* It rapidly becomes covered with small fingerlike projections, the *chorionic villi* (Fig. 32-12). Enzymes are produced by the

Fig. 32-12 Diagram of a section of an early stage of a human embryo. Three of the extra-embryonic membranes are shown: the amnion (A), yolk sac (Y), and the chorionic villi (V). The cavity of the blastula is indicated by (E). (From McEwen, Vertebrate Embryology, Holt, 1957)

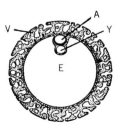

villi that enable them to sink into the mucous membrane of the uterus and make close contact with the blood supply, so as to obtain nourishment for the developing embryo. The second extraembryonic membrane is the *amnion*, which will later surround the embryo and enclose it in a fluid medium, the amniotic fluid. This fluid keeps the developing embryo moist and protects it from mechanical injury. The third extraembryonic membrane is the *yolk sac*. This has a minor part in human development but is of great importance in the forms of animals that hatch from eggs. In these, the yolk sac contains the remains of the yolk of the egg and serves as food for the embryo until it is capable of getting its own food. A fourth extraembryonic membrane that appears early in development is the *allantois*. In birds and reptiles this acts as an embryonic lung, but in man it is present for only a short time.

The nourishment of the embryo begins when the chorionic villi become embedded in the uterine wall. Prior to this, the developing cells obtain food from the yolk material and other substances in the cytoplasm of the egg. Although relatively small in quantity, they are sufficient to permit the early

stages of development to occur because there is almost no growth in size of the cells between division stages. After the chorionic villi have made their connection, capillaries in the wall of the uterus break down and form blood sinuses around the villi. This region of close association between the villi and the maternal blood supply increases greatly in size and forms a large thin membrane, the *placenta*.

At no time is there any direct nerve or blood connection between the mother and the embryo. The two blood streams run roughly parallel to each other, separated by thin membranes. Through these membranes the exchange of food and oxygen and metabolic wastes occurs by diffusion. Because there is no direct connection between the circulatory systems of mother and child, infectious diseases cannot be transmitted from the mother to the offspring unless there is some break in the membranes.

The developing child is connected to the placenta by a long ropelike structure called the *umbilical cord*. This cord develops partly from the stalk which earlier attached the yolk sac to the embryo, and partly from the allantois. Through it run two large umbilical

Fig. 32-13   A, B, and C, the same embryo 4¼ weeks after fertilization. The placenta is closed in A, cut back to expose the embryo in B, and the embryo is enlarged in C to show structure. Note the limb buds and eye beginning to form. D, 5½ weeks after fertilization. E, 10 weeks after fertilization, fetus within the uterus. F, 12½ weeks after fertilization. The placenta is present in all but C as the outermost membrane with many villi. The fluid filled amnion lies within the placental cavity surrounding the fetus, best seen in D but also visible in B and F. The yolk sac can be seen protruding from the embryo in B and its remains attached to the inside of the placenta lower right of D. The relation between the umbilical cord and the placenta is best shown in E and F. (Courtesy of the Carnegie Institution of Washington)

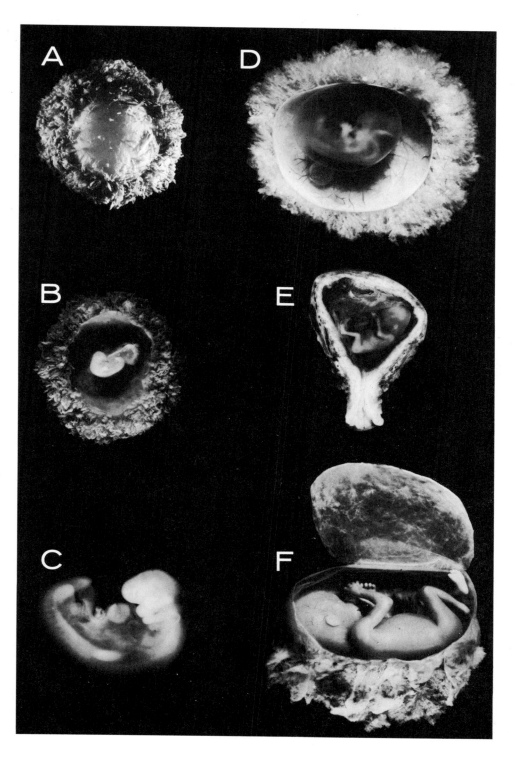

arteries and the umbilical vein. Blood flows from the child to the placenta through the arteries and returns through the veins. The relationship of the developing child, the umbilical cord, and the placenta is shown in the photographs in Fig. 32-13.

Growth and Birth

The term *embryo* is usually applied to the developing young when its general body form is such that it cannot be recognized as a member of a specific species of animal. The human embryo develops slowly. About $4\frac{1}{2}$ weeks after fertilization, the mass of cells is about 5 millimeters long. As Fig. 32-13 B and C show, at this stage there is little to indicate that the embryo is human.

Further development of the human embryo is shown in the other photographs in Fig. 32-13. As soon as the embryo becomes recognizable as human, it is called a *fetus*. The same term is used for the unborn young of other mammals.

The process of human birth, or *parturition,* occurs approximately nine calendar months after fertilization of the ovum. We do not know exactly how the birth process is started, but undoubtedly hormones play an important part. Birth begins with the appearance of *labor pains* caused by strong contractions of the muscles in the walls of the uterus. As these contractions increase in strength and regularity, the amnion surrounding the fetus breaks, discharging the amniotic fluid through the vagina. At this time the muscles of the cervix and the vagina relax to increase the size of the passage from the uterus. The pubic bones also separate slightly by the stretching of ligaments that normally hold them firmly together. As a result of the relaxation of the cervix, vagina, and pubic ligaments, the strong contractions of the uterus literally squeeze the fetus through the vagina.

Usually the child appears head first because that is the position it normally takes in the uterus (Fig. 32-14). Continued contraction of the uterus then forces the rest of the body through the birth canal. The child is still attached to the placenta by the umbilical cord, which is usually tied off and cut on the side of the tie near the mother, shortly after the birth. The *navel,* or umbilicus, is the scar left on the abdomen which marks the place where the fetus was attached to the cord. Shortly after birth, the placenta and the remains of the amnion (known as the *afterbirth)* are expelled from the uterus.

While the fetus is in the uterus, it obtains food and oxygen from its mother's blood through the placenta. It also discharges carbon dioxide and other wastes through the same membrane. The lungs of the fetus are collapsed and most of the blood that will eventually flow through them is diverted by a blood vessel that connects the pulmonary artery with the aorta. At the time of birth, this vessel closes so that blood from the right ventricle must then flow through the pulmonary arteries to the lungs before it returns to the left side of the heart. If this vessel does not close off normally, the baby's blood may not be properly oxygenated and the child is known as a *blue baby.* Fortunately, this condition can be corrected by surgery. During the process of birth, the rise in the carbon dioxide content of the baby's blood stimulates the respiratory center of the medulla. As a result of this, breathing movements begin and thereafter continue normally.

After the child is born, the mother's uterus returns rapidly to its normal condition. However, its cavity is slightly larger than it was before pregnancy, the blood vessels are straighter and more distinct, and the muscle layers are more sharply defined. The mother's mammary glands enlarge during pregnancy and begin to produce milk at the

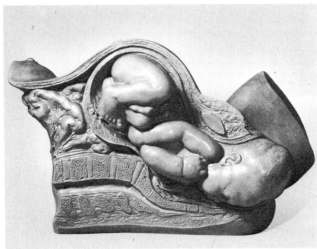

Fig. 32-14   Photographs of models showing two stages in the birth of a child. In A, the head is passing through the cervix into the vagina. The amnion usually breaks at about this stage in birth. In B, the head is emerging from the opening of the vagina. At this stage the child is still attached to the placenta by the umbilical cord (not shown in the model). (Courtesy of the Cleveland Health Museum)

time of birth. This function is controlled by the hormone prolactin, which was mentioned in Chapter 30. Many babies are fed cow's milk or formulas containing cow's milk, instead of being breast fed. This is not always satisfactory, since human milk is slightly different from cow's milk, as shown by the percentages in the following table.

	Human Milk	Cow's Milk
Calcium	0.032	0.118
Protein	1.4	3.4
Fat	3.7	3.9
Lactose	7.2	4.9
Minerals (total)	0.2	0.7
Calories per pound	307	310

VOCABULARY REVIEW

Match the phrase in the left column with the correct word in the right column. *Do not write in this book.*

1   Produced by the seminiferous tubules.
2   Surgical removel of the prepuce.
3   Occurs ten days before menstruation.
4   An extraembryonic membrane important in the nourishment of the fetus.
5   Surrounds the maturing ovum.
6   Twins resulting from separation of the two halves of the fertilized egg at the first cleavage.
7   A germ layer formed by growth of cells into the blastocoel.
8   Attachment of the blastula to the inner wall of the uterus.
9   The germ layer from which the nervous system develops.
10 Organ of mother in which the fetus develops.

a   circumcision
b   ectoderm
c   endoderm
d   follicle
e   fraternal
f   identical
g   implantation
h   mesoderm
i   ovulation
j   ovum
k   placenta
l   sperm
m   uterus

TEST YOUR KNOWLEDGE

Group A

Select the correct statement. *Do not write in this book.*

1   Vertebrates that produce live young are said to be (a) oviparous (b) vivarium (c) viviparous (d) ovarian.
2   One type of gonad is the (a) ovum (b) egg (c) ovary (d) sperm.

3   The primary sex organ of a female is the (a) oviduct (b) ovary (c) uterus (d) vagina.

4   Sperms are produced in the (a) seminal vesicle (b) prostate gland (c) testis (d) Cowper's gland.

5   The sperm and its fluid medium is called (a) prepuce (b) scrotum (c) labia (d) semen.

6   The union of the sperm and egg is called (a) implantation (b) fertilization (c) ovulation (d) maturation.

7   The union of the human sperm and egg normally takes place in the (a) ovary (b) cervix (c) oviduct (d) uterus.

8   An accessory structure of the male reproductive system is the (a) urethra (b) ureter (c) chorion (d) Fallopian tube.

9   The outer layer of cells of the gastrula is called (a) mesoderm (b) endoderm (c) ectoderm (d) coelom.

10 Of the following, the structure which is an embryonic membrane is the (a) uterus (b) vagina (c) placenta (d) corpus luteum.

Group B

1   Describe the general development of the zygote through the formation of the primary germ layers.

2   How does the formation of identical twins differ from that of fraternal twins?

3   (a) Describe the path taken by the sperm from the moment of its introduction into the vagina of the female through fertilization. (b) Relate the differences in structure between sperm and ova to their different functions.

4   Name the extraembryonic membranes. What are their functions in the human embryo?

5   (a) Briefly outline the stages in the process of birth from the appearance of labor pains. (b) What major change in the circulation of the baby's blood takes place at birth and with what result?

Chapter 33 Genetics

In Chapter 3 we learned that cells are the basic units of the body and that their structure is specialized according to their activities. We also learned that within the nucleus of each cell are the materials that control the heredity of the individual. We should remember that only the chromatin in the gametes is involved in heredity which is determined during the process of fertilization when the nuclei of the egg and sperm unite. The science that deals with the inheritance of characteristics is called genetics.

Genes

We may say that a person has inherited the color of his eyes from one parent and the color of his hair from the other. This of course, does not mean that there are small bits of eye color or hair in the gametes. However, each of us has in his cells the determinants of physical and mental traits. These are called the *genes*. They have a very definite composition, which will be discussed in the next chapter. The differences which exist among the genes are apparently due to the coding of parts of

the long chain deoxyribonucleic acid (DNA) molecule.

Some early studies by Drs. Morgan, Bridges, and Sturtevant indicated that some of the hereditary characteristics of the fruit fly, Drosophila, could be related to definite regions in the chromosomes of this animal. Later, by a careful analysis of the results from breeding flies with different features, these men were able to locate within the chromosomes the regions that determine specific hereditary characteristics.

Investigation of the structure of the chromosomes in the salivary glands of Drosophila has produced very interesting and significant results. The chromosomes in these glands are many times larger than those in the gametes or the other body cells of the animal. In fact, evidence that these large chromosomes are made up of numerous strands of chromatin material, perhaps as many as a thousand. Such chromosomes are said to be polytenic. If these giant chromosomes are properly stained, alternating bands of dark and light materials can be seen (Fig. 33-1). Through comparison of these with or-

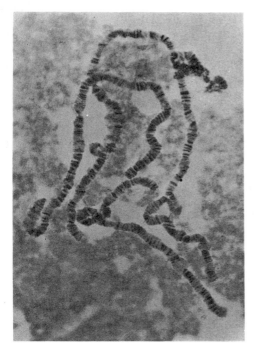

Fig. 33-1 Photomicrograph of the giant chromosomes from the salivary glands of Drosophila. Note the pronounced banding of the chromosomes. (Bausch and Lomb Inc., Rochester, N.Y.)

dinary chromosomes, it has been determined that bands are present at exactly the same points (loci) on the chromosomes at which previous methods forecast the presence of genes.

Embryo Development

Following its fertilization, the egg cell is known as a zygote (Greek, *zygosis*, joining). It then divides by mitosis to form two cells, and later, four, eight, sixteen, and so on. At first these divisions are very regular, but soon the mathematical exactness is lost since some cells start dividing more rapidly than others. As a result of this change in division rate, certain tissues develop earlier in life than others. This differentiation into various types of cells appears to be influenced by the action of materials in the cytoplasm of the egg. These serve to determine the fate of the cells into which they are eventually segregated, producing muscle or nerve or epithelium or whatever other type of tissue is required. As Fig. 33-2 shows, this has been demonstrated in the case of the frog's egg. If the colored area in the egg is destroyed, no reproductive system can develop. Other regions are responsible for the development of still other organ systems. It is the eventual separation of these regions into groups of cells that determines the development of the animal.

During the development of the reproductive organs, the cells divide by mitosis just like the other body cells. Each receives the same number and same types of chromosomes as the others. However the *germ* cells in the gonads, cells which eventually produce the mature eggs or sperm (the *gametes*), divide differently. The result is that the mature gametes receive only one of each pair of chromosomes while the cells of the rest of the body, the *somatic* cells, receive both. In man, for example, the number of chromosomes present in somatic cells is 46, but this is reduced to 23 in the mature germ cells. Until quite recently, the number 48 was generally accepted as the chromosome count for man. However, researchers using much improved methods of staining have found that the basic number is 46. Variations of 47 and 48 may appear occasionally.

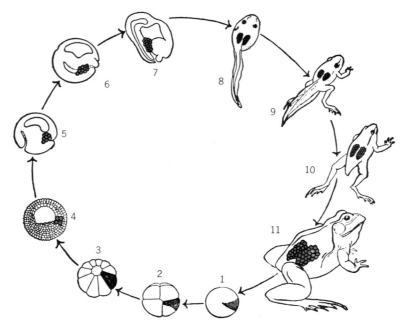

Fig. 33-2   The sequence in the development of eggs in a frog. In 1, the colored area of the fertilized egg contains materials that determine the development of the reproductive system. In 2 through 10 are shown various stages in tadpole development. In 11 we see the adult frog with mature eggs in the ovaries. (After Conklin from Guyer, Animal Biology, *permission of Harper and Bros.)*

The full number of chromosomes found in each somatic cell is called the diploid (2n) number, while that of the germ cells is known as the haploid (n) number. Each species of animal or plant has a constant diploid number of chromosomes. The number of chromosomes is not the important consideration: the quality of the genes is the principal factor in determining the characteristics of an organism. The following table shows how the number of chromosomes varies among a few selected animals and plants.

As these data indicate, the structural or physiological complexity of an ani-

Organism	Diploid Number (2n)	Haploid Number n
Honey bee	16	8
Fruit fly (Drosophila)	8	4
Crayfish	196	98
Rabbit	44	22
Pig	40	20
Cattle	60	30
Horse	66	33
Monkey (Rhesus)	48	24
Man	46	23
Corn	20	10
Wheat	82	41
Black pepper	128	64
Edible pea	14	7

mal or plant is not solely dependent on the number of chromosomes present in its cells.

Meiosis

The process by which the number of chromosomes is reduced from the diploid to haploid number is known as meiosis (Greek, *lessening*). This usually occurs in man just before the gametes become mature within the gonads. In Fig. 33-3 are shown the important steps in the process, the details of which are beyond the scope of this text. Plants undergo a similar series of nuclear changes.

The two cell divisions of meiosis result in 4 gametes each containing the haploid number of chromosomes. This process differs essentially from mitosis as shown in the table on page 444.

There are two features of the process of meiosis that should receive special attention. During the formation of the

Fig. 33-3 The process of meiosis in an animal having a diploid number of six chromosomes. Note the reduction in the number of chromosomes in the gametes. (Redrawn with modifications from Miller and Haub, General Zoology, Holt, 1956)

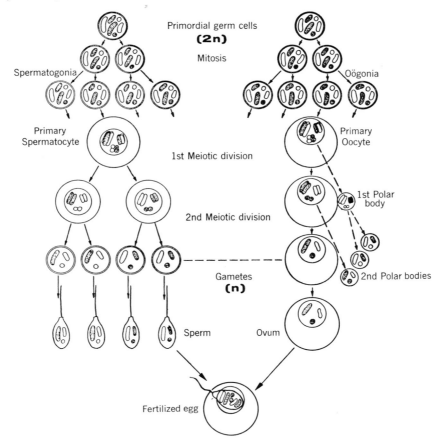

	Mitosis	Meiosis
Kind of division	Asexual	Sexual
Number of divisions	One	Two
Number of cells formed	Two	Four
Number of times chromosomes replicate	Once	Once (in first division only)
Number of chromosomes in each cell formed	Diploid (2n)	Haploid (n)
Kind of cells formed	Somatic (all identical)	Gametes (all different)

sperm *(spermatogenesis)*, most of the cytoplasm disappears from the cell, and the sperm contributes only chromatin material to the egg cell at the time of fertilization. The cytoplasm of the egg, on the other hand, supplies the early stages of the developing embryo with nourishment. It is therefore of extreme importance that the mature ovum retain as much cytoplasm as possible. During the formation of the egg *(oogenesis)*, the meiotic divisions of the chromosomes are accompanied by unequal division of cytoplasm (Fig. 33-3). As a result, miniature cells *(polar bodies)* are formed which contain the same number of chromosomes as the egg but have a minimum amount of cytoplasm (Fig. 33-4). These small polar bodies may be considered as waste products of the process of reproduction, since they cannot normally be fertilized.

During meiosis, the distribution of specific chromosomes to the mature cells is purely a matter of chance. Since they are present as pairs of homologous (Greek, *homo,* equal + *logos,* origin) chromosomes, the members of the pairs separate at the time of reduction and each passes into a different cell. For example, if we consider a cell with a diploid number of 6 chromosomes (3 pairs of homologous chromosomes) which we label Ee, Ff, and Gg, we find that the separation of the members of these pairs may take any of the eight patterns shown in the following table.

1	2	3	4	5	6	7	8
E	e	E	e	E	e	E	e
F	f	F	f	f	F	f	F
G	g	g	G	G	g	g	G

If only three pairs of chromosomes result in eight (2^3) different combinations of mature gametes, it is not difficult to see how 46 (23 pairs) of human chromosomes may result in many more possible combinations. Mathematically, the number of possible combinations of chromosomes appearing in either the mature egg or sperm cell would be 2^{23} or 8,388,608. Since fertilization consists of the union of any sperm with any ovum, the number of possible combinations of chromosomes in the zygote reaches the amazing total of 70,368,744,177,664 ($8,388,608^2$). If each chromosome contained only one gene, this alone would explain why the possibility of any two people being exactly alike is infinitesimally small. It has been estimated that the chances are about 300 trillion to 1 against the occurrence of the same gene content in two people. Of course, the possibility of similarity between closely related people would be considerably greater.

By the process of fertilization, the new individual receives specific hereditary characteristics from his parents. Fertilization also stimulates the egg cell to start dividing. As Dr. Jacques Loeb demonstrated in the early 1900's, a mature unfertilized frog egg need only be pricked by a sharp needle in order to start its division. Normally, the stimulus that starts cell division is provided by the penetration of the sperm through the cell membrane of the egg. If, however, an artificial stimulus is used to start division, the animal that develops has only the haploid number of chromosomes. This process, known as *parthenogenesis,* occurs normally among some animals, notably certain insects. In these forms, the stimulus that starts the division is presumably chemical.

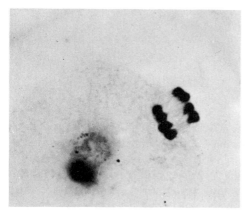

Fig. 33-4 Photomicrograph of the formation of a polar body (telophase of the first meiotic division) on a nematode egg (Ascaris megalocephala). (Carolina Biological Supply Co.)

PRINCIPLES OF HEREDITY

Studies of heredity before 1866 were not conclusive. The results obtained by earlier investigators offered little explanation of the way inheritable features are transmitted from one generation to the next. This was due either to faulty techniques or failure to interpret the results correctly. In 1866 an Augustinian monk who lived in the monastery in Brünn (Brno), Moravia, published two small treatises on the laws of heredity in the journal of the local natural science society. This monk, Johann Gregor Mendel, later became the head of the monastery. His many added duties prevented him from continuing his study of inheritance. He died in 1886 without realizing that his two small contributions would form the foundation for all the work that has since been done in this field.

It was not until 1900 that three investigators (de Vries, Correns, and Tschermak) discovered what they thought to be new facts about heredity. In searching through literature in this field, they found Mendel's earlier reports on the same subject. They realized immediately that Mendel was the pioneer in this investigation and gave him full credit for these basic discoveries. The principal reason for the delay in recognizing the value of Mendel's experiments was that Mendel's papers were published in a small journal that usually contained only natural science articles of a very generalized type. Although the publication was received by large libraries, Mendel's reports were apparently disregarded by outstanding biologists of that era.

The Monohybrid Cross

Mendel's work was concerned with the inheritance of various characteristics shown by the edible pea *(Pisa sativum)*. He grew plants of this species in the monastery garden and kept very careful notes on the results he obtained. Because peas are normally self-pollinating, Mendel could control his experiments. He concentrated on one characteristic at a time, and made his crosses between plants which had shown the trait through several generations.

In this way, Mendel discovered that when he pollinated purebred plants having long stems with pollen from those having short stems, all of the offspring were as tall as the tall parent. He also found that whenever he cross-pollinated two purebred plants with any pair of contrasting characteristics, one of these would appear in the offspring while the other did not. To the characteristic that appeared in the offspring he gave the name of *dominant characteristic,* while the characteristic of the pair that did not appear he called the *recessive characteristic.* The following table shows some of the dominant and recessive pairs discovered by Mendel in his work with garden peas.

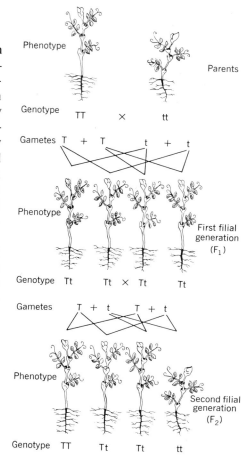

Fig. 33-5   *The inheritance of stem length in the edible pea. In the F_2 generation it is impossible to distinguish between the homozygous and heterozygous plants because the gene for tallness is dominant.*

Dominant	Recessive
Tall plants	Short plants
Round seeds	Wrinkled seeds
Green pods	Yellow pods
Gray seed coat	White seed coat

Mendel found that if he crossed two hybrid plants (the offspring of purebred parents showing contrasting characters), the following generation contained some plants showing the recessive feature. If we use the same pairing given in the preceding table, Mendel's actual results are shown in the table on the next page.

Feature	Characteristic		Ratio
	Dominant	Recessive	
Stem length	787 tall	277 short	2.84 : 1
Seed shape	5474 round	1850 wrinkled	2.96 : 1
Pod color	428 green	152 yellow	2.82 : 1
Seed coat color	705 gray	224 white	3.15 : 1
Average ratio	——	——	2.94 : 1

The diagram in Fig. 33-5 illustrates the results obtained in crossing tall and short plants. It will be noted that the letters T and t are used to designate the traits of tallness and shortness. The capital letter, T, designates the dominant feature, and the small letter, t, the recessive one. Thus the sperm and egg cells of the tall plant each contain one gene for tallness. The genes for shortness are separated in the same manner.

Since the parents in this case show pure characteristics for tallness and shortness, they are said to be *homozygous* (Greek, *homo*, equal + *zygosis*, joining). After fertilization, the offspring of these parents contain one gene for tallness and one for shortness, and are said to be *heterozygous* (Greek, *heteros*, different). The term *hybrid* is used for the heterozygous condition.

The generation that results from breeding together the pure (homozygous) parents is called the *first filial generation* (F_1). All these offspring are heterozygous. However, if these heterozygous plants are bred with each other, the *second filial generation* (F_2) will show an approximate ratio of three plants with the dominant characteristic (tallness) to every plant with the recessive characteristic (shortness). This 3 : 1 ratio is seen in the table of Mendel's results above. It is based on the probable sorting out of the genes into different gametes and their random recombination in the zygote. It should be pointed out that the 3 : 1 ratio holds only when the offspring are numerous; the data given above are based on the observation of 9,897 individual plants.

Unless you are familiar with the ancestry of an organism, it is impossible to tell by inspection whether it is heterozygous or homozygous for any *dominant* characteristic, since its appearance gives no clue to the genetic content of its cells. Because of this, we need terms to differentiate appearance from genetic makeup. The appearance of the organism is known as the *phenotype* (Greek, *phaino*, to show + *typos*, model). The assortment of genes in the cells is the *genotype* (Greek, *genos*, race). Thus, in the pea plants mentioned before, there is no way of knowing which tall offspring in the F_2 generation are heterozygous and which are homozygous. The phenotype (tallness) is the same in all of them, but they may be of two different genotypes (*TT* or *Tt*). In order to determine whether a specific individual is homozygous or heterozygous for the dominant characteristic, it is necessary to perform a *back cross*. This is done by crossing the unknown individual with one showing the *recessive* feature.

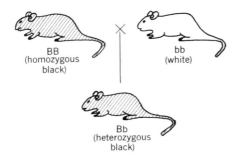

Fig. 33-6   A back cross between a homozygous black and a recessive white mouse.

The use of the back cross can be illustrated in the case of mice. If a mouse having black hair is crossed with a white mouse, the black color is dominant and the offspring in the F_1 generation are all black. If two of these mice that are heterozygous for color are crossed, the members of the F_2 generation are in the ratio of three black mice to one white mouse, the same ratio that Mendel found in his pea plants. Since all of these black mice are as black as the parents and the black grandparent, it is impossible to determine visually their genetic makeup. A

Fig. 33-7   A back cross between a heterozygous black mouse and a recessive white. Compare this with Fig. 33-6.

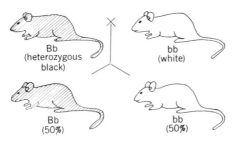

back cross, using one of the black mice and a white mouse (which is necessarily homozygous) will reveal the genotype of the black mouse. Fig. 33-6 shows the results obtained when a homozygous black mouse is crossed with a white mouse. The offspring will all be black (phenotype) and heterozygous (genotype *Bb*) as shown by the symbols in the boxes.

Genes From One Parent	Genes From the Other Parent	
	b	b
B	Bb	Bb
B	Bb	Bb

If the heterozygous mouse is crossed with a white mouse (Fig. 33-7) half of the offspring will be heterozygous black and half will be white as follows:

Genes From One Parent	Genes From the Other Parent	
	b	b
B	Bb	Bb
b	bb	bb

Sex Determination

In one instance, the *entire* chromosome, rather than the gene, controls a definite characteristic of an organism. The sex of an individual is determined by the presence of certain definite chromosomes, known as the *X* and *Y chromosomes*. In the human male there are 22 pairs of homologous chromosomes plus one pair in which the members are quite unlike each other. The larger of this pair is known as the X chromosome and the smaller the Y chromosome. Together they constitute the sex chromosomes. In females we find the pair consists of two X chromosomes.

During meiosis, the number of chromosomes is halved so that each mature gamete receives 23 chromosomes, distributed as follows:

Sperm	Ova
22 + X	22 + X
22 + Y	22 + X

Fertilization of an ovum by a sperm containing an X chromosome results in the development of a female child, while a male develops if the sperm contains a Y chromosome. Theoretically, this should result in an equal distribution between male and female births in the population as a whole. Actually, however, there is a slight tendency in favor of male births. It has been suggested that an explanation of this is the greater speed of movement of the sperm carrying the Y chromosome. Some students of population trends consider the idea that the male is the stronger sex to be a myth. Data from various life insurance companies lend weight to the idea that two X chromosomes are more effective in promoting longevity (long life) than the $X—Y$ combination. At present the average married woman has a life expectancy of six or seven years more than the average married man.

Sex-linkage

Color blindness has already been discussed briefly in Chapter 11. It is a hereditary defect that is transmitted from one generation to the next by a defective gene on the X chromosome. Hemophilia (Chapter 23) is inherited in the same way. Both of these are recessive characteristics.

As Fig. 33-8 shows, a major part of the X chromosome does not have a homologue (similar part) on the smaller Y chromosome. If, therefore, this nonhomologous region of the X chromosome contains the gene for color blindness or hemophilia and the X is combined with a Y chromosome, there is no gene for the normal condition to overcome the recessive defect. The male will therefore inherit color blindness or hemophilia. If, however, two X chromosomes join in the zygote, and one contains a defective gene and the other a gene for the normal characteristic, the female will not exhibit the defect although one of her chromosomes still carries the defective gene.

Fig. 33-8 A diagram of the X and Y chromosomes showing the homologous and nonhomologous regions.

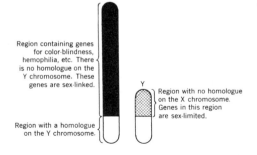

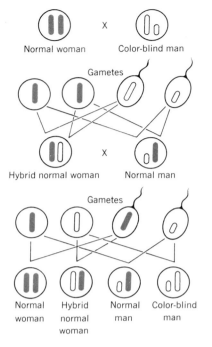

Fig. 33-9   The inheritance of color blindness. X chromosomes bearing the dominant gene for full color vision are shown red. Those bearing the Recessive gene for color blindness are shown white. The Y chromosome bears neither gene.

Characteristics that are transmitted in this fashion are known as *sex-linked* traits. In Fig. 33-9 are shown several crosses between individuals carrying the gene for color blindness. It will be noted that a woman can be color blind only if she has two genes for the condition, which means that she must inherit the characteristic from both her father and her mother.

If a gene on the *Y* chromosome determines a characteristic, it is termed *sex-limited*. Again referring to Fig.

33-8, it will be noted that there is a region of the *Y* chromosome that has no homologue on the *X*. It is in this region that the sex-limited genes are found. Only a few of these have been identified with certainty. One results in the webbing of the second and third toes, another the appearance of excess hair in and around the ears, and a third (sometimes fatal) condition characterized by a barklike skin.

The Dihybrid Cross

We have thus far discussed only the results obtained when one pair of contrasting characteristics is crossed *(monohybrid cross)*. In such a mating, one member of the pair is dominant over the other in the ratio of 3 : 1 in the F_2 generation. If we want to determine the ratios that result from crossing two pairs of characteristics, we make a *dihybrid cross*. Mendel also performed experiments of this type. He crossed plants which consistently yielded round green seeds with those yielding wrinkled yellow seeds. The resulting seeds were all round and yellow. The plants which grew from these seeds by normal self-pollination produced seeds in the following ratios: 9 yellow-round : 3 green-round : 3 yellow-wrinkled : 1 green-wrinkled. The reason for this distribution is shown diagrammatically in Fig. 33-10. If these results are analyzed in terms of dominant and recessive characteristics, it will be seen that the basic ratio of 3 dominants to one recessive still exists. There are still 3 round-seed plants for each wrinkled-seed, and 3 yellow-seed plants for each green-seed.

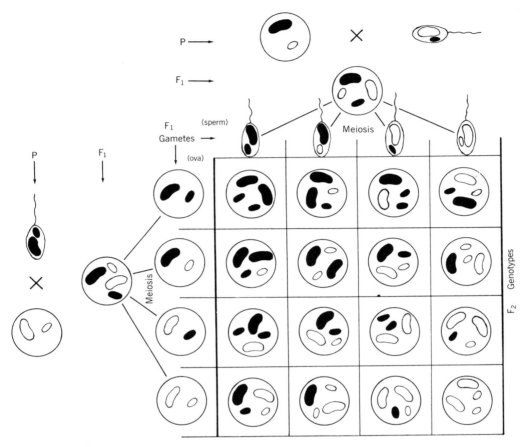

Fig. 33-10 Chart chowing the results of a dihybrid cross. The genes from one parent are shown across the top of the square, while those from the other parent are along the side. The chromosomes carrying the dominant characteristic are shown in solid color (black), while those with the recessive gene for the characteristic are shown in outline.

Mendel's Laws

Mendel's work laid the basis for the formulation of three laws, which control the inheritance of most known characteristics. These are called *Mendel's Laws* and may be stated as follows:

The Law of Dominance. The offspring of parents which are pure for contrasting characteristics will resemble one parent only. The hereditary unit (gene) for one characteristic, the dominant, prevents the other, the recessive, from expressing itself when the two are combined in the same organism.

The Law of Segregation. When hybrids are crossed with other hybrids or with individuals showing either the

Mating	Offspring
Heterozygous × Heterozygous	75% show the dominant feature: 25% the recessive This may also be shown as 25% homozygous dominants : 50% heterozygous dominants : 25% recessives.
Heterozygous × Recessive	50% heterozygous : 50% recessives
Heterozygous × Homozygous dominant	50% homozygous dominants : 50% heterozygous

dominant or recessive characteristic, the dominant and recessive both appear in the offspring, in a definite ratio. The results obtained by various types of matings are shown in the table above.

The Law of Unit Characters. Each pair of characteristics operates independently of others, following the laws of dominance and segregation.

Incomplete Dominance

Since Mendel's time, some apparent contradictions of his basic findings have been discovered. However, if these are examined critically, they will be found to follow the fundamental laws. One of these contradictions is a condition known as *incomplete dominance,* or *blending,* in which neither of the contrasting genes appears to be strictly dominant or recessive in the heterozygous condition. The result is an apparent blending of the characteristics so as to produce an intermediate condition. This is well illustrated in the case of certain flowers such as zinnias or four o'clocks. A pink hybrid results from the crossing of a red flower with a white flower in these species. The same is true of some of the colors shown by pigs (Fig. 33-11). Spotted animals are hybrids showing incomplete dominance of either color.

The inheritance of skin color in man is another example of blending, but the condition is not as simple as that shown by either flowers or cattle. There are evidently two pairs of genes involved in human skin color. Therefore, the inheritance of skin color resembles the dihybrid cross shown in Fig. 33-10. If we consider *AB* as representing the genes for dark skin color, and *ab* those for white skin color, then *ABAB* would represent the homozygous condition existing in the dark-skinned races, and *abab* that of the white-skinned races. A mating between two individuals who are homozygous for these characteristics would produce a heterozygous person with a genotype of *AaBb*. A cross between two people having this hybrid condition could produce any of sixteen possible combinations illustrated in Fig. 33-10. The occurrence of intermediate shades of color is due to differences in the total number of genes for dark or white skin color which the individual has received.

Fig. 33-11   Incomplete dominance in pigs. Above: this speckled hybrid boar resulted from a cross between a red Duroc and a black Berkshire. The family of piglets in the lower photograph resulted from a cross between the boar in the upper photograph and a hybrid white Yorkshire sow. How do you explain the resulting color combinations? Clue: the color white is dominant to both black and red, and the white sow in this photograph carries the gene for black color. (Grant Heilman)

Linkage and Crossing Over

Some characteristics do not obey Mendel's law of independent segregation. When pure bred, purple flowered sweet pea plants with long pollen grains are crossed with red flowered plants with round pollen grains the F_1 generation, as could be expected, consists entirely of plants with purple flowers and long pollen grains, showing that these are the two dominant character-istics. However, when members of the F_1 generation are crossed with one another, the expected $9:3:3:1$ ratios illustrated in Fig. 33-10 are not obtained. There is instead an excess of the two parental types (Fig. 33-12) because the characteristic for purple flower segregates with the characteristic for long pollen and that for red flowers with that for round pollen. In 1910, T. H. Morgan postulated that characteristics which do not segregate independently

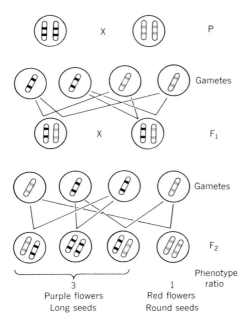

Fig. 33-12 Segregation of linked characteristics in sweet pea plants. The chromosomes bearing the genes for purple flowers and long pollen grains are shown banded with black and the chromosomes bearing the genes for red flowers and round pollen grains are banded with red.

are located on the same chromosome and he called them *linked characteristics*.

Occasionally linked characteristics do segregate independently; due to a phenomenon called *crossing over* which takes place during the first division of meiosis. At one stage in this division, chromosome pairs become entwined, one chromosome around the other. The chromosomes may break due to the strains developed and exchange material as they rejoin. Fig. 33-13 shows how the characteristic *P* for purple flowers might become separated from

L the characteristic for long pollen and become linked instead to the characteristic *l* for round pollen while *L* becomes linked to *p* the characteristic for red flowers. Crossing over and the recombination that accompanies it is the usual process by which new groupings of characteristics arise. Theoretically, such crossovers can occur in one to fifty percent of the offspring, but the actual results of this process appear much less frequently.

MUTATIONS

A new hereditary characteristic that appears in any species of animal or plant is known as a *mutation*. Any such change must, of course, be the result of alterations in the pattern of the genes, otherwise it could not be inherited. There are two general cate-

Fig. 33-13 In a crossover, the homologous chromosomes become entwined. When they separate, a new arrangement of genes results. (From Altenburg. Genetics, Holt, 1957)

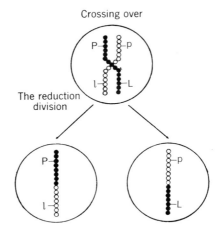

gories of mutations based on the degree of change: *chromosome mutations* and *gene mutations*. Changes in chromosome number, for example polyploid plants whose cells contain several times the diploid number of chromosomes, and changes to large portions of the chromosomes involving loss, duplication, and alteration in the order and kind of large numbers of genes are all considered chromosomal mutations. One example is known as deletion. The activity of the cytoplasm during meiosis may cause two breaks in a chromosome. The broken ends may then reunite excluding a segment of the chromosome and the accompanying genes. When a deficient chromosome such as this pairs with a normal chromosome, an unpaired loop of normal chromosome results corresponding to the deleted segment of the deficient chromosome. (See Fig. 33-14.)

Gene mutations are changes in a single gene. White eye color in the fruit fly *Drosophila*, taillessness in Manx cats and albinism in man are examples. These mutations are due to a localized change in the coded information of the DNA molecule (the chromatin). In the case of *Drosophila*, the white-eyed mutant form is unable to synthesize the red and brown eye pigments.

Most mutations are harmful but some are beneficial. A beneficial mutation increases the chances of survival and breeding success of the bearer by rendering him better adapted to his surroundings. It follows that a beneficial mutant gene will quickly spread through a population by the normal processes of breeding and Mendelian inheritance. Harmful mutations on the other hand,

Fig. 33-14 Portion of a giant chromosome pair showing the region containing a deletion. Note the unpaired loop of the right-hand chromosome of the pair corresponding to the portion deleted from the left-hand chromosome.

due to their adverse effects on survival and breeding success, do not spread. Evolution, both animal and human, has taken place through the occurrence and subsequent inheritance and spread of beneficial mutations.

The question of the effects of nuclear radiations and other penetrating radiations (X rays) on the occurrence of mutations has been the subject of much speculation during recent years. Since 1946, geneticists have been studying the effects of the nuclear bombs dropped on Hiroshima and Nagasaki. They have attempted to determine whether there has been any change in the sex ratio of children born to survivors of the bombings, the birth weight of children, their size at the age of nine months, and the rate of death during the first nine months. Thus far, little information of a really conclusive nature has been obtained. This might be expected because of the shortness of the time that has elapsed since the bombings. Whether future generations will show an increase in mutations is, of course, a question that cannot be answered now.

Experiments on fruit flies, mice, and various types of plants have been undertaken to determine the effects of radiation on germ cells. Many of these have indicated that radioactive materials and powerful X rays can produce changes in the less stable regions of the DNA molecule with the overall result that the rate of naturally occurring mutations is increased. In high doses these rays cause such large changes that sterility or unviable offspring result. In the case of nuclear radiations, the effects of radiation decrease with distance from the center of the blast area so that the chances of a survivor of an atomic attack receiving a harmful dosage are also reduced by distance. The preliminary studies from Hiroshima and Nagasaki indicate that the human genes are not unusually sensitive to atomic radiations, nor do they have any special ability to resist them. Studies of the effects of X-ray radiations on human heredity are not conclusive. Because of the possibility that such radiations may cause deleterious mutations, health authorities are now cautioning us against the indiscriminate use of X rays, particularly in the region of the sex organs. Physicians and dentists are advised to be more careful than they formerly were in shielding both themselves and their patients from excessive exposure to X rays and to avoid "routine" use of X rays where they are not necessary.

VOCABULARY REVIEW

Match the phrase in the left column with the correct word in the right column. *Do not write in this book.*

1 A structural element on a chromosome that determines a specific characteristic.	a diploid number
2 A highly complex chemical substance within the nucleus that controls the cell's activities.	b DN
	c gene
	d haploid number
	e heterozygous

3 The number of chromosomes present in a muscle cell.

4 The name given to the two final cell divisions in the production of the mature gamete.

5 A characteristic that is transmitted by the X-chromosome.

6 Another name for a hybrid.

7 A small cell that is produced and then discarded during oogenesis.

8 Pairs of similar chromosomes.

9 A term used to refer to the cells of the body, exclusive of the germ cells.

f homologous chromosomes
g homozygous
h meiosis
i mutation
j polar body
k sex-linked
l somatic

TEST YOUR KNOWLEDGE

Group A

Select the correct statement. *Do not write in this book.*

1 The sex of an individual is determined at the time of (a) spermatogenesis (b) differentiation (c) fertilization (d) oogenesis.

2 The number of chromosomes in a gamete is called (a) haploid (b) diploid (c) triploid (d) tetraploid.

3 The material in the cell nucleus that determines hereditary traits is the (a) chromatin (b) centrosome (c) mitochondria (d) Golgi body.

4 Fertilization occurs when the sperm unites with the (a) ovary (b) ovum (c) zygote (d) polar body.

5 The formation of one type of gamete is called (a) meiosis (b) mitosis (c) spermatogenesis (d) parthenogenesis.

6 A cell that contains the haploid number of chromosomes is the (a) spermatocyte (b) oocyte (c) spermatogonium (d) ootid.

7 Members of a pair of similar chromosomes are said to be (a) analogous (b) homologous (c) homogeneous (d) heterogeneous.

8 Mendel did most of his work on heredity using (a) fruit flies (b) mice (c) garden peas (d) four o'clocks.

9 The loss of a gene or group of genes from a chromosome is called (a) segregation (b) deletion (c) omission (d) blending.

10 A new hereditary characteristic that appears in a species of plant or animal is called a (a) dominant trait (b) recessive trait (c) mutation (d) phenotype.

Group B

1 In man, brown eyes are dominant over blue eyes. If two brown-eyed parents have a blue-eyed child, what are the genotypes of the parents?
2 Human heredity appears to follow the same laws that govern inheritance in animals and plants. Explain and illustrate with specific examples.
3 Why may plants that look alike produce offspring that are quite unlike each other or their parents?
4 Why is a knowledge of the laws of inheritance necessary for the identification of mutations?
5 Explain why the following statement is either true or false. "One-quarter of the sons of a man who lost his right foot in an automobile accident will be born with a right foot."

Chapter 34 Nucleic Acids. Human Genetics

THE NUCLEIC ACIDS

Deoxyribose Nucleic Acid

The chemical basis for the inheritance of characteristics is an extraordinary substance, *deoxyribonucleic acid (DNA)*. This substance is the principal constituent of the chromosomes and belongs to a group of chemical compounds known as *nucleic acids*. The chemical structure of these acids is not as complex as that of most proteins. Nucleic acids have relatively high molecular weights, between 500,000 and 2,000,000. This indicates not only their chemical complexity, but also their variation in composition.

DNA contains deoxyribose, a derivative of the 5-carbon sugar, ribose. The structure of both these sugars is shown in Fig. 34-1. In the molecule of DNA,

Fig. 34-1   Structural formulas for the chemical components of DNA and RNA. Above are the two sugars and the phosphate group and below are the five nitrogen containing bases.

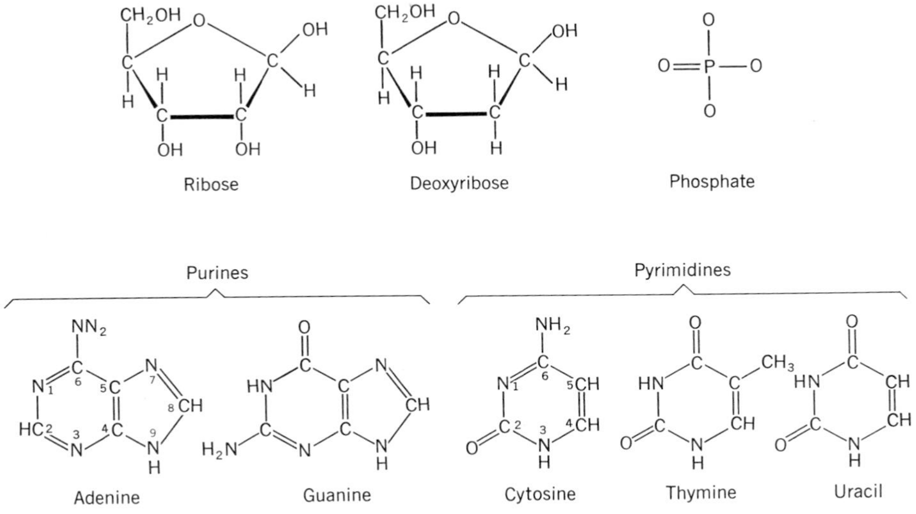

Ribose   Deoxyribose   Phosphate

Purines   Pyrimidines

Adenine   Guanine   Cytosine   Thymine   Uracil

459

deoxyribose is combined with phosphate groups and molecules of four different nitrogen containing bases: *adenine* (A), *cytosine* (C), *guanine* (G), and *thymine* (T). Adenine and guanine are purine bases while cytosine and thymine are pyrimidine bases. The structural formulas are shown for these substances in Fig. 34-1.

Studies made by F. H. C. Crick and J. D. Watson indicate that the DNA molecule is composed of two long chains of base-sugar-phosphate subunits called *nucleotides*. (See Fig. 34-2.) These apparently are not straight chains, but exist in the form of a double spiral sometimes called a *double helix*. The alternating deoxyribose and phosphate groups form two continuous helices, while the bases attached to the deoxyribose groups link the helices by each pairing up with the complimentary base on the opposite chain.

To visualize this structure, think of a straight ladder made of flexible materials. In this ladder the long rails represent the deoxyribose and phosphate groups joined end-to-end, and the rungs of the ladder represent the paired bases, as in Fig. 34-3. Now, if the ladder were twisted, the long rails would form a double helix with the rungs still joining them, as in Fig. 34-4. This analogy is oversimplified, but it serves as a working model of the general structure of DNA molecules.

The four bases are always paired in such a manner that adenine joins with thymine and guanine joins with cytosine. In Fig. 34-4 we see a representation of only a small portion of a DNA molecule with only one possible sequence of the paired bases. Since the sequence of pairs can be varied indefinitely in any double helix, a great many different DNA molecules are possible.

Fig. 34-2   The structure of one nucleotide. Both DNA and RNA helices are composed of nucleotide subunits.

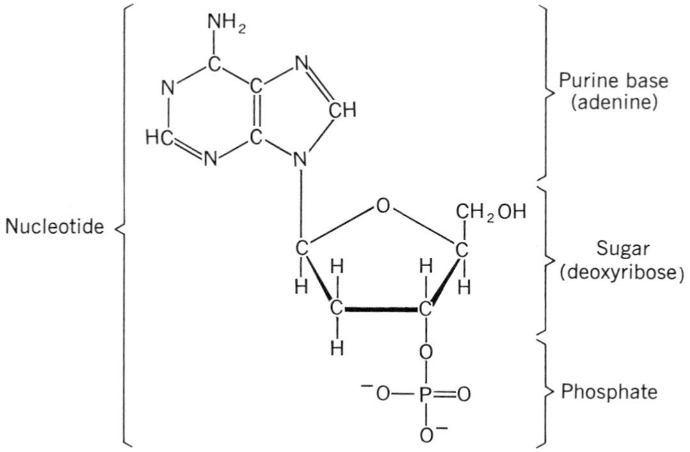

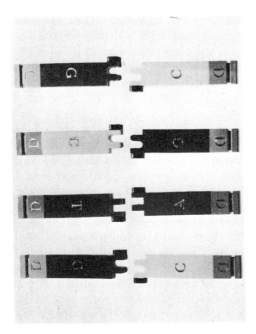

Fig. 34-3 The DNA bases with attached sugar groups represented by elements of a model. Bases: C, cytosine; G, guanine; A, adenine; T, thymine. Sugar: D, deoxyribose. Note that each base is laid opposite the base with which it pairs exclusively. (Fundamental Photographs)

Morse code, only two symbols are used, and no combination of these ever exceeds four units. If however, there are combinations of the symbols (A, C, G, T) in an infinite variety of sequences, it is evident that all of the many individual characteristics that distinguish one person from another can be obtained. The sequence on the DNA molecule of the four symbols of the DNA code determines the nature and order of the amino acids in the protein molecules synthesized by the cell. It is in fact by determining protein

Fig. 34-4 Model of portion of the DNA molecule. The base-sugar elements shown in Fig. 34-3 have been assembled here to form the characteristic double helix. The straps and their attachments to the base-sugar elements represent the phosphate groups. (Fundamental photographs)

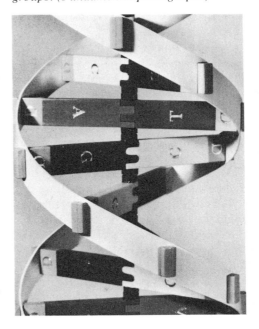

If we also consider that the total number of nucleotides in any helix may differ, it is apparent that there is an almost unlimited number of possible variations in DNA molecules.

Recent work on the structure of the DNA molecule indicates that the bases in the molecule are the real determinants of the hereditary characteristics of animals and plants. Their linear order in the molecule of DNA can be likened to the dash-and-dot system of the Morse telegraphic code. In the

synthesis that DNA controls heredity. In the different cells of the body at different stages in the life cycle different portions of the DNA are actively promoting protein synthesis. This is basically what gives cells their peculiar characteristic.

Occasionally mistakes occur in the sequence of the bases in the DNA molecule, just as they do in the transmission of a telegram. If the analogy of the telegraph code is continued, we can liken errors in inheritance to those made by a telegrapher who would transmit the word *live* as *like,* or by losing a letter make *live* read *lie.* The latter would be an example of deletion. In the same manner, the bases may occasionally become rearranged to form nonsense words. These may result in mutations, some of which might have very serious effects on the individual who receives them. Furthermore, since these changes become coded in the DNA, they are then transmitted to future generations.

Another unusual feature of the DNA molecule is that it can serve as a *template,* or mold, for another exactly similar part. As has been pointed out in Chapter 3, following mitosis each chromosome duplicates itself. Thus, before going into the interphase stage, it makes an exact copy of itself, in a process known as *replication.* The evidence suggests that, in replicating, the DNA helix comprising the chromosome unwinds and the bonds linking the base pairs together are broken splitting the DNA molecule into two mirror image chains of nucleotides. Each single chain builds a complimentary chain of nucleotides on to itself. complimentary bases linking up as in the original molecule. The result is two identical DNA helices in place of one, and each one identical to the original double helix. The mechanism of this process is represented in Fig. 34-5.

Ribonucleic Acid (RNA)

To the biochemist, a molecule of a protein is composed of long chains of amino acids held together by chemical bonds. In an average protein there may be between five hundred and one thousand amino acid units in the peptid chain, while in more complex proteins there may be several chains linked together. The primary difference between one protein and another is the sequence of the individual amino acids in the chain. As will be pointed out later in this chapter, any difference in the order of the amino acids can result in profound changes in the chemical nature of the protein. As previously mentioned, the order of the amino acid units is determined by the DNA of the nucleus. Since this material does not enter the cytoplasm where the proteins are formed, there needs to be some method of conveying the correct information to the cytoplasm.

Within the nucleus, another substance is formed that is known as ribonucleic acid (RNA). This differs from DNA in two important respects. In the first place, the desoxyribose is replaced by the 5-carbon sugar, *ribose,* and the pyrimidine base, thymine, is replaced by *uracil.* Also RNA is not self replicating.

Within the nucleus of a cell, the DNA is presumably in a very loose spiral form except during the time of cell division. When in this uncoiled

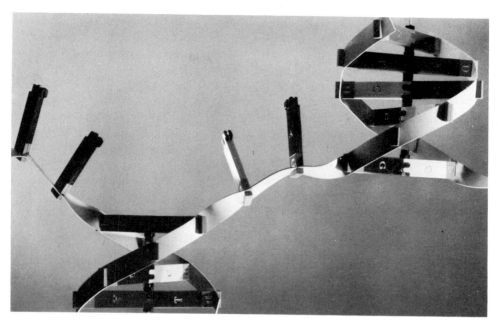

Fig. 34-5   Replication of DNA. A possible mechanism is demonstrated by the model. Bottom left: the double helix is unwinding from above downwards. Top right: a new complimentary helix is being built from top downward on to one half of the original double helix. The same is happening outside the photograph to the left. The result is two identical double helices where before there was only one. (Fundamental Photographs)

condition, it behaves like a template, or mold. The four bases become associated with the complimentary bases of the DNA that control the composition of a definite protein. These RNA bases then become part of a helical double strand of nucleotides, which can be seen under an electron microscope, and then leave the nucleus. The RNA strands have the order of the amino acids as determined by the DNA code imprinted on them also in the form of a code the letters of which are the bases adenine (A), guanine (G), cytosine (C), and uracil (U). Some very careful re-search work has shown that specific combination of only three of these bases is necessary to determine a particular amino acid. In other words, the RNA carries a triplet code. For example, the amino acid tyrosine is coded as A–U–U, tryptophan as G–G–U, aspartic acid as G–U–A, and so forth. Some of the amino acids are evidently coded by several different combinations of the letters. Thus we find that the amino acid serine has a code of U–C–U, or U–C–C, or U–C–G, while asparagine is coded as A–C–A or A–U–A. These few examples have been cited simply to give some indication of the extreme complexity of processes occurring within a cell.

The RNA that has been coded by the DNA is known as messenger RNA. This RNA leaves the nucleus and enters the channels of the endoplasmic reticulum (Chapter 3). Along the walls of these are arranged the ribosomes which are composed of an RNA-protein complex that has a very high molecular weight and is chemically quite stable. This RNA has also been shown to originate from the DNA template as also has the transfer RNA present in the cytoplasm. The different roles of these three RNA's are discussed in connection with cell metabolism in Chapter 3.

HUMAN GENETICS

The inheritance of characteristics by human beings follows the same general patterns that have been established for pea plants or fruit flies. The basic laws are applicable to all organisms, and it is simply a question of time before it will be possible to give answers to some of the problems that are currently plaguing students of human inheritance. It is easy to control the mating of plants or lower animals, but human matings cannot be controlled in the same manner. Therefore, the understanding of human inheritance presents special problems which can be solved only slowly. Solutions are usually based on a statistical study of a characteristic in a given population. This population must have a fairly uniform background, generally of a racial nature. After conclusions of a rather tentative nature have been drawn, either chemical or environmental studies must be undertaken before any definite judgment can be reached. This method is time-consuming, and results are usually incomplete, but some progress is being made and the future development of this field of genetics promises much more.

Blood Types

One human characteristic that has received serious study is the inheritance of blood types. In 1900, Dr. Carl Landsteiner of the Rockefeller Institute accurately described the principal blood types, designated as A, B, AB, and O. Their characteristics were discussed on page 318.

The gene which determines the inheritance of the agglutinogen on the erythrocytes has been called "L" in honor of Dr. Landsteiner. A conventional method of writing the symbol has been adopted so that L^A represents Type A blood, L^B is used for Type B individuals, and L^{AB} for type AB. Since type O does not contain any agglutinogen and is recessive to both A and B, it is designated as ll. In the table, the relation of the various groups that can be easily identified is shown with the genotypes that can produce them.

It will be easily realized that since L^A and L^B are both dominant to type O (ll), the heterozygotes of types L^Al and L^Bl cannot be easily distinguished from the homozygotes, L^AL^A and L^BL^B. Also, neither L^A nor L^B is dominant with respect to the other. The hererozygote, L^AL^B, can be identified as the blood type, AB, and has some charac-

Blood Group	Cell Agglu- tinogen	Serum Agglu- tinin	Genotype
A	A	b	$L^A L^A$ or $L^A l$.
B	B	a	$L^B L^B$ or $L^B l$.
AB	A and B	none	$L^A L^B$
O	none	a and b	ll

teristics of both the homozygotes. Here is an example of the inheritance pattern mentioned in the preceding chapter as incomplete dominance.

The results of this pattern are obvious. An individual of group *O (ll)* could be the result of the mating of $L^A l \times L^A l$, $L^B l \times L^B l$, $L^A l \times L^B l$, or two *O* type parents *(ll)*. He could not, however, have one *AB* type parent. As a result of an analysis of this type, it is sometimes possible to determine the parentage of children whose identity may be in doubt because of mixups in hospital procedures.

Rh Factor

Another hereditary factor which is present in the serum of the blood is the Rh factor. Early investigation seemed to indicate that the inheritance of this factor was a case of simple Mendelian dominance. With the refinement of techniques, it became obvious that the original methods of identifying the factor were not sufficiently exact. It is now known that there are several classes of the Rh factor, each evidently determined by a separate gene. From the behavior of the Rh factor in its inheritance pattern, the idea has arisen that the genes governing the various classes are arranged in clusters on the chromosomes. While one gene does not necessarily control the action of a neighboring gene, it does confuse the picture of the exact method of inheritance of the factor. It is quite possible that this clustering of slightly different genes around a central point explains why the number of cases of erythroblastosis fetalis is somewhat less than the theoretical expectation mentioned on page 447.

There are other inherited factors in the serum which can be identified by appropriate tests. Two of these, *M* and *N,* are quite well known. To identify these, however, requires the injection of human serum into a rabbit to force it to form agglutinins against the human proteins. These are then taken from the rabbit and subjected to rather complicated chemical procedures. The serum that is produced in this manner can then be used to distinguish between *M* and *N* human individuals. A test for these groups has relatively little significance for the average person. It is, however, very useful in determining the correct parentage of a child when used in combination with other blood-type tests.

Abnormal Hemoglobins

In Chapter 22 there is a brief mention of an abnormal type of hemoglobin which results in sickle cell anemia. Under the microscope, the blood of a person suffering from this disease shows some red corpuscles that are curved so that they resemble the blade of a sickle. If the blood of a person showing this characteristic is kept in a tightly sealed bottle for several days, most of the cells will assume this shape. A normal person's cells kept under the same condition will not show the sickling. It has been known for some time that this is an inherited condition, but the chemistry of the abnormal hemoglobin has only been discovered recently. This knowledge offers new evidence that the precise sequence of the amino acids in a protein molecule is controlled by the genes.

In normal hemoglobin, the molecule is composed of chains of peptids. Each chain consists of eight amino acids that are always found in the same order. The amino acids are *valine* (Val), *histidine* (His), *leucine* (Leu), *threonine* (Thr), *proline* (Pro), *glutamic acid* (Glu), and *lysine* (Lys), arranged as follows:

$$+ \quad + \quad \quad \quad - \quad - \quad +$$
—Val–His–Leu–Thr–Pro–Glu–Glu–Lys—

Some of these have positive electrical charges, others have negative charges, while some of the acids are electrically neutral.

In sickle cell anemia, there is a change in the sequence of the amino acids in the peptid chain in one place. The first *glutamic acid* molecule is replaced by a molecule of the amino acid, *valine*, resulting in a peptid chain of the following type:

$$+ \quad + \quad \quad \quad + \quad - \quad +$$
—Val–His–Leu–Thr–Pro–Val–Glu–Lys—

As a result of the difference in the electric charge of valine as compared with that of the original glutamic acid, the molecule of hemoglobin becomes elongated and relatively inflexible. This rigid shape then distorts the shape of the corpuscle and results in sickling.

The inheritance of this characteristic is due to the presence of a gene, S. If a person is heterozygous for this condition (Ss), he does not usually show any severe anemia, and the number of corpuscles that have a sickle-shape is relatively small. If however, he is homozygous for the condition (SS), many of the erythrocytes show sickling and he rarely reaches maturity because of the effects of the disease. Thus the homozygous condition may be considered to be lethal, while those with the heterozygous condition remain to continue the trait.

An interesting side-light on the condition of sickling has been found by a statistical study of the populations in which it occurs. It has been observed that peoples showing this trait are usually found in areas where the disease of malaria is common. "Sicklers" are seldom found living in malaria-free districts. In a study made of 288 children in a malaria-infested district of Africa, 43 (14.9 percent) were found to be heterozygous for sickling. Of these, only 12 (28 percent) showed evidence of infection by the malaria parasite. Of the remaining 245 children who lacked the S gene, 113 (46 percent) had contracted malaria. Although evidence of this type is too incomplete at present to serve as definite proof of the relation of the S gene to immunity to malaria, it suggests a possible explanation of why some racial groups are more likely to survive in a region that proves fatal to others.

From the standpoint of genetics, the human organism is highly complex and individual. The study of human heredity is further complicated by the undirected matings of people. Any study of human inheritance must, therefore, be slow. Geneticists frequently speak of the "tendency" to inherit various traits. This is a conservative way of saying that there is some evidence that a characteristic may be inherited, but there is as yet no definite proof of the dominance or recessiveness of the characteristic.

At present there appears to be a tendency for the inheritance of *diabetes mellitus*, but the exact type of transmission is not definitely known. Is it dominant, recessive, sex-linked, or incompletely dominant? There are many views on this. The same may be said for several types of cancer, as for example, cancer of the breast. For many years strains of mice have been bred which transmit cancer of the breast (mammary glands) as a dominant characteristic. Other strains show a high resistance to this trait. Among the human population, the current data appear to follow this observation, but heredity of this type of cancer is still considered as a "tendency." Other conditions are better known. *Huntington's chorea*, a nervous disorder which appears in middle-life; *polydactyly*, the presence of extra fingers or toes; and *syndactyly*, the webbing of fingers and/or toes, are all recognized defects of man, and their dominance or recessiveness is known.

A few of the human characteristics for which the inheritance pattern has been determined are shown below.

Dominant	Recessive
Black hair	All other colors
Dark brown hair	Light browns and reds
Curly hair	Straight hair
Brown eyes	Blue eyes
Astigmatism	Normal vision
Double-jointedness	Normal condition of joints
Normal pigmentation	Albinism
Cleft palate	Normal palate
Lack of musical ability	Musical ability
Normal hearing	Deaf mutism
Tongue-curling	Lack of ability to curl tongue
Free ear lobes	Attached ear lobes
Ability to taste PTC (phenyl-thiocarbamide)	Inability to taste PTC

Sex-linked Characteristics

Broad terminal joints of the digits
Red-green color blindness
Hemophilia
White forelock of hair
Baldness
Icthyosis simplex (scaly skin)

Mutations

Mutations occur in members of the human race just as they do in other living organisms. Like those in lower

9 In a normal human, the number of chromosomes in a cell is (a) always an odd number (b) usually 46 (c) twice as great in a sperm as in a muscle cell (d) half as great in a sperm as in an egg.

10 The genotype of an individual (a) is always the same as the phenotype (b) refers to his visible traits (c) refers to the composition of his chromosomes (d) is indicated by the presence of either an X or Y chromosome.

Group B

1 Why may two parents that have A-type blood have a child with O-type blood?

2 What is the general shape of a DNA molecule? What groups of chemical compounds are present in it?

3 Briefly explain what occurs during the replication of a chromosome.

4 How do the abnormal hemoglobins illustrate the ability of DNA to determine characteristics that are based on minute variations?

5 (a) Explain what you understand by "the genetic code" in relation to the formation of RNA. (b) Name three differences between RNA and DNA.

APPENDIX

Units of Measurement

1 Ångstrom unit (Å) = 3.937×10^{-9} inch
             = 1×10^{-10} meter
             = 1×10^{-4} micron
1 Atmosphere = 33.899 feet of water at 39.1° Fahrenheit
             = 760 millimeters of mercury at 0° Centigrade
             = 14.696 pounds per square inch
             = 10,333 newtons
1 centimeter (cm) = 0.3937 inch
1 gram (gm) = 0.03527 ounces (avoirdupois)
1 kilogram (kg) = 1000 grams = 10^6 milligrams
             = 2.2046 pounds (avoirdupois)
1 kilometer (km) = 0.62137 miles
1 liter (l) = 1.056710 U.S. quarts (fluid)
             = 1000 milliliters
1 meter (m) = 0.001 kilometers = 100 centimeters = 1000 millimeters
             = 1×10^{10} Ångstrom units
             = 3.2808 feet = 39.37 inches
1 micron (μ) = 10^{-6} meters
             = 10^4 Ångstrom units
             = 3.937×10^{-5} inches
1 millimeter (mm) = 0.03937 inch
1 milligram (mg) = 3.352739 ounces (avoirdupois)
1 milliliter (ml) = 0.0338147 U.S. ounces (fluid)
1 newton = the force required to accelerate one kilogram of mass one meter per second per second
1 U.S. foot = 30.48 centimeters exactly
1 U.S. inch (in) = 2.54 centimeters exactly
             = 2.54×10^8 Ångstrom units
1 ounce (avoirdupois) = 28.349527 grams
1 pound (avoirdupois) = 453.592 grams
1 U.S. quart (fluid) = 0.946333 liter

Temperature Conversion Factors

1° Fahrenheit (° F) = 0.556° Centigrade (° C)
1° Centigrade = 1.8° Fahrenheit
0° Kelvin = $-273°$ C = $-459.4°$ F = Absolute Zero

To convert Fahrenheit or Centigrade readings: $\dfrac{°\,C}{°\,F - 32} = \dfrac{5}{9}$

GLOSSARY

Abdomen (ab-*doh*-men). The lower part of the trunk below the level of the ninth pair of ribs. It contains the major parts of the digestive and excretory systems as well as many endocrine glands.

Abduction. Withdrawal of a part of the body from the axis of the body, or of an extremity.

Absorption. The passage of materials through a cell membrane by either diffusion or osmosis.

Acetylcholine (as-eh-til-*koh*-leen). A substance produced at the junction of nerve and muscle fibers which results in the contraction of the muscle.

Actin (*ak*-tin). One of the two proteins concerned in muscle contraction. See myosin.

Action current. The electric current accompanying the activity of any reacting tissue.

Adduction. Any movement resulting in one part of the body or a limb being brought toward another or the median line of the body.

Adenosinediphosphate (*ADP*). A compound of one molecule of adenine, one of ribose, and two of phosphoric acid.

Adenosinetriphosphate (*ATP*). A compound consisting of one molecule of adenine, one of ribose, and three of phosphoric acid.

Adipose (*ad*-ih-pohz) *tissue*. A type of tissue in which each cell contains a single large vacuole filled with oil.

Afferent (*af*-er-ent). The passage of a nervous impulse toward the CNS or blood toward the heart.

Afterimage (*af*-ter-im-ij). The retention of a visual image on the retina of the eye after the light stimulus has been cut off.

Agglutination (ah-gloo-tih-*nay*-shun). The clumping together of cells, caused by the chemical action of agglutinins on their surfaces.

Agglutinin (ah-*gloo*-tih-nin). A chemical substance which affects the outer layer of a cell's membrane and makes it sticky.

Agglutinogen (ag-loo-*tin*-oh-jen). A chemical substance in the cell's membrane toward which the agglutinin is antagonistic.

Albumin (al-*boo*-min). A relatively simple type of protein.

Albuminuria (al-boo-mih-*noo*-ree-a). The excretion of albumin in the urine.

All-or-none. Occurring completely or not at all, as the response of a nerve or a muscle bundle to a stimulus.

Amino acid. A chemical compound that contains both the amine (NH_2) and the organic acid (COOH) radicals.

Anabolism (a-*nab*-oh-liz'm). Those chemical processes occurring during the life of a cell that result in the formation of new protoplasm or the replacement of old.

Anaphase (*an*-a-fayz). A stage in mitosis that is characterized by the migration of chromosome halves to opposite poles of the cell.

Anatomy. The branch of the biological sciences dealing with visible body structures.

Anemia. A condition resulting from a lowering of the amount of hemoglobin in the body.

Anoxia (an-*ok*-see-a). The condition arising from a lack of oxygen.

473

Antagonistic action of muscles. The opposing action of the members of a pair of skeletal muscles.

Anterior. Referring to the front of the human body. Also at or toward the head end of an animal.

Antibodies. Substances which oppose specific foreign bodies.

Antigen. A substance, foreign to the body, which stimulates the formation of a specific antibody.

Anvil. See *incus.*

Appendage. Any structure having one end attached to the main part of the body and the other end more or less free.

Appendicular skeleton. The skeletal structures composing and supporting the appendages. These include the shoulder and hip girdles as well as the arms and legs.

Arachnoid (a-*rak*-noid). The middle layer of the tissues covering the brain and spinal cord.

Arteriole (ahr-*tee*-ree-ohl). A fine branch of an artery.

Artery. A muscular blood vessel carrying blood away from the heart.

Arthritis (ahr-*thry*-tis). An inflammation of the joints, sometimes resulting in fusion of the bones.

Articulation. 1, the enunciation of speech. 2, a joint; the junction of two or more bones.

Asphyxia (as-*fik*-see-a). A condition whereby the oxygen in the body is reduced and the carbon dioxide is increased, leading to suffocation.

Assimilation. The formation of new protoplasm from nonliving materials within cells.

Aster. A series of cytoplasmic fibers which radiate from the centriole during mitosis.

Astigmatism (a-*stig*-ma-tiz'm). A defect of vision caused by abnormal curvature of the surface of either the cornea or the lens.

Atom. The smallest unit of matter.

Atrio– (*ay*-tree-oh–). A prefix referring to the atrium of the heart.

Atrioventricular (ay-tree-oh-ven-*trik*-yoo-lar) *bundle.* A bundle of nerve fibers that transmits impulses from the atria to the ventricles.

Atrium (*ay*-tree-um). A chamber of the heart that receives blood.

Atrophy (*at*-roh-fee). A decrease in size of some organ or structure of the body.

Auricle (*aw*-rih-k'l). 1, the pinna of the ear; the flap of cartilage and skin on the outer ear. 2, an atrium of the heart.

Autonomic (aw-toh-*nom*-ik) *nervous system.* A portion of the nervous system formed of two chains of ganglia lying outside of the central nervous system but connected to it by branches (rami).

Axial (*ak*-see-al) *skeleton.* The central, supporting portion of the skeleton composed of the skull, vertebral column, ribs, and breast bone.

Axon (*ak*-son). That portion of the neuron that carries impulses away from the cell body.

Back cross. A cross between either a homozygous dominant or a heterozygous dominant of the F_2 generation and a homozygous recessive of the P generation.

Bainbridge reflex. Acceleration of the heart and a rise in arterial pressure resulting from a rise in pressure in the right atrium and the roots of the great veins.

Barometer (ba-*rom*-uh-ter). A device for measuring air pressure.

Basal (*bays*-al) *metabolic* (met-a-*bol*-ik) *rate.* The basic rate of the body's metabolism.

Basophil (*bay*-soh-fil). A type of white blood corpuscle in which the cytoplasmic granules stain blue with methylene blue.

Bel. A unit used to measure the intensity of sound.

Bilateral symmetry. A type of body structure in which bisection through the vertebral column and the breast bone results in two halves which are mirror images of each other.

Binocular vision. The ability to see an object with both eyes simultaneously.

Biochemistry. The chemistry of the tissues; physiological chemistry.

Biological sciences. The sciences that deal with living organisms.

Biophysics. A study of life processes by the methods and instrumentation of physics.

Blending. See *incomplete dominance.*

Blind spot. The region of the retina that is

insensitive to light because it has no rods and cones.

Blood. The circulating tissue of the body.

Blood platelets (*playt*-lets). The smallest of the blood solids, which aid in the formation of a blood clot.

Blood pressure. The pressure exerted by flowing blood on the walls of an artery.

Bolus (*boh*-lus). A ball of food which passes along the digestive canal.

Bone. A relatively hard, porous substance containing a high percentage of calcium and phosphorous salts. It forms the major part of the skeleton.

Bone marrow. The fatty tissue lying within the central cavity of many bones. See *red bone marrow* and *yellow bone marrow.*

Boyle's Law. The gas law which deals with the relation of pressure to the volume of a gas.

Brain. The part of the central nervous system contained within the cranium of the skull.

Brain stem. Structurally, the lowest part of the brain. It is composed of the pons, midbrain (thalamus and hypothalamus) and the medulla oblongata.

Brightness. The amount of light emitted by a luminous body.

Brownian movement. Movement of small particles which results from their bombardment by smaller invisible particles.

Buffering. An action made by any substance in a fluid tending to decrease the change in hydrogen ion concentration.

Bursa (*bur*-sa). A sac of fluid between the bones of a joint.

Bursitis (bur-*sy*-tis). An inflammation of a bursa, frequently accompanied by the formation of calcium salts in the cavity.

Callus. Any thickening of a tissue resulting from the normal growth of cells.

Calorie (*kal*-oh-ree). The amount of heat required to raise the temperature of 1,000 grams of water one degree Centigrade.

Calorimeter (kal-oh-*rim*-uh-ter). A device to measure heat production.

Capillary (*kap*-uh-ler-ee). A minute blood vessel that arises from an artery and joins with other small vessels to form a vein.

Carbohydrate (kahr-boh-*hy*-drayt). A chemical compound composed of carbon, hydrogen, and oxygen, with hydrogen and oxygen in the ratio of 2:1.

Cardiac (*kahr*-dee-ak) *muscle.* A branching, lightly striated type of muscle found only in the heart. It contracts rhythmically.

Cartilage (*kahr*-tih-lij). A flexible supporting tissue composed of a nonliving matrix within which are the living cartilage cells.

Catabolism (ka-*tab*-oh-liz'm). Those processes occurring with living cells that result in the breaking down of protoplasm and the release of energy.

Cell. The unit of structure and function in a living body.

Cell membrane. A membrane surrounding a cell. It is highly selective in controlling the passage of materials into and out of the cell.

Cell theory. All living things are composed of cells and the products of cells.

Central nervous system. The brain and spinal cord.

Centriole (*sen*-tree-ohl). Two small bodies lying within the centrosome, which can be stained by hemotoxylin.

Centromere (*sen*-troh-meer). A small centrally located body which attaches the two chromosome halves to each other.

Centrosome (*sen*-troh-some). A granular area lying near the nucleus of an animal cell, containing the centrioles.

Cephalin (*sef*-ah-lin). A chemical produced by blood platelets.

Cerebellum (*ser*-uh-bel-um). The region of the brain that coordinates muscular activities.

Cerebral (*ser*-uh-bral) *hemisphere.* One of the two principal divisions of the cerebrum.

Cerebrum (*ser*-uh-brum). The largest of the three principal parts of the brain.

Cerumen (suh-*roo*-men). The waxy secretion formed by glands in the external acoustic meatus.

Cervical (*sur*-vih-kal). Referring to the neck region.

Charles' Law. The gas law dealing with the relation between the volume of a gas and its temperature.

Chemical compound. A substance composed of two or more unlike atoms which are chemically united in definite percentage.

Chemical formula. A method of stating the composition of a substance by the use of chemical symbols and numbers.

Cholinesterase (koh-lin-*es*-ter-ayz). A substance produced at the junction of a nerve and muscle fiber which neutralizes the action of acetylcholine.

Choroid (*koh*-roid) *layer*. A pigmented, highly vascular layer of the eye that lies between the retina and the sclera.

Chromatic (kroh-*mat*-ik) *aberration* (ab-er-*ay*-shun). The scattering of light rays at the edges of a lens so that rings of color appear to surround the object that is being viewed.

Chromatid. One of the pair of filamentous spiral threads that make up a chromosome.

Chromatin (*kroh*-ma-tin). The material within the nucleus of a cell that is easily stained by basic dyes.

Chromosome (*kroh*-muh-sohm). One of the rodlike bodies appearing during the mitotic division of a cell.

Chyle (kyl). Lymph containing considerable amounts of emulsified fats.

Chyme (kym). The fluid form of food within the stomach and intestine.

Cilium (*sil*-ee-um). A small, whiplike projection from a cell.

Coccyx (*kok*-siks). The end of the vertebral column beyond the sacrum.

Cochlea (*kok*-lee-a). The spiral bony canal of the inner ear in which are located the auditory receptors.

Coenzyme. A substance associated with an enzyme and activating it.

Cold-blooded. A term used to describe animals which cannot maintain a constant body temperature.

Collagen. The albumin-like substance in connective tissue, cartilage, and bone. It is the source of gelatin.

Colloid (*kol*-oid). Matter in a very finely divided state. Colloidal particles vary in size from large molecules to the threshold of visibility with a high power optical microscope.

Colon (*koh*-luhn). The large intestine.

Cone. A receptor in the retina of the eye that is sensitive to light of high intensity and color.

Conjunctiva (kon-jungk-*ty*-va). A delicate layer of epithelium that covers the outer surface of the cornea and is continued as the lining of the eyelids.

Connective tissue. Tissue that supports and holds together various parts of the body.

Convolution of the brain. The raised area between two sulci.

Cornea (*kawr*-nee-a). The transparent layer in the front of the eye, the outer surface of which is covered by the conjunctiva.

Coronary (*kawr*-uh-ner-ee) *circulation*. The arteries, capillaries, and veins of the heart.

Corpus callosum (kah-*loh*-sum). A band of white matter that connects the two cerebral hemispheres. It is present only in mammals.

Cortex. A general term applied to the outer region of an organ.

Cranial nerve. A nerve that arises from the brain and passes to some region of the body. There are twelve pairs of cranial nerves.

Crossing over. The recombination of genes resulting from the twining of chromosomes during meiosis and their later separation.

Cytology. The branch of the biological sciences that deals with the minute structure of cells.

Cytoplasm (*sy*-toh-plaz'm). The material lying within the cell membrane and outside of the nucleus. It contains both living and nonliving substances.

Dalton's Law of Partial Pressures. The gas law which describes the effect of the pressure exerted individually by each of the gases in the atmosphere.

Deamination. Removal of the amine ($-NH_2$) group from an organic substance, especially an animo acid.

Decibel. A unit used to measure sound intensity. One-tenth of a bel.

Deciduous (dee-*sid*-yoo-us). Describing any structure which is lost during the normal growth period of an animal or plant. It is applied to the milk teeth or the leaves shed by a tree.

Deglutition (deg-loo-*tish*-un). The act of swallowing.

Deletion (duh-*lee*-shun). In genetics: the loss of a part of a chromosome during meiosis.

Dendrite. The nerve cell process that carries impulses toward the cell body. There may be many dendrites on one nerve cell.

Dentine. The hard, bonelike material that forms the major part of a tooth.

Deoxyribonucleic (de-ok-see-rih-boh-*noo*-klee-ik) *acid* (*DNA*). A nucleic acid; basis for inheritance of characteristics.

Detoxification. Removal of the poisonous property from a substance.

Diabetes (dy-a-*bee*-teez) *mellitus* (meh-*ly*-tus). A disease resulting from a failure by the pancreas to produce enough insulin.

Dialysis (dy-*al*-ih-sis). The separation of a colloidal material from other substances through a semipermeable membrane.

Diaphragm (*Dy*-a-fram). A wall of muscle separating the thoracic and abdominal cavities. It is present only in mammals.

Diaphysis (dy-*af*-ih-sis). The main part, or shaft, of a bone.

Diastole (dy-*as*-toh-lee). The period during the beat of the heart when the atria and the ventricles fill with blood.

Diffusion. The movement of molecules of a substance from a region where it is in relatively high concentration to an area of lower concentration.

Diploid (*dip*-loid) *number*. The number of chromosomes present in the somatic (body) cells of an individual. It is usually twice that of the haploid number.

Disaccharide (dy-*sak*-a-ryd). A type of carbohydrate formed by the chemical union of two hexoses.

Dislocation. The forcing of a bone out of its normal position in a joint.

Distal (*dis*-tal). Toward the free end of an appendage. The hand, for example, is distal to the arm.

Division of labor. The division of the activities of the body among specialized organs and tissues.

DNA. See *deoxyribonucleic acid*.

Dominant characteristic. The member of a pair of contrasting characteristics which appears in the phenotype.

Dorsal. Referring to the back of an animal; the side along which the backbone passes; sometimes called the posterior side.

Dura (*doo*-ra) *mater* (*may*-ter). The outermost of the meninges.

Ear drum. See *tympanum*.

Edema (ee-*dee*-ma). Swelling caused by the collection of fluids in a tissue.

Effector. A muscle, gland, or other type of tissue that responds to a definite stimulus.

Efferent. Referring to any nerve or blood vessel that carries an impulse or blood away from the CNS or heart.

Electrocardiogram (ee-lek-troh-*kahr*-dee-oh-gram). A record of the electric currents produced by a beating heart.

Electroencephalogram (ee-lek-troh-en-*sef*-ah-loh-gram). A record of the electric currents produced by the brain.

Electron. A negatively charged particle of an atom.

Elements. The basic forms of matter.

Embryology (em-bree-*ol*-oh-jee). A study of the early stages in the development of the individual. The embryonic period extends up to the time when the individual assumes the general characteristics of the race.

Emulsifying agent. A substance used to assist in forming an emulsion.

Emulsion. The suspension of an insoluble liquid in another liquid in the form of small droplets.

Enamel. The hard, outer layer of a tooth.

Endocrine (*en*-doh-kryn) *gland*. A gland lacking ducts and secreting directly into the blood stream.

Endoplasmic reticulum. A network of fine channels passing through the cytoplasm of a cell and connecting the plasma and nuclear membranes; ribosomes lie along the walls.

Endothelium (en-doh-*thee*-lee-um). A type of tissue resembling squamous epithelium.

Energy. The ability to do work.

Enzyme (*en*-zym). A chemical compound which affects the rate of a chemical reaction.

Eosin (*ee*-oh-sin). A commonly used biological stain that is bright red in color.

Eosinophil (ee-oh-*sin*-oh-fil). A type of white

corpuscle in which the cytoplasmic granules stain deeply with eosin.

Epiphyseal (ep-i-*feez*-ee-uhl) *line.* The line between the epiphysis and the diaphysis (shaft) of a bone.

Epiphysis (ee-*pif*-ih-sis). A bony process attached to another bone by a layer of cartilage.

Epithelium (ep-i-*thee*-lee-um). The type of tissue which covers all body surfaces.

Erythrocyte (ee-*rith*-roh-syt). See *red blood corpuscle.*

Excitability. The readiness of a tissue to respond to a stimulus.

Excretion. The elimination of liquid wastes.

Exocrine (ek-soh-kryn) *gland.* A gland with a duct or ducts.

Extension. The movement of one part of the body away from another.

External acoustic meatus (mee-*ay*-tus). A slightly curved canal in the temporal bone which conducts sound waves from the exterior to the tympanum.

Exteroceptors. A nerve ending in the skin or a mucus membrane that receives stimuli from the external world.

Extrinsic muscles of the eye. The muscles attached to the outer surface of the eyeball.

Farsightedness. See *hyperopia.*

Fasiculus. 1, a bundle of nerve, muscle, or tendon fibers separated by connective tissue. 2, a bundle or tract of nerve fibers presumably having common connections and functions.

Fat. A simple type of lipid composed of one molecule of glycerine and three molecules of fatty acid.

Fatigue. 1, weariness from physical exertion. 2, the inability of nerves and tissues to respond to a stimulus following over-activity.

Fatty acid. A weak organic acid present in lipids.

Feces. Solid intestinal excretions.

Fertilization. The union of egg and sperm nuclei.

Fetus. The unborn offspring of mammals in the later stages of development.

Fibrinogen (fy-*brin*-oh-jen). A plasma protein required for the formation of a blood clot.

Fissure. A deep indentation on the surface of the cerebrum.

Fistula. A narrow tube or canal in the walls of an organ or the body surface resulting from the failure of the region to heal properly following a disease process or a wound.

Flexion. The bending of one part of the body on another.

Focal point. The point at which the rays of light passing through a lens are focused.

Food. Any substance which may supply energy to the body, increase or repair protoplasm, or be stored for future use.

Foramen (foh-*ray*-men). A normal opening in a tissue through which fluids, nerves, or blood vessels pass.

Fovea (*foh*-vee-a) *centralis* (sen-*tray*-lis). The most sensitive region of the retina. It contains only cones.

Frontal (*coronal*) *plane.* An axis which divides the body into anterior and posterior halves.

Galvanometer (gal-va-*nom*-uh-ter). A device used to measure the flow of an electric current.

Gamete (*gam*-eet). A general term applied to either egg or sperm cells.

Gametogenesis (gam-eh-toh-*jen*-ee-sis). The processes involved in the maturation of the gametes.

Gas. A state of matter characterized by an indefinite volume and shape.

Gel. A colloidal system which is a solid or semisolid mass consisting of solid and liquid phases.

Gene (jeen). The determiner of a heritable characteristic.

Genetics (juh-*net*-iks). The study of the inheritance of characteristics.

Genotype (*jen*-oh-typ). The genetic makeup of an individual.

Gland. A specialized group of cells that manufactures materials which are released in the form of secretions.

Globulin (*glob*-yoo-lin). A type of protein.

Glycosuria (gly-koh-*soo*-ree-a). The excretion of sugar in the urine.

Golgi apparatus. Small bodies in the cytoplasm of most animal cells, concerned with formation of secretions.

Gonad. A general term applied to a reproductive organ.

Gray matter. Those parts of the central nervous system having nerve tissue lacking a medullary sheath.

*Gyrus (jy-*rus). See *convolution of the brain.*

Hammer. See *malleus.*

Haploid number. The number of chromosomes present in the mature gametes.

Haversian (ha-*ver*-shan) *canal.* A small canal in bone tissue, occupied by a blood vessel.

Heart muscle. See *cardiac muscle.*

Hematin (*hem*-a-tin). The iron compound that is combined with globin to form hemoglobin.

Hemoglobin (hee-moh-*gloh*-bin). A chemical compound of iron and protein that is responsible for the transport of oxygen around the body.

Hemophilia (hee-moh-*feel*-ee-a). An inherited defect in which the blood does not clot normally.

Hemorrhage (*hem*-o-rij). The loss of blood from the circulatory system.

Henry's Law. The statement of the solubility of gases in a liquid.

Heterozygous (het-er-oh-*zy*-gus). The condition in which an individual possesses both genes for a pair of contrasting characteristics. The hybrid condition.

Hexose (*hek*-sohs). A form of carbohydrate in which there are six carbon atoms.

HHb. See *hemoglobin.*

Histology. The branch of the biological sciences dealing with the microscopic structure of tissues.

Homeostasis. The maintenance of a steady state by the coordinated activities of various organ systems.

Homologous (hoh-*mol*-oh-gus). A term used in describing parts of the body that are similar in structure and origin.

Homoiothermic. A term used to describe warm-blooded animals in which the internal temperature remains more or less uniform and is not affected by the temperature of the surroundings.

Homozygous (hoh-moh-*zy*-gus). Referring to the condition in which an individual possesses only one type of gene for a given characteristic. The pure condition.

Hormone. A chemical messenger produced by an endocrine gland.

Hue. A characteristic of the color of an object; the difference between the color of the object and gray of the same brilliance.

Hydrogen ion concentration. The degree of acidity or alkalinity of a substance measured in terms of the concentration of hydrogen ions.

Hydrolysis (hy-*drol*-ih-sis). A chemical process in which a molecule of water combines with another compound, reducing it to a simpler form.

Hyper–. A prefix denoting an excess of a substance.

Hyperopia (hy-per-*oh*-pee-a) (*farsightedness*). A defect of vision in which the focal point of the lens falls behind the retina due to a shortening of the eyeball.

Hypertrophy (hy-*pur*-troh-fee). An increase in size of some organ or region of the body.

Hypo–. A prefix denoting a subnormal amount of a substance or the lower position of one object in relation to another.

Incomplete dominance. A type of inheritance in which neither of a pair of genes for contrasting characteristics is completely dominant or recessive.

Incus (*ing*-kus) (*anvil*). The middle of the three small bones of the middle ear.

Infrared (in-fra-*red*) *radiations.* The wave lengths of light lying just beyond the red end of the visible spectrum.

Ingestion. The intake of solid or liquid food materials through the mouth.

Inhibit (in-*hib*-it). An action which slows down an activity or prevents it from occurring.

Inorganic. Not organic; a substance which lacks the element carbon in its molecule (carbon dioxide and its compounds are exceptions).

Integument (in-*teg*-yoo-ment). The skin and its appendages.

Interoceptors. Receptors situated in the viscera which receive stimuli connected with digestion, excretion, circulation, etc.

Interphase. The period in the life of a cell when it is not dividing.

Invertebrate (in-*vur*-tuh-brayt). An animal lacking a backbone.

Inverted retina (*ret*-ih-na). The type of

retina in which the rods and cones point toward the choroid layer.

Iodopsin (ey-oh-*dop*-sin). A pigment in cones which is sensitive to color differences.

Ion (*ey*-on). An electrically charged atom, or group of atoms.

Iris (*ey*-ris). The colored portion of the eye.

Irritability (ir-ih-ta-*bil*-ih-tee). The ability of protoplasm to respond to a stimulus.

Islets of Langerhans (*Lahng*-er-hahns). Small groups of isolated cells embedded in the pancreas. They form insulin.

Isotopes (*ey*-soh-tohps). Forms of the same element whose atoms differ in the number of neutrons they contain.

Jaundice (*jawn*-dis). Yellow color of skin due to blockage of bile ducts causing excess bile pigments to accumulate in blood.

Joint. The point of union between two bones.

Karyolymph (*kar*-ee-oh-limf). The fluid within the nucleus of a cell.

Kinetic energy. The energy of motion or activity.

Kyphosis (ky-*foh*-sis). The abnormal curvature of the vertebral column or the sternum in an anterior-posterior plane.

Labyrinth. An intricate system of connecting passages; a maze.

Lacteal (*lak*-tee-uhl). A lymph duct in a villus in which fat accumulates.

Lacuna. A small depression or space.

Lamella. A thin scale or plate.

Larynx (*lar*-ingks). The voice box, or Adam's apple.

Lateral. Pertaining to the side of the body; situated on either side of the median vertical plane.

Lens. An optical device that refracts light so that its rays either diverge (biconcave lens) or converge (biconvex lens).

Leukemia (loo-*kee*-mee-a). A disease of the blood characterized by an abnormally large number of white blood corpuscles.

Leukocyte (*loo*-koh-syt). See *white blood corpuscle.*

Leukocytosis (loo-koh-sy-*toh*-sis). An increase in the number of white blood corpuscles during certain types of infection.

Leukopenia (loo-koh-*pee*-nee-a). A decrease in the normal number of white blood corpuscles due to bone marrow damage or certain infections.

Ligament (*lig*-a-ment). A tough band of connective tissue which connects bones to each other.

Lipids. Chemical compounds usually composed of the elements carbon, hydrogen, and oxygen, in no special relation to each other. The simplest forms are the fats and oils.

Liquid. The state of matter that is characterized by a definite volume but indefinite shape.

Lordosis (lawr-*doh*-sis). An abnormal curvature of the lumbar vertebrae in an anterior direction resulting in a swayback condition.

Lumbar. Referring to the region of the back extending from the thorax to the sacrum.

Lymph (limf). A colorless fluid within the lymphatics, derived from the extracellular fluid.

Lymphocyte (*lim*-foh-syt). A type of white blood corpuscle formed in lymph tissue.

Malleus (*mal*-ee-us). One of the small bones in the middle ear, one end of which is attached to the tympanum and the other to the incus.

Mammals. A group of vertebrates possessing hair, giving birth to living young, and capable of producing milk.

Marrow. The soft tissues within the medullary canals of long bones.

Marsupial. A group of mammals in which the young are born in a very immature state and later develop within the pouch, or marsupium, of the female.

Mass. Quantity of matter.

Mastication (mas-tih-*kay*-shun). The act of chewing food.

Matrix. See *cartilage.*

Matter. Any substance having weight and occupying space. It may be in the form of liquid, solid, or gas.

Medial (*mesial*). Toward the midline of the body as opposed to lateral.

Medulla (muh-*dul*-a). A general term applied to the internal region of a solid organ.

Medulla oblongata (ob-long-*gay*-ta). A part of the brain stem.

Megakaryocyte (meg-a-*kar*-ee-oh-syt). A primitive type of blood cell formed in the red marrow. It breaks up to form blood platelets.

Meiosis (my-*oh*-sis). The type of cell division which occurs during maturation of gametes. It results in the formation of sperm and ova containing the haploid number of chromosomes.

Mendelian (Men-*dee*-lee-an) *laws*. The laws formulated by Gregor Mendel to explain the inheritance of characteristics.

Meninges (mee-*nin*-jeez). The coverings of the brain and spinal cord.

Mesentery (*mes*-en-ter-ee). A fold of the peritoneum that supports parts of the digestive system.

Metabolism (meh-*tab*-oh-liz'm). The chemical processes occurring within cells.

Metaphase (*met*-a-fayz). A stage in mitosis in which the halves of the chromosomes separate.

Metaplasm (*met*-a-plaz'm). Nonliving materials formed by the cytoplasm of a cell.

Micella (mih-*sel*-a). A subdivision of a myofibril.

Midsaggital plane. A plane which passes through the skull and spinal cord thus dividing the body into right and left halves.

Mitochondria (mit-oh-*kon*-dree-a). Small living structures lying within the cytoplasm of a cell, associated with energy transformations.

Mitosis (my-*toh*-sis). The process of cell division which results in the chromatin material of a cell being divided into equal halves.

Mixed gland. A gland possessing both exocrine and endocrine parts; for example, the pancreas.

Mixed nerve. A nerve containing both efferent and afferent fibers.

Molecule. The smallest particle of an element or a compound that can exist free and still show all of the properties of the substance.

Monosaccharide (mon-oh-*sak*-a-ryd). A form of carbohydrate containing a single saccharide group. These may be either pentoses or hexoses.

Mucosa (myoo-*koh*-sa). The mucous membrane lining an organ.

Mucus. A complex protein-carbohydrate compound that is formed by specialized epithelial cells. It lubricates internal body surfaces.

Muscle tissue. A type of tissue that has the ability to contract.

Muscle tone. The constant state of partial contraction of a muscle resulting from a subthreshold flow of nervous impulses to it.

Mutation (myoo-*tay*-shun). A suddenly appearing new characteristic that can be inherited.

Myasthenia (my-as-*thee*-nee-a) *gravis*. A condition of easy fatigability of voluntary muscles thought to be caused by an excess of a chemical which blocks acetylcholine at the myoneural junction.

Myelin (*my*-uh-lin) (*medullary*) *sheath*. A covering of fatty material over a nerve fiber or process.

Myofibril (my-oh-*fy*-bril). A subdivision of a striated muscle fiber.

Myopia (my-*oh*-pee-a) (*nearsightedness*). A defect of vision resulting from an elongation of the eyeball and the failure of the lens to focus the image on the retina.

Myosin (*my*-oh-sin). One of the two proteins concerned in muscle contraction.

Nearsightedness. See *myopia*.

Negative afterimage. The appearance of the image in complementary colors after the stimulus has ceased to act.

Nephron (*nef*-ron). A tubule in the kidneys.

Nerve. A cablelike structure composed of axons, dendrites, or both.

Nerve impulse. A wavelike impulse passing over a nerve cell process that is characterized by both electrical and chemical effects.

Nerve tissue. A specialized tissue that is adapted to respond to a stimulus.

Neural (*noo*-ral). Referring to the nervous system.

Neurofibrils (noo-roh-*fy*-brils). Strands of cytoplasm in the nerve cell body that connect the dendrites and the axons.

Neuron (*noo*-ron). The unit of structure in the nervous system. A nerve cell.

Neuroplasm. The protoplasm in the cell body of a neuron and filling the spaces between the neurofibrils.

Neutron (*noo*-tron). An electrically neutral particle in the nucleus of an atom.

Neutrophil (*noo*-troh-fil). A type of white blood corpuscle in which the cytoplasm does not stain with either hematoxylin or eosin.

Nissl (*nis'*l) *body*. One of many small cytoplasmic inclusions in the nerve cell body.

Nuclear (*noo*-clee-er) *membrane*. A semipermeable membrane surrounding the nucleus of a cell and separating it from the cytoplasm.

Nucleic (*noo*-*klee*-ik) *acid*. A complex type of organic acid found in chromatin.

Nucleolus (*noo*-*klee*-oh-lus). One or more bodies within the nucleus that appear to be associated with the formation of nucleoproteins.

Neucleoplasm (*noo*-klee-oh-plaz'm). A general term used to identify the living material of the nucleus.

Nucleoprotein (*noo*-klee-oh-*proh*-tee-in). A type of protein found in the nucleus of a cell.

Nucleotide. The combination of a purine or pyrimidine base with a sugar and phosphoric acid.

Nucleus (*noo*-klee-us). 1, the controlling center of a cell's activities. 2, the center of an atom, composed of protons and neutrons. 3, a group of specialized nerve cells lying within the white matter of the central nervous system.

Nutrient (*noo*-trih-ent). See *food*.

Oil. A type of relatively simple lipid that is fluid at ordinary temperatures.

Olfaction. The process of smelling.

Optic nerve. The nerve passing from the eye to the occipital lobe of the cerebrum.

Orbit (*orbital cavity*). A cavity in the upper half of the anterior surface of the skull; the eye cavity.

Organ. A group of tissues which function together in the performance of a definite body function.

Organ of Corti. The structure in the inner ear which contains the sense organs of hearing.

Organ system. A group of organs with related functions.

Organelle. A structure lying within the protoplasm of a cell and performing a specific function.

Organic. A chemical compound containing the element carbon. See *inorganic*.

Organic nutrient. Any carbohydrate, lipid, or protein.

Orifice. An opening; an entrance to a cavity or tube.

Osmosis (os-*moh*-sis). Diffusion through a semipermeable membrane.

Osmotic (os-*mot*-ik) *pressure*. The pressure that is established on one side of a semipermeable membrane as a result of the passage of a material through it.

Ossicle. A small bone; especially one of the three bones of the middle ear.

Ossification (os-ih-fee-*kay*-shun). The process of bone formation.

Osteoblast. A cell concerned with the formation of bone.

Osteoclast (*os*-tee-oh-klast). A cell which destroys bone to form channels through it for the passage of blood vessels and nerves.

Osteocyte (*os*-tee-oh-syt). A bone-forming cell after it has left a Haversian canal.

Osteomyelitis (os-tee-oh-my-ee-*ly*-tis). An infection of bone tissue.

ovary (*oh*-va-ree). The egg-producing organ.

Oviparous. An animal that lays eggs.

Ovoviviparous. Animals which produce eggs that are not laid but the developing young do not have a placental connection with the female.

Ovum (*oh*-vum). A female reproductive cell; an egg cell.

Oxyhemoglobin (ok-sih-hee-mah-*gloh*-bin). Hemoglobin containing oxygen. It is bright red in color.

Papilla (pa-*pil*-a). Any projection of tissue above a normal surface.

Parthenogenesis. The development of an egg which has not been fertilized.

Pathology (pa-*thol*-oh-jee). The study of disease.

PBI. See *protein bound iodine*.

Pentose. A sugar containing five carbon atoms, such as ribose.

Pericardium (*per*-ih-kahr-dee-um). A double-walled sac of endothelial tissue surrounding the heart.

Periosteum (per-ih-*os*-tee-um). The tissue which covers the outside of a bone.

Peripheral (puh-*rif*-er-al) *nervous system.* The part of the nervous system lying outside of the central nervous system and containing the receptors and effectors.

Peristalsis (per-ih-*stal*-sis). A rhythmic, wavelike series of contractions moving over the walls of a tube and resulting in the passage of materials along the tube.

Peritoneum (per-ih-toh-*nee*-um). A layer of tissue that lines the abdominal cavity and surrounds many of the abdominal organs.

pH. The symbol for hydrogen ion concentration.

Phagocyte (*fag*-oh-syt). A type of white blood corpuscle that devours bacteria and other foreign bodies.

Pharynx (*far*-ingks). The cavity in back of the mouth which is a common passageway to both the respiratory and digestive systems.

Phenotype (*fee*-noh-typ). The visible characteristics of an individual as determined by the genes.

Phonation (foh-*nay*-shun). The ability to produce sounds by means of vocal cords.

Photopic (foh-*top*-ik) *vision.* Vision in a bright light resulting from the activities of the cones.

Physiology. The study of the functions or activities of the various parts of the body.

Pia (*pee*-a) *mater* (*may*-ter). The innermost of the meninges.

Pinna. See *auricle.*

Placenta. The organ on the wall of the uterus to which the embryo is attached by the umbilical cord and through which it receives its nourishment.

Plasma. The fluid part of the blood.

Plasma membrane. The external covering of an animal cell.

Pleura (*ploor*-a). A two-layered sac of endothelial tissue surrounding each lung.

Plexus. A network of interlacing nerves, blood vessels, or lymphatics.

Poikilothermic. See *cold-blooded.*

Polar body. A small cell that is formed during the maturation of the ovum. It cannot be fertilized by a sperm.

Polysaccharide (pol-ee-*sak*-a-ryd). A complex form of carbohydrate composed of many hexose groups.

Polytene. Refers to the giant chromosomes found in cells of the salivary glands and intestinal wall of some insects.

Portal vein. A large vein, formed from capillaries in the walls of the digestive system, that passes to the liver.

Positive afterimage. The retention of an afterimage in its original color or shape after the stimulus has ceased.

Posterior. Referring to the back of the human body. Also at or toward the tail end of an animal.

Potential energy. The energy of position, as a raised weight or a coiled spring.

Presbyopia (prez-bee-*oh*-pee-a). A condition of old age in which the lens of the eye becomes fixed at a universal focal distance.

Prone. Lying face downward; the opposite of supine.

Prophase (*proh*-fayz). A stage in the mitotic division of a cell in which the chromatin network becomes thicker and individual chromosomes appear.

Proprioceptors. A receptor located in a muscle, joint, tendon, or the vestibule of the ear whose reflex is connected with locomotion or posture.

Protein (*proh*-tee-in). An extremely complex chemical compound composed of carbon, hydrogen, oxygen, nitrogen, and other elements. It forms the basis of protoplasm.

Protein bound iodine (*PBI*). Iodine present in the blood in the form of part of a protein molecule produced by the thyroid gland.

Proton. A positively charged particle within the nucleus of an atom.

Protoplasm. The living material in cells.

Proximal (*prok*-sih-mal). The part of an appendage that is toward the main part of the body.

Psychology (sy-*kol*-oh-jee). The study of an organism's responses to its environment (behavior).

Psychosomatic. Pertaining to the interrelationship between the mind and the body.

Pulmonary circulation. The path of the blood through the lungs.

Pulse. A wave of blood flowing along the arteries.

Pulse wave. A wave of muscular contraction that passes along the walls of the arteries.

Pupil. The opening in the iris of the eye.

Receptor. The ending of a nerve fiber that is specialized to receive one particular type of stimulus.

Recessive characteristic. The member of a pair of opposing characteristics which does not appear in the phenotype when the dominant is present.

Reciprocal innervation. A condition in which the members of a pair of antagonistic muscles are stimulated and inhibited simultaneously by nerve impulses.

Red blood corpuscle. A nonnucleated, biconcave disk containing hemoglobin.

Red bone marrow. The principal site of the formation of blood corpuscles.

Referred pain. A feeling of pain in one part of the body resulting from the stimulation of nerves in another region.

Refractory period, absolute. Said of a muscle or nerve resisting stimulation immediately after it has responded to a previous stimulus.

Refraction. The bending of a beam of light as it passes from one medium into another.

Replication. The process in which a molecule of DNA makes an exact duplicate of itself.

Reproduction. The ability of a living organism to produce similar living things.

Respiration. The exchange of gases between cells and their surroundings.

Retina (*ret*-ih-na). The innermost layer of the eye composed of nerve endings that are sensitive to light.

Rh factor. Hereditary factor present in the serum of the blood.

Rhodopsin (roh-*dop*-sin) (*visual purple*). A reddish-blue pigment present in the choroid layer and in the rods in the retina.

Ribonucleic (ry-boh-*noo*-klee-ik) *acid* (*RNA*). A nucleic acid found principally in the cytoplasm of a cell.

Ribose (*ry*-bohs). A sugar found in ribonucleic acid.

Ribosomes (*Palade granules*). Term used for granules containing RNA.

RNA. See *ribonucleic acid.*

Rod. A light receptor in the retina that is sensitive to low intensities of light but not color.

Sacrum (*say*-krum). The region of the vertebral column composed of fused vertebrae that forms part of the hip girdle.

Sarcolemma. The delicate sheath surrounding a muscle fiber.

Saturation. The color or hue of an object depending on the amount of white light that is present.

Sclera (*sklee*-ra). The tough outer layer of the eye; the white of the eye.

Scoliosis (skoh-lih-*oh*-sis). A sideward curvature of the vertebral column.

Scotopic (skoh-*top*-ik) *vision.* Vision in a dim light due to the activities of the rods.

Secretagogue (see-*kree*-ah-gog). An endocrine secretion that stimulates another gland to activity.

Secretion. The act of releasing a product of a gland from the cells that formed it. Also, the product of the gland.

Semicircular canals. Three bony canals in the inner ear, concerned with maintenance of balance.

Semipermeable (sem-ey-*pur*-mee-a-b'l) *membrane.* A type of membrane that will permit the passage of certain materials but prevent the movement of others through it.

Septicemia (sep-tih-*see*-mee-a). A generalized type of infection caused by bacteria or other pathogenic organisms in the blood.

Septum. A wall of tissue dividing a cavity.

Serous fluid. Resembling serum; the fluid produced by a serous gland.

Serum. Plasma lacking fibrinogen.

Serum albumin (al-*boo*-min). A type of protein present in plasma and concerned with the maintenance of osmotic pressure.

Serum globulin (*glob*-yoo-lin). A type of plasma protein that contains antibodies.

Sesamoid bone. A small bone developed in the tendon of a muscle.

Sex-limited characteristic. A characteristic which is determined by a gene that is present on the nonhomologous portion of the Y chromosome.

Sex linkage. The type of inheritance in which the gene is carried on the nonhomologous portion of the X chromosome.

Sinoatrial (sy-noh-*ay*-tree-al) *node.* A bundle of nerve fibers in the wall of the right atrium that serves as the pacemaker for the heart beat.

Sinus (*sy*-nus). A normal cavity within a bone or other organ.

Skeletal muscle. See *striated muscle.*

Skeleton. The bony and cartilaginous framework of the body.

Smooth muscle. Elongated, thin, spindle-shaped muscle fibers that are involuntary in their action.

Sociology (soh-see-*ol*-oh-jee). A study of the relation of human beings to each other and their environment.

Sol. The phase of a colloidal suspension in which the dispersed phase is uniformly scattered through the dispersing phase; it is quite fluid, as the white of an egg.

Solid. The state of matter that is characterized by definite shape and volume.

Sperm. A male reproductive cell.

Spherical aberration. The scattering of beams of light at the edges of a lens so that they are not focused sharply.

Sphincter muscle. A layer of smooth muscle arranged in a circular manner around a tube or opening. It behaves like a drawstring in controlling the size of the opening.

Sphygmomanometer (sfig-moh-ma-*nom*-uh-ter). A device for measuring blood pressure.

Spike potential. See *action current.*

Spinal nerve. One of thirty-one pairs of nerves arising in the spinal cord.

Spindle. A series of elongated cytoplasmic fibers connecting the two centrosomes during mitosis.

Spirometer (spy-*rom*-uh-ter). A mechanical device used to measure the capacity of the lungs.

Sprain. The result of tearing or straining of tendons and ligaments at a joint.

Stapes (*stay*-peez). The innermost of the three small bones of the middle ear. It articulates with the incus at one end and the membrane of the oval window at the other.

State of matter. The form taken by matter; solid, liquid, or gas.

Stereoscopic (ster-ee-oh-*skop*-ik) *vision.* The ability to see in three dimensions.

Stimulus. Any change in the environment that brings about a response.

Stirrup. See *stapes.*

Striated (*stry*-ayt-ed) *muscle.* Tissue composed of spindle-shaped fibers with striations; its action is voluntary.

Stroke volume. Referring to the output of the heart; this amount is arrived at by dividing the output of *either* ventricle by the number of beats per minute.

Subarachnoid (sub-ah-*rak*-noid) *space.* The space between the arachnoid and pia mater containing the cerebrospinal fluid.

Substrate (*sub*-strayt). The material affected by an enzyme.

Sulcus (*sul*-kus). A relatively shallow indentation on the surface of the brain.

Summation. The effect produced by the accumulation of several subthreshold stimuli.

Supine. Lying on the back with the face upward, or extending the arm with the palm upward; opposite of prone.

Suspension, colloidal. The suspension of finely divided particles of one substance in another.

Synapse (*sin*-aps). The meeting place of two nerve cell processes.

Synovium. A clear fluid that is normally present in joint cavities.

Systemic (sis-*tem*-ik) *circulation.* The path followed by the blood in flowing through all organs of the body except the lungs and the tissues of the heart.

Systole (*sis*-toh-lee). The period of active contraction of the atria and ventricles.

Telereceptor. Any receptor that can be stimulated from a distance, such as the eye.

Telophase (*tel*-oh-fayz). The final stage in mitosis in which a new cell membrane appears and the cytoplasm of the parent cell divides into two halves.

Tendon. A tough band of connective tissue that attaches a muscle to a bone.

Testis. The male sperm-producing organ.

Tetanus (*tet*-a-nus). A state of sustained contraction of a muscle caused by the repeated flow of nerve impulses to the muscle. Also, a bacterial disease in which the muscles contract and fail to relax.

Thoracic (thoh-*ras*-ik). Referring to the region of the thorax.

Thorax (*thoh*-raks). The upper part of the trunk.

Threshold. The minimum stimulus required to bring about a response.

Thrombocyte (*throm*-boh-syt). See *blood platelet*.

Thrombus (*throm*-bus). A clot of blood that forms within a blood vessel.

Tissue. A group of similar cells performing a specific function.

Tourniquet (*toor*-ni-ket). A mechanical device to prevent excessive loss of blood.

Trabecula (tra-*bek*-yoo-la). A wall of firmer and more compact tissue lying within a bar.

Transformation of energy, law of. The conversion of energy from a quiet (potential) form to an active (kinetic) state.

Transverse plane. A plane passing at right angles to both the frontal and saggital planes and dividing the body into two parts; a cross-section.

Trunk. The main part of the body.

Tympanum (*tim*-pa-num). A thin membrane covering the internal end of the external acoustic meatus. It vibrates when struck by sound waves.

Tyndall effect. The scattering of a beam of light by the particles of a colloidal suspension.

Ultraviolet radiations. That region of the spectrum lying just beyond the shortest visible violet rays.

Uterus. The cavity that receives the fertilized egg and in which the fetus develops.

Vacuole (*vak*-yoo-ohl). A space within a cell that is filled with a fluid.

Vaso– (*vas*-oh–). A prefix referring to a blood vessel.

Vein. A blood vessel carrying blood toward the heart.

Ventral (*ven*-tral). Referring to the side of an animal opposite to that along which the backbone passes; sometimes called the anterior side in human beings.

Ventricle (*ven*-tri-k'l). 1, a chamber of the heart from which blood flows to various parts of the body. 2, one of several cavities in the brain.

Venule (*ven*-yool). A small vein.

Vertebral column. The backbone.

Vertebrate. An animal possessing a backbone.

Villus (*vil*-us). A small, fingerlike projection from the mucosa of the small intestine, through the walls of which digested foods are absorbed by the blood.

Viscera. (sing. viscus). The organs contained within the cranial, thoracic, and abdominal cavities.

Visual acuity. The ability to see objects clearly with the naked eye.

Visual purple. See *rhodopsin*.

Vitamin. An essential accessory food substance.

Voluntary muscle. See *striated muscle*.

Warm blooded. See *homoiothermic*.

White blood corpuscle. A relatively large, nucleated blood cell having the power of independent movement.

White matter. That part of the central nervous system characterized by neurons having medullary sheaths.

Womb. See *uterus*.

Yellow bone marrow. Marrow in the medullary cavities of long bones, containing many fat cells.

Yolk. Nonliving material in the cytoplasm of an ovum. It serves as food for the developing embryo.

Zygote (*zy*-goht). A fertilized egg cell.

INDEX

Page references for illustrations are in *boldface* type.

487